The Post-Imperative Negative Variation in Schizophrenic Patients and Healthy Subjects

Psychophysiologie in Labor und Feld

Herausgegeben von Michael Myrtek und Jochen Fahrenberg

Band 3

PETER LANG

Frankfurt am Main · Berlin · Bern · New York · Paris · Wien

Christoph Klein

The Post-Imperative Negative Variation in Schizophrenic Patients and Healthy Subjects

PETER LANG
Europäischer Verlag der Wissenschaften

Die Deutsche Bibliothek - CIP-Einheitsaufnahme

Klein, Christoph:

The post-imperative negative variation in schizophrenic patients and healthy subjects / Christoph Klein. - Frankfurt am Main ; Berlin ; Bern ; New York ; Paris ; Wien : Lang, 1997
(Psychophysiologie in Labor und Feld ; Bd. 3)
Zugl.: Konstanz, Univ., Diss., 1995
ISBN 3-631-31078-1

NE: GT

D 352
ISSN 1430-8169
ISBN 3-631-31078-1
US-ISBN 0-8204-3233-4

For Burghard

„ERP research from both patho- and psychophysiologists has definitely not lived up to the unrealistic expectations of twenty and more years ago: It did not lead to a better distinction between schizophrenic and other psychiatric patients (...). On the other hand, there is a remarkable set of well established findings, sensitive to disorders of schizophrenic patients, which is worth following up in order to advance our understanding of or at least the precesion of our thinking about their dysfunctions in the course of schizophrenic disorders."

Rudolf Cohen (1990)

Preface

This dissertation ist the product of three years of teamwork. Many persons participated in this teamwork, whom I would like to thank.

I would like to thank all subjects who were willing to participate in the experimental sessions and the staff of the research ward for their assistance.

Jutta Keppner, Andrea Heinz, and Hansjörg Hautkappe assisted in data collection, as did Marc Dressel and Heiko Krüger as part of their diploma works. Without their aid and support the experiments of this dissertation would have been impossible to conduct.

I would also like to thank the members of the research groups of Prof. Dr. Rudolf Cohen and Prof. Dr. Brigitte Rockstroh at the University of Konstanz for their continuous preparedness for discussing research plans and research results, most noteably among them Dr. Michael Wagner, Dr. Matthias Müller, Dr. Petra Werther and Dr. Regine Schlenker.

Thanks also to Prof. Dr. Thomas Elbert for the idea to investigate the topography of the PINV in schizophrenic patients.

Furthermore, I would like to thank Dr. Patrick Berg for his patient and competent aid in many technical and theoretical problems and for his excellent ERP analysis programs.

I am especially grateful to Prof. Dr. Rudolf Cohen and Prof. Dr. Brigitte Rockstroh for their advice during numerous discussions, for their encouragement, and for all helpful support.

Christoph Klein

Contents

Chapter 1 Introduction into Schizophrenia—The DSM-III-R Diagnostic Concept

Summary of chapter 1 The introductory chapter shall provide an impression of the clinical features of the patients participating in the experiments of this dissertation. Although these features do not play a prominent role in the investigations reported here, it appears necessary to provide at least a rough impression of what is meant with "schizophrenic patients". Patients diagnosed as schizophrenic by DSM-III-R criteria are characterized by a set of florid and prodromal or residual symptoms which together are present for at least six months, and exhibit no schizoaffective or affective symptoms. Since chronicity of the disease is a defining symptom, a chronical future course can be expected in many of the patients. The multiplicity of schizophrenic symptoms can be reduced to two to three factors, but subgrouping of patients has proven unsuccesful in many regards. Although an organic etiology is denied by the DSM-III-R schizophrenia diagnosis, growing evidence for structural or physiological abnormalities in the brain of schizophrenic patients suggest the abandonment of this exclusion criterion.

To give a *brief* introduction into schizophrenia is only possible if most of the issues that are pertinent to an understanding of this disease are *omitted*. For the present purpose I will focus on the diagnostic system chosen to recruit patients for the experiments decribed in later chapters. This is done in order to give the reader an impression of the disease the patients who participated in the experiments of this dissertation were suffering from.

The schizophrenia concept chosen is the DSM-III-R system published by the American Psychiatric Association in 1987 (Wittchen et al., 1991). DSM-III-R provides an operational diagnosis of schizophrenia. The diagnosis of a schizophrenic disorder *includes* the presence of at least two of the following symptoms for a duration of at least one week: (a) delusions, (b) hallucinations, (c) disorganization or associative loosening, (d) catatonic behavior or (e) flat or inadequate affect. Furthermore, a cleary deteriorated (f) social adjustment (concerning work, relationships or self-management) has to be present. These symptoms have to be present for at least 6 months, including at least one florid phase (characterized by symptoms (a)-(e)) and a six months prodromal or residual phase (both characterized by at least two of the symptoms social isolation, social maladjustment, eccentric behavior, deficient hygiene, flat or inadequate affect, eccentric or poor speech, magical beliefs, recurrent illusions or loss of initiative and drive)). Diagnostic criteria that *exclude* a diagnosis of a schizophrenic disorder are (g) an etiological organic factor and (h) the presence of a schizo-affective or a major depressive disorder. Finally, in case of a past autistic disorder a schizophrenia diagnosis may be given only if delusional or hallucinatory symptoms are most prominent. Together these criteria constitute a concept of schizophrenia which is, compared to other diagnostic systems, rather „narrow". This means, that some of the patients that would have been diagnosed as schizophrenic by other diagnostic systems would not have received the same diagnosis according to DSM-III-R[1].

One of the most important aspects of defining a narrow diagnosis was the definition of affective symptoms as an exclusion criterion for the diagnosis of a schizophrenic disorder, which resulted in an improvement of diagnostic validity in terms of (poor) prognosis, (no lithium) treatment response and (less affective) family history, as thoroughly reviewed by Pope & Lipinsky (1978). Another narrowing aspect of the DSM-III-R diagnosis is the presence of florid and prodromal or residual symptoms for at least 6 months, resulting in an inclusion of chronically ill patients only, thereby exploiting the feature that is most predictive as to a chronic, „Kraepelinian", course of the disease: chronicity.

There is a huge phenomenological heterogeneity between *symptoms* of prodromal, florid or residual phases. Consider, for example, the differences between symptoms such as delusions or hallucinations on the one hand, and symptoms such as flat affect or loss of initiative and drive on the other hand. The heterogeneity of symptoms provoked attempts to group them empirically into more homogeneous syndromal —and patient— subgroups, most noteably by classical (e.g. Arndt et al., 1991; Malla et al., 1993; Silver et al., 1993) and confirmatory (e.g. Lenzenweger, 1989) factor analytic techniques or multidimensional scaling (Minas et al., 1992). This explains most of the symptom variance with two or three factor solutions, for example „reality distortion" (delusion and hallucinations), „psychomotor poverty" (poverty of speech, flat affect, decreased spontaneous movement), and „disorganization" (inappropriate affect, formal thought disorders) in the studies of Liddle and colleagues (Liddle, 1987; Liddle et al., 1992). Other distinctions, like „positive" versus „negative" or „type 1" versus „type 2", have been suggested by, e.g., Andreasen & Olson (1982).

Some of the symptoms have been related to brain abnormalities of schizophrenic patients. This issue will be presented in greater detail in *chapter 2*. These results are pertinent to the last exclusion criterion of schizophrenia, the organic etiology. On the one hand, if schizophrenia is considered an „endogeneous" psychosis, an exclusion of an organic etiology is straightforward.

[1] For a comparison between DSM-III, the precursor of DSM-III-R, and other diagnostic systems (Feighner, RDC and others) see Coryell & Zimmerman (1987), Endicott et al. (1982), Klein (1982), McGlashan (1984), Overall & Holister (1979), Stephens et al. (1980), Strauss & Gift (1977).

On the other hand, as will be reported in *chapter 2*, there is growing evidence for structural or physiological abnormalities in the brain of schizophrenic patients.

In sum: the patients investigated in the present experiments received a diagnosis of a schizophrenic disorder since they exhibited a certain set of symptoms, defined by the DSM-III-R criteria. Structural and physiological brain abnormalities may be present in these patients. The next chapter will examine these findings and their possible relation to cognitive impairments in greater detail.

Chapter 2 Neuroetiological Facts and Theories About Schizophrenia

Summary of Chapter 2 In his neuropsychological theory of schizophrenia, Frith (1992) makes a case for the linking of schizophrenic symptoms with underlying brain abnormalities through deviant cognitive processes. He considers schizophrenia a disorder of willed actions, of self-monitoring, and of monitoring of the intentions of others. Each of these disorders is related to certain cognitive or physiological abnormalities, and each of these abnormalities is related to a set of putatively afflicted brain regions, especially frontal, striatal and temporal brain structures. In linking symptoms, underlying cognitive or physiological mechanisms, and the aforementioned brain regions, Frith (1992) provides a comprehensive picture of the neuropsychology of schizophrenia. However, except for the account of some positive symptoms with disorders of self-monitoring, most of the arguments for the pairwise linkage of symptoms, cognitive and physiological processes, and brain regions are rather indirect and do not sufficiently refer to the already existing literature on structural and physiological abnormalities in schizophrenia. Therefore, primary studies on physiological or structural deficits in schizophrenia are surveyed, which, in sum, favor hypotheses of physiological prefrontal cortex and structural temporo-limbic abnormalities in schizophrenia. Weinberger (1991) assumes the prefrontal hypofunctions to result from a deficient communication with temporo-limbic structures, especially when the integrity of prefrontal functions depend upon "information networking" with the temporo-limbic system. The results on the neurophysiology of schizophrenic patients during the execution of the Wisconsin Card Sorting Test provide some evidence for this hypothesis. Nonetheless, while these particular instances of deficient "networking" exist, it is not convincing to reduce prefrontal dysfunction to abnormalities in the temporo-limbic system or the communication between both brain structures.

2.1 The Neuropsychological Theory of Schizophrenia by Frith (1992)[2]

Although DSM III-R denies an organic etiology of schizophrenia, both structural and physiological abnormalities in schizophrenic patients have been observed and linked to schizophrenic sings and symptoms either directly via empirical correlational or related methods, or indirectly via theoretical accounts of schizophrenic symptoms by typical cognitive processes linked to disorders of functions of certain brain regions. One example of the latter class of approaches is the neuropsychological theory of Frith (1992; 1987; Frith & Done, 1988), which will be presented in the following paragraphs.

Frith (1992) suggests schizophrenia to be a three-fold disorder: (1) of "willed actions", (2) of "self-monitoring", and (3) of "monitoring of the intentions of others". The next paragraphs will describe these types of disorder in greater detail.

2.1.1 Schizophrenia as a Disorder of Willed Action

In describing the gap between (intact) responsive and (deficient) spontaneous speech behavior in schizophrenic patients with negative symptoms, Frith (1992) introduces the theoretical distinction between responsive and self-initiated behavior (p. 43) suggesting schizophrenic patients to be deficient in the latter category[3] of the initiation of actions. This deficit leads to *poverty of action*[4], *perseverative behavior*[5] and *inappropriate behavior*[6]. According to Frith (1992) such negative signs can occur in different brain disorders, as, for example, psychomotoric retardation after frontal lobe lesions and Parkinson's disease (p. 53), but with specific underlying cognitive impairments in each of these disorders (p. 55). Referring to the known neuro-etiologies of disorders with negative signs, Frith assumes that the "limbic" frontal lobe (orbito-frontal cortex, cingulate cortex, supplementary motor area (SMA)), and the striatum, which are both dopaminergically innervated, are afflicted brain regions in patients with negative signs (p. 55). Frontal lobe lesions can result in a kind of stereotyped behavior

---

[2] The present text focuses on the monograph "Neuropsychology of Schizophrenia" by Frith (1992), which still provides the most detailed presentation of his theory. Arguments and statements of earlier (Frith, 1987; Frith & Done, 1988) or later (Frith, 1994) publications which differ from this text will be cited explicitly.

[3] For a similar differentiation see Callaway & Naghdi (1982)

[4] Poverty of action is described as a difficulty in generating actions spontaneously (p. 46), as in psychomotor retardation, flattening of affect (p. 51), and word or design fluency impairments (pp. 47).

[5] Perseverative behavior is defined as the predictability of current behavior by previous behavior (p. 49).

[6] Behavior that is guided by irrelevant stimuli (p. 51) as in incoherence of speech and incongruity of affect (p. 50)

which Luria (1973, cited after Frith, 1992) called "inert stereotypy" and which Frith labels "stimulus-driven" behavior, a behavior completely determined by environmental stimulus configurations. As Frith argues, stimulus driven behavior resembles Shallice's (1988) behavior due to contention scheduling without modulation by the Supervisory Attentional System. Stereotyped behavior can be elicited by amphetamine in experimental animals, an effect which Frith (1992) assumes to result from the repetitive initiation of one and the same contention scheduled response (Frith, 1992, p. 59). Amphetamine doses sufficiently small enough *not* to induce overt behavioral abnormalities impair reversal learning in experimental animals, presumably due to a deficient inhibition of the no longer rewarded response. These effects are similar to those produced by experimental frontal lobe lesions in animals and suggest that amphetamine results in a stimulus-driven response selection (p. 60), as is the case of frontal lobe lesions. According to Frith (1992), the effects of amphetamine in experimental animals —like the effects of certain frontal lobe lesions— can explain negative signs of schizophrenia such as stereotyped behavior, incoherence, and incongruity. Since amphetamine in man predominantly produces positive symptoms, Frith assumes that positive and negative features of schizophrenia share etiological factors (p. 61).

As outlined above, Frith (1992) distinguishes stimulus driven and willed actions, and assumes only the latter to be deficient in schizophrenia. Accordingly, Frith assumes the dorsolateral prefrontal cortex (DLPFC), the anterior cingulate cortex (ACC), the supplementary motor area (SMA), and the basal ganglia (BG) to be involved in positive and negative symptoms of schizophrenia. These brain structures are functio-anatomically linked (Alexander et al., 1986) and have been suggested to pertain to a system responsible for willed actions (Goldberg, 1985). The connection between the frontal ACC, SMA, and DLPFC (as putative anatomical substrates of the Supervisory Attentional System and responsible for plan and goal setting) on the one side and the basal ganglia (as putative anatomical substrates of the contention scheduling system and response execution) on the other side, is "lesioned" in schizophrenia, according to Frith (1992, p. 62), with positive and negative symptoms as consequences of this "lesion". With the PET technology, Frith et al. (1991) could demonstrate the involvement of PFC, SMA, and the Basal Ganglia in the execution of willed actions, supporting the relationship between these brain structures and negative symptomatology.

2.1.2 Schizophrenia as a Disorder of Self-Monitoring

While Frith (1992) introduces disorders of willed action by referring to clinical experiences, he introduces *disorders of self-monitoring* by referring to input versus output theories of auditory hallucinations in schizophrenia. Auditory hallucinations, which are the most prevalent hallucations in Schizophrenia, have been explained by so-called output theories, which consider auditory hallucinations to reflect inner speech which is misperceived by the patient "as coming from somewhere else" (p .71). Some studies cited by Frith (1992; Gould, 1949; McGuigan, 1966; Green & Preston, 1981; Bick & Kinsbourne, 1987; Green & Kinsbourne, 1989) are compatible with this hypothesis and may be theoretically explained by abnormalities in central speech processes of working memory functions (inner voice/inner ear), as conceptualized by Baddeley (1990).

As Frith (1992) argues, the reason for the occurence of auditory hallucinations in schizophrenia is not the occurence of inner speech (which is a normal working memory process), but the putative failure of schizophrenic patients "to recognize that this activity is self-initiated" (p. 73). The distinction between the execution of an action and the perception of its intention is suggested by the "corollary discharge" (Sperry, 1950) and "re-afference copy" (Von Holst & Mittelstaedt, 1950) principles of motor control. These principles state that in parallel to efferent muscle commands, messages about the *intention* to exhibit these efferent commands are sent to the perceputal system in order to distinguish self-initiated from externally initiated movements, which constitutes internal monitoring. Thinking has been conceptualized as "internalized action" (Piaget, 1936/53, cited after Frith, 1992), and internal monitoring systems for thoughts have been postulated (Feinberg, 1978) which suggest the generalization from the motor system control to "all voluntary actions" (Frith & Done, 1988). However, while there is some evidence of impaired self-monitoring in thought-disordered schizophrenic patients (Harvey, 1985), associations with auditory hallucinations have not yet been established (Frith, 1992, p. 76).

Frith points out that the misattribution "of self-generated actions to an external agent" (p. 73) cannot only explain passivity experiences occurring within the context of hallucinations, but may also explain the occurence of delusions. Delusions of alien control, thought insertion, and other Schneiderian "First Rank Symptoms" result from a loss of "awareness of the sense of effort that reflects central monitoring" (Frith, 1992, p. 81). As a result, confronted with "alien" thoughts, the patient experiences delusions with passivity feelings.

Some studies on fast visuomotor error correction (which in the absence of external feedback is considered to reflect central, corollary discharge mediated monitoring) showed impairments in this task in schizophrenic patients, especially those with passivity experiences (Malenka et al., 1982; Frith & Done, 1989[7]), supporting Frith´s hypothesis which considers a lack of corollary discharge[8] to underlie positive symptoms.

---

[7] These studies will be discussed in greater detail in *chapter 6*.

[8] "Positive symptoms occur because the brain structures responsible for willed actions no longer send corollary discharges to the posterior parts of the brain concerned with perception." (p. 93). This has not always been Frith´s argument. Frith (1987) assumed adequate production of corollary discharges in schizophrenia (" (...) I would suggest that it is not corollary discharge, but some other information which is missing.", p. 638).

By analyzing symptoms of certain neuropsychological damages, Frith attempts to delineate brain systems involved in positive symptoms of schizophrenia. The "alien hand" sign, a kind of "utilization behavior" (i.e. execution of unintended actions; Lhermitte, 1983), bears some complementarity with positive symptoms in that due to a loss of monitoring, stimulus-driven actions, which are normally not associated with a feeling of effort or intention, are "released" without awareness by the subject. The "alien hand" sign is usually associated with lesions of the medial frontal lobe (SMA; Goldberg et al., 1981), the brain region which normally monitors stimulus-elicited actions (Frith, 1992, p. 86).

Other brain damages associated with actions that are executed without self-awareness are blindsight (occipital lobe), amnesia (temporal lobe), and delusional misidentification (right parietal lobe; Capgras' syndrome; Frith, 1992). As Frith (1992) argues, all these syndromes have in common that the execution of actions is not accompanied by self-awareness. As mentioned before, concerning positive symptoms of schizophrenia, self-awareness means self-awareness of the "sense of effort". Frith (1992, p. 91) assumes sense of effort to be generated in the frontal eye fields (Brodmann's area 8), referring to neurophysiological studies on eye movements in the rhesus monkey by Robinson & Wurtz (1981), or in the anterior cingulate cortex as the source of willed vocalizations, referring to studies on the vocalization behavior in squirrel monkeys by Müller-Preuss et al. (1980).

In a recent formulation of his theory, Frith (1994) has suggested that a "failure of central monitoring is the consequence of an inability to represent our own mental states, including our intentions" (p. 154).

2.1.3 Schizophrenia as a Disorder of the Monitoring of the Intentions of Others

Not only does schizophrenic *speech* reflect disorders of willed action[9] and disorders of self-monitoring[10], but also the third disorder proposed by Frith (1992), *disorders in the monitoring of the intentions of others*. In addition to speech abnormalities in schizophrenia[11], Frith (1992) illustrates disorders in the monitoring of the intentions of others with the deficiencies in mentalizing in childhood autism. Mentalizing is both the "belief that other people have minds different from our own" and the "ability to infer the beliefs, wishes, and intentions of other people in order to predict their behavior" (p. 118). While both

---

[9] e.g. poverty of speech, poverty of content of speech; incoherent speech; neologisms (Frith, 1992, pp 101-103)

[10] e.g. impaired self-editing (Cohen, 1976), impaired self-repair of speech errors (Leudar et al., submitted), and disorganized discourse (Hoffmann, 1986) (Frith, 1992, pp 104-106)

[11] e.g. lack of references and cohesive ties (Rochester & Martin, 1979; reduced predictability of discourses (Rutter, 1985); impaired apprehension of metaphorical meanings (Cutting & Murphy, 1990) concerning schizophrenic speech (Frith, 1992, pp 106-108, p. 115)

components of mentalizing are considered to be deficient in autistic children, only the latter is assumed by Frith (1992) to be deficient in schizophrenic patients *after* becoming ill (p. 121). Therefore, although being aware of the existence of other people´s minds, schizophrenic patients cannot infer their beliefs and intentions, with the consequence of paranoid delusions and delusions of reference (p. 125).

2.1.4 Schizophrenia as a Disorder of Meta-Representations

According to Frith (1992) the three aforementioned "mechanisms are all special cases of a more general mechanism" (p. 116) of conscious experience which Frith calls *metarepresentations*. With the perception of, e.g., objects of the outer world *representations* emerge. Becoming aware of this perceptual act amounts to the representation of a representation, i.e. a metarepresentation, as "the crucial mechanism that underlies this self-awareness" (p. 116). Metarepresentations (i.e. awarenesses) of own goals, own intentions, and intentions of others are missing in schizophrenic patients, with the aforementioned symptoms as consequences of this disorder (Frith, 1992, p. 125).

Table 2.1 Tabular Survey of Key Concepts of the Neuropsychological Theory of Schizophrenia by Frith (1992)

Type of Disorder	Clinical Parallel	Associated Symptoms	Mechanism	Afflicted Brain Region
Willed Actions	frontal lobe damage	negative symptoms (poverty of action; perseverative and inappropriate behavior), positive symptoms	stimulus-driven behavior; lack of self-initiated behavior	OFC, DLPFC, ACC, SMA; striatum
Self-Monitoring	no direct clinical parallel given	positive symptoms, some hallucinations, some delusions	lack of corollary discharges	ACC, FEF
Monitoring of the Intentions of Others	childhood autism	paranoid delusions and delusions of reference	impaired mentalizing ability	TC, A; PFC. (in social contexts)

Abbreviations: PFC= Prefrontal Cortex; OFC=Orbital PFC; DLPFC=Dorsolateral PFC; ACC=Anterior Cingulate Cortex; SMA=Supplementary Motor Area; FEF= Frontal Eye Fields; TC=Temporal Cortex; A=Amygdala; S=Symptoms

Since the concept of metarepresentations is rather new, brain systems involved are not yet experimentally delineated. However, referring to brain systems that have been suggested to be involved in abilities that may require metarepresentations, Frith (1992) assumes that the interaction of the temporal cortex and the amygdala on the one hand, and the frontal cortex on the other hand, is required for metarepresentations in social contexts (Frith, 1992, p. 129).

Before discussing the neuropsychological theory of Frith, a summary of the key concepts of that theory shall be given in **table 2.1**.

2.1.5 Discussion of the Theory of Frith (1992)

Frith´s neuropsychological theory of schizophrenia attempts to delineate cognitive processes underlying some schizophrenic symptoms and to infer afflicted brain regions from the deviant cognitive processes.

Explicitness is not only one of the major strengths of Frith´s theory, but the basis for some concrete comments. I think four major points seem worth discussion: *Firstly*, not all schizophrenic symptoms appear explainable by Frith´s theory. Although it is Frith´s explicit intention to explain *only some* of the symptoms of schizophrenia, the clinical relatedness of symptoms which are associated with symptoms that are not discussed by Frith and do not appear to be explainable by his theory renders his explanations somewhat "patchy". Although visual, olfactory, and gustatory hallucinations in schizophrenic patients may be less prevalent than auditory hallucinations, they do exist and frequently coexist with auditory hallucinations. Since Frith´s output theory of (auditory) hallucinations is intrinsically modality-specific, these coexistent symptoms cannot be explained by his theory. Furthermore, Frith explains delusions of reference or paranoid delusions with deficient mentalizing processes. Since mentalizing refers exclusively to social interactions, it cannot explain non-social components of these delusions (e.g. delusional misinterpretations of real world objects) which occur alone as well as intermingled with social components or occur intermingled with hallucinatory experiences (e.g. being influenced by rays).

Secondly, the relationship between (deviant) cognitive processes and resultant symptoms is not always unambiguous. According to Frith, mentalizing is necessary in order to predict the behavior of others. Frith assumes that deficient mentalizing can result in paranoid or referential delusions. This is, however, *but one* plausible causal chain. For example, it is at least equally plausible that disappointed social behavioral predictions due to deficient mentalizing results in social withdrawal, which is yet explained by different processes (willed actions) within the framework of Frith´s theory.

Thirdly, there appear to exist some intrinsic contradictions: Frith´s account of negative symptoms by disorders of willed action and positive symptoms by disorders of self-monitoring suggest *inverse* relations between the occurence of both symptom groups. This becomes apparent when Frith discusses the effects of neuroleptic medication on positive symptoms (Frith 1992, pp. 89-90). Positive symptoms occur "because patients act without being aware of any intention to act" (p. 90). (By neuroleptics) they can be ameliorated "by reducing the patient´s spontaneous activity" (p. 90). Spontaneous activity, however, is self-initiated activity in Frith´s terminology and its lacking his *exclusive* account of negative symptoms. Hence, the existence of negative symptoms (i.e. reduced self-initiated behavior) should reduce the occurence of positive symptoms. On the other hand, Frith (1992) assumes that positive and negative symptoms share etiological factors (p. 61). Furthermore, the conclusion to be drawn from the preceding argument is at odds with cross-sectional as well as longitudinal data in schizophrenia. In most schizophrenic patients positive and negative symptoms do co-exist cross-sectionally, and anamnestic data suggest that positive symptoms are frequently preceded by negative symptoms in the longitudinal course (e.g. Häfner et al., 1992; see also Huber & Zerbin-Rüdin, 1979, for a discussion of schizophrenic prodromes).

Fourthly, the existing evidence on functio-anatomical abnormalities in Schizophrenia is not sufficiently discussed or integrated into his theory: One of Frith´s (1992) primary concerns is "to consider how to relate the various signs and symptoms of schizophrenia to disturbances in particular brain systems" (p. 24). He argues that the search for direct relationships between symptoms and specific locations of brain damages is less promising than the search for associations between symptoms and underlying cognitive processes (p. 85). Since he explains schizophrenic symptoms by abnormal cognitive processes which he associates with locations of brain lesions in patients with *different* neuropsychological disorders who exhibt these abnormal cognitive processes, he explicitly relates schizophrenic symptoms to putatively afflicted brain regions (see **table 2.1**). However, in my opinion, this linkage is too indirect, for the reported relations between abnormal cognitive processes and brain abnormalities in patient groups with different brain damages are relevant to schizophrenic disorders only if the location of a brain abnormality can unambiguously and independently of the investigated clinical group be infered from the knowledge of the abnormal cognitive process, which is obviously not possible. Therefore, evidence of relations between symptoms and cognitive processes on the one hand and brain abnormalities on the other hand *in schizophrenia itself* should *in addition* be considered. Evidence in this respect has already been discussed. For example, reduced prefrontal metabolism has been associated both with negative symptoms and with deficiencies in neuropsychological functions which are also seen in patients with PFC damage.

It is therefore appropriate to examine which of the structures implicated in Frith´s theory have indeed been found to function abnormally in schizophrenia, and if there are some hints to link these abnormal brain processes or structures to symptomatological or cognitive processes in schizophrenic patients. The following presentation will focus on structural and physiological findings of the prefrontal cortex and of the temporo-limbic system in schizophrenic patients[12].

2.2 Structural and Physiological Brain Findings in Schizophrenia

The following survey of studies on brain findings in schizophrenia will —for the sake of conciseness— focus on structural and physiological findings in prefrontal and temporo-limbic systems. Results on subcortical structures will be mentioned only briefly, due to the relative paucity of results.

2.2.1 Structural Findings in the Prefrontal Cortex

Bilaterally reduced frontal lobe volumes have been found by some authors with *Magnetic Resonance Imaging* (MRI; Andreasen et al., 1986; Raine et al., 1992; Woods & Yurgelun-Todd, 1991) or *Computed Tomography* (CT; Keilp et al., 1988; Morihisa & McAnulty, 1985; Rosse et al., 1991; Yates et al., 1990) techniques. Some authors have found frontal volume reductions primarily in the left hemisphere (Golden et al., 1981; DeMyer et al., 1988). In addition to these morphological results, postmortem histological abnormalities have been found in the cingulate cortex of schizophrenic patients (Benes et al., 1986).

Only few studies have related these structural deviancies to neuropsychological test performance (Rosse et al., 1991; Yates et al., 1990; Keilp et al., 1988; Seidman et al., 1994) in schizophrenic patients.

2.2.2 Physiological Findings in the Prefrontal Cortex

Beginning with the pioneering work on rCBF of Ingvar & Franzén (1974; Franzén & Ingvar, 1975), schizophrenic patients have been shown to exhibit normal frontal blood flow during resting, but a relatively reduced increase of frontal metabolism during activating conditions (including Raven´s Progressive Matrices). Other investigators have replicated this *condition-dependent* activational pattern with neuropsychological tests considered to involve mainly PFC functions, such as the Tower of London (Andreasen et al., 1992), the Chicago Word (Lewis et al., 1992) and other verbal fluency tests (Warkentin et al., 1989; Frith et al., 1991), the auditory Continous Performance Test (Cohen et al., 1987; Buchsbaum et al., 1990), and the Wisconsin Card Sorting Test (e.g. Weinberger et al., 1986; Berman et al., 1986, 1987, 1993; Paulman et al., 1990; Rubin et al., 1991). In some of these studies, reduced PFC activation was related

[12] Results on ventricular enlargement in schizophrenia will not be discussed here because their relation to the theoretical background of the present experiments is too indirect.

to WCST performance (Weinberger et al., 1986; Paulman et al., 1990), or resting rCBF was correlated with subsequent WCST performance (Sagawa et al., 1990).

During resting conditions reduced frontal metabolism has also been found in schizophrenic patients by e.g. Farkas et al.(1984), Géraud et al. (1987), Jernigan et al. (1985), Kishimoto et al. (1987), Sagawa et al. (1990), and Volkow et al. (1987). This reduced metabolism was independent of brain atrophy (Wolkin et al., 1988). Some studies, however, indicate that reduced frontal blood flow under resting conditions is not seen in acute patients with positive or exacerbating symptomatology (Cleghorn et al., 1989; Kling et al., 1986; Warkentin et al., 1990). These results suggest that resting frontal blood flow may be increased during acute exacerbation of the illness, a conclusion which was empirically supported by the intraindividual comparisons in the study of Géraud et al. (1987).

As was the case for structural findings, particularly *left-sided* regions of the PFC may be dysfunctional (Andreasen et al., 1986; Kawasaki et al., 1993; Lewis et al., 1992; Rubin et al., 1991; Wolkin et al., 1985). It is interesting to note that with the exception of Wolkin et al. (1985), all studies reporting a particular *left* PFC involvement refer to metabolic dysfunctions during the execution of prefrontal tests. Furthermore, only left-frontal metabolic reductions were related to subsequent WCST errors in the study of Sagawa et al. (1990).

While structural frontal deficits are related to the performance in neuropsychological tests, physiological aspects of frontal lobe functions seem related to negative symptomatology (Wolkin et al., 1985; Andreasen et al., 1986; Volkow et al., 1987; Kawasaki et al., 1992; Lewis et al., 1992). In the study of Warkentin et al. (1990) resting rCBF and negative symptomatology were unrelated; only patients with florid symptoms exhibited anterior-to-posterior flow gradients which were related to their behavioral distortions. In the study of Devous et al. (1985) reduced frontal flow was more prevalent in paranoid (40%) compared to non-paranoid (11%) patients.

It has been argued that reduced frontal metabolism might be an effect of neuroleptic medication (e.g. Sheppard et al., 1983). This topic may be adressed by comparing neuroleptic-naive with neuroleptic-experienced patients and by investigating the effects of neuroleptic medication itself. Some studies have shown that neuroleptics either have no effect on frontal metabolism (e.g. DeLisi et al., 1985; Gur et al., 1987; Resnick et al., 1988) or have an increasing effect (e.g. Berman et al., 1986; Buchsbaum et al., 1987). While this group of results does *not* support the hypothesis of a neuroleptically induced hypofrontality, studies investigating this dysfunction in neuroleptic-naive schizophrenic patients have yielded conflicting results. Some authors observed reduced prefrontal metabolism in neuroleptic-naive schizophrenic patients (Andreasen et al., 1992; Chabrol et al., 1986; Geraud et al., 1987), but others did not (Sheppard et al., 1983; Volkow et al., 1986). With the exception of the study by Andreasen et al. (1992) who investigated 13 patients, as few as 6 (Sheppard et al., 1983) or 4 (Volkow et al., 1986) patients were recruited, rendering any generalization problematic.

In conclusion, it appears that if neuroleptics have any effect on prefrontal metabolism, it is an increasing, but not a decreasing (cf. Weinberger & Berman, 1988) effect.

To summarize: the aforementioned results indicate that schizophrenic patients exhibit reduced frontal metabolism under conditions which increase frontal metabolism in healthy subjects, especially in the left hemisphere. Reduced frontal metabolism under resting conditions may not necessarily be present in schizophrenic patients, especially during acute exacerbation. An association between PFC hypometabolism and negative symtpoms has repeatedly been found.

2.2.3 Structural Findings in the Temporo-Limbic System

Compared to PFC results, a left-hemisphere affliction in Schizophrenia is indicated even more strongly by studies on the temporo-limbic system (Crow, 1990). Reduced size of the left temporal lobe (Bogerts et al., 1985; Dauphinais et al., 1990) and the left planum temporale (Barta et al., 1990; Falkai et al., 1992; Hoff et al., 1992; Rossi et al., 1994), which is the posterior part of the superior temporal gyrus, have been found in schizophrenic patients. Also, bilaterally reduced volumes of amygdala and hippocampus were descriptively more prominent in the left hemisphere (Rossi et al., 1994). Furthermore, postmortem (Crow et al., 1989; Bogerts et al., 1990) and MRI (Degreef et al., 1992; DeLisi et al., 1994) studies have observed more left than right hemisphere temporo-limbic pathology in schizophrenia.

Nevertheless, evidence for left-hemisphere pathology is not completely consistent. Weinberger et al. (1980), Bogerts et al. (1985), Altshuler et al. (1990), and Christison et al. (1989) failed to find left-hemisphere hippocampus reduction, and Kleinschmidt et al. (1994) did not find a planum temporale asymmetry. Other authors reported *bi*laterally reduced temporal lobes (Rossi et al., 1994), or *bi*laterally reduced volumes of the hippocampus and the amygdala (Suddath et al., 1990; Shenton et al., 1992). Furthermore, *bi*lateral reductions in parahippocampal gyri (Brown et al., 1986), cytoarchitectural abnormalities (Jakob & Beckman, 1986), gliosis of the hippocampus (Stevens, 1982), and reduced entorhinal volume (Falkai & Bogerts, 1988) were found in schizophrenic patients, and reduced (normal) asymmetries in temporal lobe volumes were confined to female schizophrenic patients in the study of DeLisi et al. (1994).

2.2.4 Physiological Findings in the Temporo-Limbic System

In addition to the structural findings cited in the preceding paragraph, physiological deviances such as heightened dopamine content in the left amygdala (Reynolds, 1983; Deakin et al., 1989) were observed. Compared to healthy subjects, deviant metabolism under resting or activating conditions have been reported and point mainly to an overactivation of the left hemisphere, e.g. left-hippocampal (Kawasaki et al., 1992, 1993) or left inferior-temporal (DeLisi et al., 1985) metabolic enhancement under resting conditions, as well as increased left superior temporal gyrus metabolism during the Chicago Word Fluency Test (Lewis et al., 1992) and enhanced left temporal lobe metabolism during degraded CPT (Siegel et al., 1993).

Left temporal lobe metabolism was positively related to substance abuse (Cleghorn et al., 1992) and to positive symptomatolgy (Kurachi et al., 1985). The latter finding points to a relationship between positive symptoms and temporal lobe pathology, a relationship further supported by the reported association between left-hemisphere temporal lobe epilepsy and schizophrenic symptoms (e.g. Marshall et al., 1993).

In sum, results on temporo-limbic pathology in schizophrenia appear more anatomical than physiological in nature, and are in this sense complementary to the PFC findings. Left-hemisphere temporo-limbic structures seem especially involved in schizophrenia and may clinically be associated with positive symptomatology and metabolic overactivity during rest or activation conditions.

2.2.5 Findings in Subcortical Structures

A third group of results points to anatomic or metabolic findings in subcortical structures of schizophrenic brains. Reduced thalami were found by Andreasen et al. (1990) with the MRI technique, albeit only in male patients. Structural (Stevens, 1982; Jernigan et al., 1991; Swayze et al., 1992; Elkashef et al., 1994) as well as metabolic reductions (Sevall et al., 1984; Sheppard et al., 1983) were found in the basal ganglia in schizophrenia.

Again, some studies specifically point to left-sided metabolic increases (Buchsbaum et al., 1987; Kawasaki et al., 1992). Early et al. (1987) observed an augmented ratio of region-to whole brain blood flow (rCBF) in the left globus pallidus of neuroleptic naive schizophrenic patients, a result that could be replicated (cf. Early et al., 1994). This hyperactivity, however, may be a consequence of abnormalities of the amygdala or the hippocampus according to Early et al. (1994). In line with this reasoning is the observation of hyperresponsivity for dopaminergic drugs after puberty in animals with perinatal lesions of the ventral hippocampus (Lipska et al., 1993). In addition, lesions of the prefrontal cortex of rats have been shown to be followed with a certain delay by hyperactivity of the basal ganglia (Pycock et al., 1980). Inverse activational relations between frontal cortex (hypometabolism) and basal ganglia (hypermetabolism) were reported for schizophrenic patients by Musalek et al. (1988).

According to Pilowsky et al. (1994) deviant D2-receptor-binding in schizophrenia is still questionable since binding-studies with neuroleptic-naive patients have yielded equivocal results.

2.2.6 Integrating the Diversity of Brain Findings—The Neuro-Developmental Theory of Schizophrenia by Weinberger (1991)

Given the diversity of structural and/or physiological results one may ask, whether schizophrenia is a neuroetiologically heterogeneous disease, comprising, e.g., prefrontal, temporo-limbic, or subcortical subgroups, or whether these abnormalities are interrelated as a consequence of another cause or due to the relatedness of the brain structures in question.

Recent reviews and studies of brain pathology in schizophrenia emphasize the functional dependency of the putatively afflicted brain regions. Buchsbaum (1990) reviews his own evidence of substantial intercorrelations between metabolic parameters from frontal, temporal and subcortical structures. Furthermore, Rubin et al. (1991) observed *both* reduced left inferior PFC metabolism *and* reduced striatal suppression. Siegel et al. (1993) detected reduced frontal *and* medial temporal, *as well as* augmented left temporal activity in schizophrenic patients.

Studies which intercorrelate the brain processes in the prefrontal, temporal and subcortical structures within large groups of schizophrenic patients in order to establish patterns of metabolic dysfunctions are scarce (Buchsbaum et al., 1990). Nonetheless, envisaging the abundant interconnectedness of the structures in question, such interdependencies of physiological processes should come as no surprise.

Some of the anatomical relations between prefrontal, temporo-limbic, and subcortical structures are outlined in the next paragraph in order to illustrate the neuronal basis of the putative functional interdependences. While many of the neuroanatomical results cited below have been found in experimental animals of different species, including non-human primates, some of them have also been investigated in human brains. Unless other authors are cited, the following presentation refers to the monograph on the PFC by Fuster (1989).

The PFC receives afferent input from cortical and subcortical structures. Among the latter group the mediodorsal thalamic nucleus relays along its medial ("magnocellular") component information from the mesencephalic reticular formation, substantia nigra, amygdala and inferior temporal cortex to the orbital and medial PFC in a topographically organized way. The lateral component of the mediodorsal thalamic nucleus projects to the dorsolateral prefrontal convexity and receives afferents only from the PFC itself and from the substantia nigra. Finally, the laterally situated pars paralamellaris projects to the frontal eye fields (Brodmann's area 8). In addition to these "thalamically relayed" afferents, the PFC receives direct input especially from brainstem tegmentum, pons, and limbic structures such as hypothalamus, amygdala, hippocampus and cingulate cortex. Neocortical afferents to the PFC stem from all nonprimary areas of the somatic (e.g. parietal area 7), auditory (e.g. temporal area 22), and visual (e.g. inferotemporal area 21) sensory systems and suggest the PFC, and area 8 in particular, to be a region for the convergence of polysensory information.

Brain regions receiving efferents from the PFC include pons and mesencephalon, and the limbic hypothalamus, septum, cingulate and entorhinal cortex, amygdala, and hippocampus. In the primate brain the orbital PFC is connected with the amygdala complex, while the dorsolateral PFC sends efferents to the hippocampus (mainly via the entorhinal cortex) and the parahippocampal cortex. The latter reciprocal connectivity seems to be especially relevant for cognitive functions and will be discussed in greater detail below. In addition, the PFC sends efferents to the basal ganglia, i.e. caudate nucleus, putamen, globus pallidus, claustrum, and substantia nigra. However, except for substantia nigra, these connections are not reciprocal, suggesting that there is no *direct* connection between basal ganglia and PFC[13]. Therefore, the nucleus caudatus, which seems to be involved in neuropsychological „prefrontal" tasks (Mendez et al., 1989), communicates with the PFC only via the putamen, the basal ganglia output region, and the substantia nigra.

Efferents from the PFC to posteriorly located frontal regions, including the supplementary motor area (SMA), also pass the basal ganglia (posterior putamen, globus pallidus) and the venrolateral thalamic nucleus.

To sum: all of the superordinate brain regions that have been suggested to be involved in schizophrenic disorders are profusely interconnected. It is therefore highly plausible that any primary pathology in one of these regions will also affect the functions of the other regions. This is certainly no new insight into functional brain organization, and neuro-etiological conceptions of schizophrenia are adopting such integrative perspectives (e.g. Weinberger, 1987; Frith & Done, 1988; Buchsbaum, 1990; Weinberger, 1991).

One attempt to integrate anatomical and physiological results into a neuro-etiological theory has been put forward by Weinberger (1991). Weinberger assumes prefrontal dysfunctions in schizophrenia to be secondary to a structural pathology in the anterior temporal lobe. Empirical support for this primary process includes evidence of structural or anatomical abnormalities in the amygdala, hippocampus, and parahippocampal gyrus (including the entorhinal cortex, which relays input to and output from the hippocampus) in schizophrenia, as reviewed in *paragraph 2.2.3*. As a consequence, information transmission between anterior temporal and prefrontal regions is deteriorated. According to Weinberger (1991) this putative deficit should become important, firstly, after the ontogenetical maturation of the PFC and, secondly, in situations in which PFC functions depend upon the "information networking", especially with the anterior temporal cortex. It is important to note that Weinberger (1991, 1994) presents arguments in favor

[13] *Indirect* connections are given via the thalamus.

of the assumption that the neuropathology leading eventually to schizophrenia occurs early in life and does not "progress" in adulthood.

As to the first argument, orbital PFC seems to mature earlier than dorsolateral PFC (cf. Fuster, 1989), and in the monkey puberty has been suggested to be the earliest developmental stage of dorsolateral PFC maturity (Alexander & Goldman, 1978). Since it has been demonstrated in experimental monkeys that lesions of the PFC become behaviorally relevant only with PFC maturity (Goldman, 1971), PFC dysfunctions (secondary to anatomical changes in the temporal lobes) in schizophrenia could also become relevant at that stage. This could plausibly explain, why schizophrenia occurs with high probability in early adulthood (Weinberger, 1991, 1994). The hypothesis put forward by Weinberger (1991, 1994) seems to be supported, firstly, by the appearingly static pathoanatomy of schizophrenia, which has frequently been considered as occurring early in life (cf. Weinberger, 1994), and, secondly, by the relatively high incidence of schizophrenia-like disorders in cases of metachromatic leukodystrophy (MLD), a metabolic neurological disorder affecting primary white matter and glia, i.e. the communication links between processing stations (cf. Weinberger, 1991, 1994).

As to the second argument, it has been argued that the execution of the WCST depends upon the intactness of working memory functions (Goldman-Rakic, 1987), since each choice during this task depends upon the accessibility of information in "representational" or "working" memory—a feature the WCST has in common with Delay Tasks (Goldman-Rakic, 1991). Although all relevant stimulus features (color, form, number, etc.) are present prior to each response during the execution of the WCST, they bear no response-relevant information. Instead, the instruction, or the concept of color or shape etc. must guide the response. Goldman-Rakic (1991, 1994) assumes schizophrenia to be associated with a loss of *working memory functions*, resulting in a "deficit in the regulation of behavior by interalized schemata, symbolic representations, and ideas". The term "working memory" refers to the "ability to bring to mind events without direct stimulation". Working memory is therefore necessary in order to guide behavior by representations rather than stimulations. These working memory functions are hypothetically subserved by the joined activity of a network including the PFC and the hippocampus (Friedman et al., 1990). For these reasons, behavioral impairment and reduced metabolism of schizophrenic patients when taking the WCST may be a consequence of a deteriorated accessibility of informations from the anteromedial temporal system (Weinberger, 1991). In a recent study, Weinberger et al. (1992) have presented evidence in favor of this hypothesis. In a group of 9 monocygotic twin pairs, discordant for schizophrenia, differences between the affected and the unaffected twin in hippocampal volume were significantly and specifically correlated with differences in PFC activation during the WCST.

With the PET technique Frith et al. (1991) observed during a verbal fluency task, which is assumed to measure prefrontal functions, metabolic augmentations in dorsolateral prefrontal, parahippocampal, and parietal cortices of the left hemisphere and in bilateral anterior cingulate cortices on the one hand, and bilateral metabolic decreases in the superior temporal gyri on the other hand. This result reminds of vocalization results in monkeys which revealed that stimulation of the cingulate area was accompanied by reduced discharge rates in the superior temporal gyrus (Müller-Preuss et al., 1980), suggesting an inhibitory influence of the former on the latter structure. Since response selection during verbal fluency is considered to depend on internal representations (e.g. of the initial letter (words beginning with "S"), keeping track of the already produced words), this is another example of a networking function of the PFC with other brain regions, especially the temporal lobes. Concerning the integrity of these brain networks in schizophrenia, it is highly interesting to note, that metabolism during the Chicago Word Fluency Test in schizophrenic patients was decreased in anterior cingulate ("mesial frontal") and increased in left superior temporal gyri (Lewis et al., 1992). This suggests that reduced prefrontal metabolism in schizophrenic patients during this task leads to increased superior temporal gyrus disinhibition. Some other, albeit non-physiological, results seem to confirm this assumption. As Frith et al. (1991) argue, it is hypothesized that words are represented in a distributed fashion in the superior temporal gyrus, with a spreading activation following the uptake of a word. This may explain why this region showed increased metabolism during the lexical decision task, with reduced metabolism in the posterior cingulate. A lexical decision task was also provided by Spitzer (1993) as a semantic priming task with schizophrenic patients, and revealed an *augmented* semantic priming in this group.

Word fluency and semantic priming results with schizophrenic patients may be interpreted as revealing distortions in a functional brain network comprising at least frontal and temporal brain regions. These results support one of the conclusions drawn from experiments with the WCST, namely that hypofrontality may occur only under conditions which require the networking of the PFC with the temporo-limbic system.

In sum, the results on WCST, word fluency, and semantic priming support part of the hypothesis of Weinberger (1991), according to which there may be a deficient "information networking" between PFC and temporo-limbic structures.

2.2.7 Discussion of the Neuro-Developmental Theory of Schizophrenia by Weinberger (1991)

One of the crucial points in the model of Weinberger (1991), in my opinion, is that he assumes temporal lobe pathology to be insufficient to explain the etiology of schizophrenia but considers PFC dysfunctions to be dependent upon this "primary" deficit. This basic argument, however, has two weaknesses. Firstly, if schizophrenic psychopathology is at least partially related to the deficient "information networking" described above, which is itself a consequence of a primary temporo-limbic disorder, *each* temporo-limbic disorder should be associated with a schizophrenic symptomatology. Superficially, this may be true of positive symptoms (e.g. in temporal lobe epilepsy), but it does not hold for negative symptoms, which appear more closely related to PFC hypofunction. Secondly, Weinberger's argument can be true only to the *extent* that PFC functions indeed depend upon „information networking" with the temporo-limbic system. For some psychological functions this information networking seems to take place, as discussed in the preceding paragraph. However, not only, as Weinberger (1991) argues, is schizophrenia *not* associated with temporal lobe disorders such as the amnesic syndrome, but, conversely, are amnesic syndromes or other temporal lobe disorders regularly *not* associated with schizophrenic symptoms, especially negative symptoms which, as recent anamnestic data suggest (Häfner et al., 1992), may frequently demarcate the dawning of schizophrenic disorders. Furthermore, if negative symptoms were a "byproduct" of a PFC dysfunction that is secondary to a temporo-limbic pathology, negative symptoms should be related to this primary pathology, as was the case for the PFC hypoactivation during WCST execution in the monocygotic twin study of Weinberger et al. (1992). However, this is, to my knowledge, not the case. For these reasons not all PFC dysfunctions and not all schizophrenic symptoms can be explained as secondary to the well established pathoanatomy of the (left hemisphere) temporo-limbic system.

On the other hand, language analysis in patients with frontal lobe damage reveals that these patients rarely give inappropriate answers (Milner, 1984) as is frequently the case for schizophrenic patients, suggesting that PFC pathology per se will not explain all deficiencies in schizophrenic patients.

Finally, Weinberger (1991) assumes deficiencies in the *communicatory links* rather than structural abnormalities in the "processing stations" to be responsible for schizophrenic psychopathology, including working memory deficiences. For this reason Weinberger (1991) refers to metachromatic leukodystrophy (MLD) as a possible neurological model of schizophrenia (see above). Some animal results, however, appear to contradict Weinberger's hypothesis: Squire & Zola-Morgan (1991) report deficient delayed nonmatching-to-sample performance after lesions of hippocampus and amygdala, but not after lesions to white matter connecting lateral temporal neocortex to diencephalic and subcortical targets.

To sum: The neuropsychological and neurodevelopmental theories of Frith (1992) and Weinberger (1991) "converge" in the assumption of an impaired communication between prefrontal and temporo-limbic (hippocampal) sites in schizophrenia. Physiological abnormalities such as reduced prefrontal metabolism, are most reliably observed in task situations that "stress" working memory functions.

Chapter 3 Event-Related Potentials in Schizophrenia

Summary of chapter 3 After brief introductory statements about the definition of an ERP component, some of the literature on the Contingent Negative Variation (CNV) is reviewed. The cited studies suggest that the CNV is composed of two different components: the initital CNV, associated with contingency formation, and the terminal CNV, which presumably reflects preparation for a motor response and for stimulus evaluation. The frontal lobes are critical for the development of a CNV, but other cortical and subcortical structures are involved as well. The CNV has repeatedly been found to be diminished in schizophrenic patients, especially if they are acutely ill and not treated with neuroleptics. Although the topography of the CNV in schizophrenic patients has not been thoroughly investigated so far, some results indicate a relative frontal augmentation of the CNV in schizophrenic patients. Under experimental manipulations of the two-stimulus paradigm which include the loss of control over an aversive IS, a Post-Imperative Negative Variation (PINV) can be observed in healthy subjects. This negativity following IS and response can be observed even under standard conditions of this task in schizophrenic patients. The PINV can be highly prevalent especially in acute or early schizophrenia. Although this component seems to be associated more with schizophrenic than affective symptomatology, it is not nosologically specific. Following Callaway (1975) and Cohen (1990) PINV may reflect response uncertainty of schizophrenic patients, resulting in an inability to terminate the task even after responding. Following the reasoning that a close relationship can be assumed between the "stressing" of dysfunctional (cortical) brain regions and the occurence of neurophysiological abnormalities in schizophrenic patients, the working hypothesis is stated that the occurence of a PINV in schizophrenic patients might be related to the "stressing" of prefrontal cortex functions during the two-stimulus paradigm, since this task is an elementary form of the delay tasks which are among the best established tests of prefrontal cortex functions. Further arguments are presented that may support the notion of a relationship between PINV and the frontal lobes. The last paragraphs of this chapter give a brief survey on theories of the P300 and on P300 findings in schizophrenic patients.

The linkage between cognitive and neurophysiological processes in schizophrenic patients as described in the last chapter was based mainly on physiological or structural neuroimaging techniques such as rCBF, PET, and MRI. Other neurophysiological measurement techniques frequently used in psychiatric research are EEG and ERP. These techniques traditionally provide lower localizatory power than, e.g., MRI. While this is a clear disadvantage of traditional EEG/ERP, these techniques are clearly superior concerning the temporal resolution of the investigated processes. This advantage of EEG/ERP even over "dynamic" methods like PET or functional MRI, renders them especially suitable for the investigation of cognitive processes in psychiatric patients. Some of the ERP components that depict differences between schizophrenic patients and healthy subjects in stimulus evaluation, preparation, and response-related processes will be discussed in this chapter.

3.1 Some Theoretical and Methodological Aspects of ERP Research

Since Event-Related Potentials (ERP) have low amplitudes they are difficult to detect within the large amplitude spontaneous EEG on a single-trial basis. By repeating the experimental task several times and averaging the resulting number of trials around a marker that identifies the times when a certain event occured ("averaging technique"), systematic, i.e. event-*related*, cortical activity can be separated from non-systematic, i.e. not event-related or spontaneous, activity by enhancing the signal-to-noise ratio.

Several criteria have been proposed to differentiate ERP components. Since this is not the place for a detailed discussion of these suggestions, they will be mentioned only. Gaillard (1988) has distinguished 4 criteria. The first criterion are dependent measures such as polarity, latency, and scalp distribution. The first dependent measure is the *polarity* of the ERP component, which can be negative or positive. Since the polarity of an ERP component is reference-dependent, and since the choice of a recording reference is ambiguous, polarity is itself an ambiguous criterion. The polarity of an ERP component is regularly expressed by the letters N or P (usually assuming an earlobe or mastoid reference). The second dependent measure is the *latency* of the ERP component. The latency criterion is as ambiguous as the polarity criterion, although not for "technical" reasons, since the latencies of ERP components that reflect certain stages of cognitive processing can co-vary in a task-dependent way with the latencies of the reflected cognitive processes. The third dependent measure is the *topography* of the ERP component or the distribution of its positive and negative proportions over the scalp. By referring e.g. to the locations of positive or negative maxima, ERP components can be distinguished. However, minor differences between conditions in the topography of an ERP component must be allowed in order not to have as many components as there are experimental conditions (take for instance the topographical differences between RP topographies preceding finger versus foot movements). The second criterion according to Gaillard (1988) is the physical characteristics of the stimuli. In the present studies the experimental stimuli are visual stimuli of a certain size, brightness, and color. The third criterion for the definition of ERP components is the state of the organism or constitutional variabels. In the present experiments experimental and control groups will be distinguished on the basis of a clinical diagnosis (experiments A to C) or a personality questionnaire (experiment D). In both instances constitutional variables are assumed to differ between the groups. The fourth criterion according to Gaillard (1988) are task or behavioral variables. The present experiments use a certain kind of delay task, presented within the "framework" of the two-stimulus paradigm (including WS, ISI, IS, and response), in order to observe (or evoke) the ERP components of interest. It is assumed that this task and the task of the standard two-stimulus paradigm are but examples of the general type of delay tasks. As a consequence, the ERP components observed here are assumed to be comparable with the ERP components reported for experiments that used other examples of delay tasks.

Several methods for the identification of ERP components have been suggested, most of them relying on Principle Component Analysis (PCA) techniques. They shall not be discussed here[14].

3.2 Contingent Negative Variation (CNV)

In the so-called two-stimulus paradigm, which is a standard paradigm of exploring slow event-related potentials, a first stimulus which is called the Warning Stimulus (WS) predicts that with a certain probability (contingency) and temporal proximity (contiguity) a second stimulus, the Imperative Stimulus (IS), will occur, which is associated with a certain experimental task (motivational value of the IS), for example a button press. Walter (1964) discovered that a negative potential shift develops between the contingent WS and IS and called this potential *Contingent Negative Variation (CNV)*. While in the early studies on CNV intervals between WS and IS of about 1 s were used, subsequent investigations using *i*nter*s*timulus *i*ntervals (ISI) of 3 s or more revealed a biphasic shift of the CNV. The two different CNV components, initial and terminal CNV, can be separated by means of PCA (Lutzenberger et al., 1981). The putative processes underlying these two components will be discussed in the following paragraphs[15].

Rohrbaugh & Gaillard (1983) have argued that the complex of ERP components following the WS in a CNV task, including the iCNV, resembles the complex of late waves observed during the "Oddball" paradigm, comprising a first positive wave after about 350 ms, followed by a frontally negative and centro-parietally positive wave (slow negative wave 1, (SNW1) and slow positive

[14] Brief introductions into the neurophysiological and bio-electrical bases of ERP are given in the introductory paragraphs of *chapters 7* and *8*. Detailed accounts can be found in the monograph of Rockstroh et al. (1989).

[15] A discussion of global concepts of the CNV, using terms like distraction, arousal, motivation etc. will not be given in this chapter for reasons of conciseness.

wave (SPW)), followed by a second broadly distributed negative wave, the slow negative wave 2 (SNW2). Rohrbaugh & Gaillard (1983), acknowledging the lack of better nouns, identify this complex with an orienting response, the O-wave. Since both complexes, evoked by paired (CNV) or unpaired task stimuli, are affected by the same experimental factors, they seem to reflect the same underlying processes. These factors include the sensitivity for novelty and probability aspects of the evoking stimulus, stimulus modality, stimulus intensity, stimulus duration, and stimulation rate (Rohrbaugh & Gaillard, 1983). In addition, while the execution of a motor response interferes with the development of the O-wave especially at frontal leads, reduced CNV amplitudes can be observed when the WS has to be responded by a motor response; finally, the PCA factorial structures of the O-wave and the iCNV are similar according to Rohrbaugh & Gaillard (1983).

However, reported sensitivity of the iCNV to the information value of the WS (McCarthy & Donchin, 1978), the sensitity of the iCNV to the probability of IS occurence (Courchesne et al., 1987) and the complexity of IS discrimination (Gopher & Donchin, 1985) may argue against an identification of the iCNV with an Orienting Response. Furthermore, iCNV may be related to response variables as demonstrated by Van Boxtel et al. (1993) and Werther (1995). Instead, iCNV is now regarded as indicative of "more complex processes of evaluation of stimulus contingencies, comprising the storage and recall of conditional probabilities as well as temporal processing of the S1-S2 interval" (Rockstroh et al., 1989).

As is the case for the iCNV, the nature of the tCNV is still a matter of controversy, since the relationship between tCNV and the Readiness Potential (RP) has not been established conclusively. While some authors identify the tCNV with the RP (e.g. Rohrbaugh & Gaillard, 1983), others consider the tCNV to be comprised of an RP *plus* a *S*timulus *P*receding *N*egativity (SPN; e.g. Brunia, 1988; Rösler, 1991), or suggest the RP not to be a main part of the tCNV (Ikeda et al., 1994). Briefly, the main arguments of these 3 positions, which appear to result from different sets of defining criteria of ERP components, are as follows:

Considering that the RP reflects motor preparation, Rohrbaugh & Gaillard (1983) and Gaillard (1988) review experimental evidence pertaining to the relatedness of the tCNV to variables of motor preparation. The main results reviewed by Rohrbaugh & Gaillard (1983) are (1) tCNV amplitude reduction when no motor response on IS is required, (2) general sensitivity of the tCNV to variables that have an impact on motor preparation, (3) larger tCNV amplitudes with increased required muscular effort or speed of the response, and (4) the topographical similarities of tCNV and BP.

In line with this reasoning, some variables such as the muscular force level (Becker & Kristeva, 1980; Freude & Ullsperger, 1990), response outcome (Elbert et al., 1986), distraction (Nageishi & Shimokuchi, 1990; Rockstroh et al., 1986), or motivation (cf. Rockstroh et al., 1989) have been shown to influence both the RP and the tCNV.

If the tCNV were identical with the RP, the brain structures responsible for the RP should underlie the tCNV, too. Three different components can be distinguished for (pre-) movement related potentials: the RP, the negative slope and the motor potential. Brain structures associated with these components include the Supplementary Motor Area (SMA), the premotor cortex, MI, SI, and the cerebellum.

While rCBF studies have demonstrated that the SMA is active during any kind of movement and even during the imagination of motor sequences (Roland et al., 1987), EEG current source density analyses reveal a succession of a mesio-frontocentral and a lateral central current sink during the initiation of simple finger movements (Lang et al., 1991). Results of Schreiber et al. (1983) confirm that RP onset, which gradually becomes earlier from scribbling to drawing to writing movements, starts at leads over the SMA areas. Prior to unilateral movement onset slow magnetic field changes can be oberserved in *both* hemispheres, although with a preponderance contralateral to the movement side (Lang et al., 1991). In patients with unilateral lesions of the SMA, RP without the normal midline predominance could be observed with finger movements of either hand, confirming primate and human results of an involvement of both SMA in movements of either body side (Deecke et al., 1985).

While these studies suggest voluntary movement initiation to begin in the SMA, simultaneous scalp and subdural electrode measurements in epileptic patients suggest that a broad, vertex predominant scalp distribution of the RP can be reduced to an extremely localized subdural distribution confined to bilateral sensorimotor areas (Neshige et al., 1988); only the later negative slope component is associated with SMA and a strictly contralateral sensorimotor subdural activity, accompanied by contralateral centroparietally predominant slopes in the scalp recordings. Accordingly, Neshige et al. (1988) conclude that the RP preceding finger movements is generated in bilateral hand sensorimotor areas and the later occurring negative slope in contralateral sensorimotor and supplementary motor areas.

If the tCNV is identical with the RP, the aformentioned brain structures must be involved in tCNV generation, too. Some authors (e.g. Brunia 1988; Rösler, 1991), however, argue that tCNV is not completely reducable to motor preparation but includes preparation for the (nonmotoric) evaluation of the IS. According to this position, tCNV includes an RP as well as a stimulus preceding negativity (SPN).

Although the existence of a non-motoric anticipatory negativity has been demonstrated with the right-hemispheric predominant SPN which precedes "knowledge of result" feedback stimuli (e.g. Chwilla & Brunia, 1991; Weinberg, 1973), these results bear no implications for the tCNV composition since tCNV precedes instructional, and not feedback, stimuli.

Unfortunately, most of the studies designed to demonstrate the existence of an SPN prior to instructional stimuli have, for one or the other reason, yielded equivocal results. Some examples will be given.

Simons et al. (1979) observed a tCNV preceding interesting pictures. Although no response was required, a preparation for explorative saccadic eye movements cannot be ruled out as an explanation for the occurence of a tCNV (cf. Rohrbaugh & Gaillard, 1983).

Van Boxtel & Brunia (1990) found a negativity preceding an instructional cue stimulus which differed with the informational value of the stimulus. This instructional stimulus was preceded by a WS by 1 s. Although the WS was not experimentally varied, it already contained task relevant information due to the blockwise presentation of the experimental conditions (cf. Brunia, 1993). The negativity following the WS might have overlapped the negativity preceding the instructional stimulus. This argument may apply also to the double priming experiment 1 explained in Rösler (1991), where the experimental variations of S1 and S2 where confounded.

In the experiment of Brunia & Van Hooff (in preparation, cited after Brunia, 1993), which used a double priming paradigm similar to that of Rösler (1991), uncertainty about the quality of the instruction cue (delivered blockwise either as the first or the second cue) affected the preceding negativity only when the instruction cue came as the second cue. Since the experimental conditions were presented in a blockwise manner, possible effects of the certainty about the instruction cue may have already affected the evaluation of the preceding "diagonal" cue, rendering the effects on the negativity preceding the instructional cue a secondary phenomenon. This interpretation is supported by the fact that certainty about the instruction cue had no effect when this cue was delivered first. Furthermore, the comparison between instruction and diagonal cue when both are delivered as the second cue appear equivocal, since the larger negativity preceding the instruction cue (compared to the negativity preceding the diagonal cue) was itself preceded by a reduced positivity (centro-parietal leads) or an increased negativity (centro-frontal leads) following the preceding stimulus (see fig. 5 in Brunia, 1993).

Finally, Damen & Brunia (in preparation, cited after Brunia, 1993) found no negativity preceding an instructional stimulus that *exceeded* a negativity explainable by the preceding self-initiated movement.

Although by means of experimental variation the existence of negativities preceding instructional stimuli such as the IS in a CNV standard task has not been established unequivocally, other arguments can be given in favor of a difference between tCNV and RP. Firstly, tCNV amplitudes are generally larger than RP amplitudes (McCallum, 1986). Secondly, topographies of tCNV and RP differ. The study of Böcker & Brunia (in press, cited after Brunia, 1993) reveals, that although the scalp amplitude distributions of ERP components preceding movement onset (RP) or the IS (tCNV) may appear similar, scalp current source density fields differ. Parietal sinks could be observed only during the tCNV, but not during the RP task, suggesting differences in the underlying generator structures. This conclusion may be supported by the results of Sasaki & Gemba (1991). In their primate experiments with subdural needle electrodes the authors revealed surface negative and depth positive cortical potentials similar in shape to the RP in bilateral premotor cortices and in motor and somatosensory cortices contralateral to the moving hand, induced by the contralateral SMA. The amplitudes of these potentials varied with the muscular force associated with the movement (Hashimoto et al., 1980). No significant potentials were found in parietal and occipital regions (Sasaki & Gemba, 1991). In contrast with self-paced movements, warned movements were associated with slow surface negative/depth positive potentials bilaterally in the prefrontal and premotor cortices and the SMA as well as in the motor and sensorymotor cortices contralateral to the moving hand. These results suggest that while CNV like potentials can be observed in the PFC, no RPs are found in this region. These results may suggest that different brain structures are involved in RP versus CNV generation. Unfortunately, they do not allow a disentangling of the brain structures underlying early and late CNV components, since the interval between WS and IS was only about 1 s. In neuronal unit recordings of PFC cells during delay tasks, an elementary form of which the task of the two-stimulus paradigm is, a majority of cells exhibited a change in firing rate with presentation of the first and the second stimulus, while a subset of 50% of all investigated cells show a sustained increase (in a few cases a decrease) in firing, spanning the whole intertrial interval; delay discharge patterns could best be observed for delays of about 10 s (Fuster, 1984). Since the task structure of all delay tasks includes all defining elements of the task structure of the two-stimulus paradigm, and since *both* unit recordings (Fuster, 1984) and CNV-like cortical surface potentials (Nakamura et al., 1993) have been observed in non-human species under such conditions, it is at least plausible to assume contingent, delay-spanning activity during CNV tasks which is not reducable to the either iCNV or tCNV also in human subjects.

Still other results support a differentiation of RP and tCNV. In the case report of a patient suffering from a midbrain infarction with decussation of the superior cerebellar peduncle (Ikeda et al, 1994) the tCNV was preserved while the RP was absent. Dependency of the RP upon cerebellar functions have also been demonstrated in monkeys (Sasaki et al., 1979; Ikeda et al., 1994) and humans (Shibasaki et al., 1986; Ikeda et al., 1994).

Recordings from the human cortical surface (Walter, 1968) suggest that the frontal cortex is involved in the generation of the CNV. Zappoli et al. (1978, 1992) could observe a CNV in patients with decussations of the thalamo-prefrontal pathways.

The aforementioned arguments can be summarized as follows. Firstly, early and late components of the CNV may share some modulating experimental factors but probably have different generator structures. Secondly, the RP and the tCNV are affected by similar experimental factors. Their topographies differ, and common as well as different brain structures can be assumed to be involved in their generation. Unequivocal experimental evidence supporting the hypothesis that tCNV includes both RP and SPN has not yet been given.

The ERP complex following the WS during a CNV task, including the iCNV, bears some similarities with the ERP complex following unpaired task-relevant stimuli. The identification of the early CNV complex with an O-wave, however, does not take into account that the iCNV, which would be a component of this complex, is also sensitive to IS or response features. Therefore both orienting and contingency formation functions may underlie the iCNV.

The previous discussion of the two components of the CNV focused on the possible functional meaning of the iCNV and the tCNV, and the possible relationship between both. Inherent in this discussion is the notion that CNV is an ERP composed of two components *which both can be observed in isolation*. This position is vigorously proposed by Gaillard (1983) and refuses the existence of a "true" *contingent* negative variation which is not reducable to those components that can be evoked in isolation, but can be observed only if a task-relevant contingency between two successively stimuli exists. During delay tasks, which evoke a CNV-like slow negative potential, exactly such contingent, delay-bridging activity has been observed in PFC cells of experimental animals and with delays of several seconds duration (Fuster, 1989). This result indicates that the CNV during delay tasks cannot be reduced to two independent and non-contingent components like iCNV and tCNV (see *chapter 5*). Rather, the processing of cross-temporal contingencies characterizes the activity of the prefrontal cortex during delay tasks, which contributes to the raising of the CNV.

3.3 Contingent Negative Variation in Schizophrenic Patients

Reduced amplitudes of the Contingent Negative Variation (CNV) have been observed in a variety of geriatric, neurologic, or psychiatric diseases including schizophrenia (Timsit-Berthier et al., 1984, 1986) and must therefore be considered as nosologically non-specific (Cohen, 1991; Shagass et al., 1978). None the less, CNV in schizophrenic patients may even be smaller than CNV in other psychotic groups (Abraham et al., 1976; Timsit-Berthier et al., 1984; Van den Bosch, 1983)[16]. While CNV appears unrelated to age and sex (e.g. Rizzo et al., 1983; Roth et al., 1979; Timsit-Berthier et al., 1983), it has been demonstrated to correlate with clinical variables such as severity of psychopathology (Abraham et al., 1976; Dongier et al., 1974; Timsit-Berthier et al., 1976), and florid or positive symptomatology (Abraham & McCallum, 1977; Abraham et al., 1976, 1980; Bachneff & Engelsmann, 1983; McCallum & Abraham, 1973; Roth et al. 1979; Timsit-Berthier et al., 1984; Verhey et al., 1986). Associations with negative symptomatology have less frequently been reported, but weak correlations of vertex CNV with affective flattening, premorbid asociality, and the EPQ scales P, N, and L have been reported (Van den Bosch, 1983). Furthermore, the flatness of the amplitude gradient of the CNV

[16] Larger CNV in schizophrenic patients compared to healthy controls were reported, but could not be replicated in the same laboratory (cf. Cohen, 1990).

from Cz to Fz correlated with negative symptomatology in the study of Eikmeier et al. (1992).

It has been discussed whether the reduced CNV is a "state" or a "trait" phenomenon in schizophrenia, but results pertaining to this question are scarce and somewhat inconclusive. In support of a "state" interpretation amplitude increases of the CNV from acute to remitted stages of the disease have been reported (McCallum & Abraham, 1973). Furthermore, CNV increased with neuroleptic treatment and remission in the study of Knott et al. (1976). The same longitudinal pattern was found as a statistical trend by Rizzo et al. (1983). Tecce & Cole (1976) reported that administration of fluphenazine in schizophrenic patients with florid symptomatology increased their CNV amplitudes and alleviated florid symptoms. Such observations correspond to results of research with animals. Marczynski (1978) reported on reduced negative slow potentials over the frontal cortex in animals treated with amphetamine, an effect that was reversible by haloperidol or barbiturates (Pirch, 1978).

Since distraction may reduce CNV in healthy subjects (Tecce et al., 1978), and since heightened distractibility is a clinical feature especially in acute schizophrenia, attentional distractibility has been suspected to underlie reduced CNV in schizophrenic patients (cf. Pritchard, 1986). Reduced CNV in hallucinating compared to non-hallucinating patients (Roth et al., 1979) and increased CNV reduction due to a secondary distraction task (McCallum & Walter, 1968) may be seen to support this hypothesis.

Others, however, failed to replicate the clinical results (Abraham & McCallum, 1977; Abraham et al., 1976). Furthermore, Rizzo et al. (1984) observed during a 5 year follow-up period a decrease of CNV amplitudes in 6 schizophrenic patients developing from acute to chronified stages of their illness.

Thus, there is some evidence of an inverse longitudinal and cross-sectional association between CNV reduction and acute or florid symptomatology, possibly due to heightended distractability during acute stages of the illness. Conflicting evidence, however, has been reported and restricts the generalizability of this phenomenon.

One reason for conflicting results could be different cognitive processes required by the experimental conditions. Hence, it would be highly important to discover which of the psychological processes, to which CNV has been shown to be sensitive, is deviant in schizophrenic patients. Unfortunately, only few studies have attempted to experimentally investigate the cognitive abnormalities underlying reduced CNV amplitudes in schizophrenic patients. Experiments by Cohen and co-workers demonstrated that neither impaired time estimation during the ISI nor disregard of the task impact of the WS can account for the CNV reduction in schizophrenic patients (cf. Cohen, 1990).

Most of the studies on CNV reduction in schizophrenia used short ISI of about 1 s. Further insight into the processes underlying these CNV reduction in schizophrenic patients might arise from the differential analysis of the components of the CNV which have been shown to reflect different psychological processes. Currently, there are only a few studies which allow this differentiation.

In the studies of Verhey et al. (1984, 1986) iCNV proved especially sensitive to florid symptomatology. While both iCNV and tCNV at Cz correlated moderately with the severity of the psychiatric diagnosis, iCNV but not tCNV discriminated schizophrenic and nonschizophrenic psychotic patients, this effect being stronger for Cz than Fz (Van den Bosch, 1983). In the study of Van den Bosch (1984) the amplitude at Cz and the diminished Cz-Fz amplitude gradient of the late CNV discriminated several groups of psychiatric patients, including schizophrenic patients, and healthy controls best. Other studies also revealed better discrimination with the late, compared to the early, CNV (Conolly et al., 1983; Borst & Cohen, 1987). Werther (1995) found both iCNV and tCNV to discriminate between schizophrenic patients and healthy controls. Klein et al. (1994) observed neither significant iCNV nor significant tCNV differences between schizophrenic and healthy subjects.

Timsit-Berthier et al. (1973) and Tecce (1978) distinguished "Type-A" and "Type-B" CNV shapes. Type-A CNV is a fast-rising negativity suggestive of a well developed iCNV component which appears to be missing in Type-B CNV. Although low amplitude CNV (<5 µV) was relatively prevalent in (affective and schizophrenic) psychotic patients, distributions of Type-A versus Type-B CNV did not differ between psychotic, neurotic and healthy subjects.

The paucity of studies differentiating iCNV and tCNV does not allow for any firm conclusions. Seemingly both CNV components may be reduced in schizophrenic patients. Some authors report reduced RP amplitudes, with the RP starting earlier in schizophrenic patients than controls (Chiarenza et al. 1985; Timsit-Berthier et al., 1973; Westphal et al., 1986). Due to the assumed overlap of tCNV and RP, these results are in agreement with the reported tCNV reduction in schizophrenic patients and suggest reduced CNV in schizophrenic patients to be at least partially due to impaired motor preparation or response initiation.

Physiological rather than cognitive informations about deviancies in CNV generation may arise from comparisons of the CNV topography in schizophrenic patients and selected control groups. Due to the restriction of the electrode montage to the sagittal midline in most clinical studies, the topography of the CNV in schizophrenic patients is currently only insufficiently known. Some studies suggest a less reduced (or relatively pronounced) frontal CNV in schizophrenic patients (Klein et al., 1994; Rockstroh et al., 1993; Tecce & Cole, 1976; Van den Bosch, 1984; Wagner et al., 1993). Differences between the CNV at Cz and Fz leads discriminated several psychiatric groups (Van den Bosch, 1984). CATEGO$^+$ schizophrenic and healthy subjects were best discriminated by the CNV differences at Cz and Fz (Verhey et al., 1986).

These results suggest that beyond reduced amplitudes deviant cortical patterns of CNV generation may be present in schizophrenic patients, recommending thorough topographical investigations. Caution in interpretation is warranted if such topographical deviations amount to shallower amplitude gradients between leads, i.e. less spatial information, since this may reflect reduced signal-to-noise ratio rather than true topographical differences (Cohen, 1990).

To summarize: It is difficult to answer which of the two CNV components discriminates schizophrenic patients from controls best, since most studies used an ISI too short to differentiate the two components and since those studies that allowed their disentangling yielded inconclusive results. So far, the topography of the CNV in schizophrenic patients is not thoroughly investigated since in most studies the electrode montage was restricted to the sagittal midline. Within this limited topographical assessment a relative augmentation of the frontal CNV has been replicated in several studies.

3.4 Post-Imperative Negative Variation (PINV)

Under standard conditions of the two-stimulus paradigm no negativity can be observed in healthy subjects following the IS. If, however, the IS is an aversive auditory stimulus that can be terminated by the subject with a fast response (contingency between response and response outcome), and if subjects are unexpectedly experimentally deprived of this control, a negativity with fronto-central preponderance called Post-Imperative Negative Variation (PINV) can occur (Rockstroh et al., 1979). Such "loss of control" over an aversive IS has repeatedly been demonstrated to induce a PINV in healthy subjects (e.g. Elbert et al., 1982; Kathmann et al., 1990). With an experimental design similar to that of Rockstroh et al. (1979), Delaunoy et al. (1978) observed a PINV in healthy subjects after withdrawal of control over the termination of an *non*-aversive IS (light flashes). Interestingly, only those subjects that did not find an explanation for the loss of control (e.g. by assuming an equipment malfunction or an interference of the experimenter) developed a PINV. Referring to these "loss of control" experiments, PINV has been suggested to reflect preparation for the "appraisal of an expected contingency" (Birbaumer et al., 1986)[17].

3.5 Post-Imperative Negative Variation in Schizophrenic Patients

While in healthy subjects experimental manipulations such as the aforementioned are necessary in order to induce a PINV, schizophrenic patients exhibit a pronounced fronto-central PINV even under standard conditions of the

---

[17] There are also other conditions, e.g. hypnosis (Rizzo et al., 1980) or nitrous oxide inhalation (Timsit-Berthier et al., 1987), under which healthy subjects may develop a PINV. These conditions will not be discussed here.
A further discussion of the possible meaning of the "loss of control" experiments will be given in *chapter 9*.

two-stimulus paradigm, i.e. when a fast motor response, which terminates, for instance, light flashes or has no further consequences, is required by the IS.

Although PINV is highly prevalent in adult schizophrenic patients (see below) and may be seen in schizophrenic children (Strandburg et al., 1984), it is by no means nosologically specific to schizophrenic disorders, since PINV can be found in depressive (Timsit-Berthier et al., 1987) and neurotic patients (e.g. Idiman & Oykay, 1987) as well as psychopaths (Paty et al., 1978) under standard CNV conditions, in phobic patients after phobogenic stimuli when these were expected to occur (Dubrovski et al., 1978), and in manic patients after a discrimination task (Rizzo et al., 1979). Furthermore, some patients with traumatic head injuries (Rizzo et al., 1978) and some patients with prefrontal lobectomy (Zappoli et al., 1992) exhibit a PINV. Only a few studies did *not* find a PINV in schizophrenic patients (Roth et al., 1979; Van den Bosch, 1983).

Different ways to *quantify* PINV have been used. Some authors define the PINV as a negativity of at least 2 s duration following the IS (e.g. Chouinard et al., 1977; Dubrovsky & Dongier, 1976; Dubrovsky et al., 1976; Rizzo et al., 1979, 1980, 1983, 1984), others measured the latency until negativity after the IS crossed the pretrial baseline (e.g. Bolz & Giedke, 1981; Delaunoy et al., 1978; Rizzo et al., 1978). Most studies, however, used one or the other amplitude measure for PINV parametrization (e.g. Mantanus et al., 1981; Timsit-Berthier et al., 1987; Van den Bosch, 1983, 1984).

Furthermore, a *qualitative* PINV typology, distinguishing different PINV shapes, has been suggested (Timsit-Berthier et al., 1973). While Type I PINV is the "normal" immediate resolution of negativity following the IS, Type II-IV PINVs are different "abnormal" shapes of negativity prolongation after the IS (see **figure 3.2**). Especially Type III and Type IV PINVs were found to be most prevalent in psychotic patients, with Type III PINV being more prevalent in schizophrenic as compared to affective psychotic patients (Timsit-Berthier et al., 1973). One to two years retest-reliability of the PINV shape was sufficient only in symptomatically stable patients (Timsit-Berthier et al., 1978).

In a large sample assessment including 146 schizophrenic and 166 neurotic patients and 96 healthy control subjects, Dubrovsky & Dongier (1976) found a PINV (defined as a negativity of more than 2 s after the IS) in 75% schizophrenic patients, 37% neurotic patients, and 15% controls.

A negativity persisting for more than 2 s was reported by Chouinard et al. (1977) in 65% of schizophrenic patients, and by Rizzo et al. (1983) in 86% acutely ill and not neuroleptically treated schizophrenic patients. Idiman & Okyay (1987) found Type IV PINV in 58% of their schizophrenic patients. It is not decidable, whether such differences in prevalence rates depend upon clinical features of the patient samples or on the definition of the PINV (duration, type etc.).

Some results, however, suggest that PINV may not be equally prevalent in all clinically distinguishable subgroups, providing some hints as to the psychopathological conditions of PINV occurence. Subdividing their sample into "early" (defined by the presence of hallucinations and/or delusions for 1 to 6 months) and "chronic" schizophrenic patients, Dubrovsky & Dongier (1976) found a PINV in 94% of the early and 68% of the chronic schizophrenic patients.

While Rizzo et al. (1983) observed a PINV in 86% acute patients, only 14% of the residual or chronic schizophrenic patients exhibited a PINV. These comparisons suggest that PINV is a state variable in schizophrenic patients (cf. Pritchard, 1986).

Figure 3.2 PINV-Typology According to Timsit-Berthier et al. (1978)

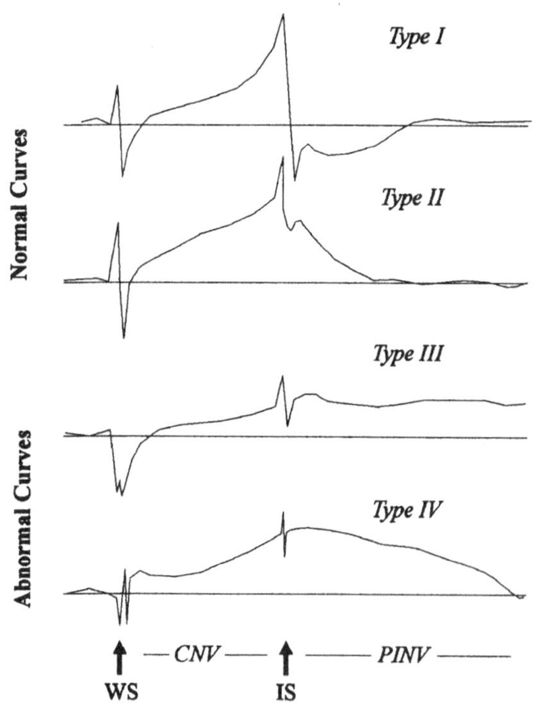

Incomprehensibility of speech (Abraham et al., 1974), ideas of reference, somaesthetic and auditory hallucinations (Bachneff & Engelsmann, 1980) were *positively* associated with PINV, but also emotional withdrawal (Chouinard et al., 1977), incongruity of affect and gloomy thoughts (Bachneff & Engelsmann, 1980). Furthermore, anxiety was *negatively* associated with PINV in the studies of Abraham et al. (1974) and Chouinard et al. (1977), as were guilt feelings, anxious depression (Chouinard et al., 1977), neurasthenic complaints, obsessive thoughts and affective and manic symptoms (Bachneff & Engelsmann, 1980).

Finally, affective flattening, social isolation, premorbid asociality (Van den Bosch, 1983) and conceptual disorganization (Chouinard et al., 1977) were *not* related to PINV.

In sum, PINV appears positively related to schizophrenic and negatively related to affective symptomatology (cf. Pritchard, 1986; Cohen, 1990) and seems to be more prevalent in acute than in chronic or residual patients. These conclusions should be read cautiously, since the studies cannot be taken as systematic clinical assessments of psychopathological correlates of the PINV. Furthermore, up to 62% of depressive patients may show a PINV (Timsit-Berthier et al., 1986) and schizophrenic and affective patients may not be separable by this component (Timsit-Berthier et al., 1984).

The topography of PINV in schizophrenic patients still awaits thorough investigation, (as does the topography of the CNV) and shall be considered in greater in detail in *chapter 8. Both* components, however, were observed with frontal and central midline recordings. This means, that at least a topographical overlap exists. Presumably due to seemingly topographical communalities, but certainly also due to CNV/PINV shapes of type III or IV, in which frequently no positivity following the IS can be observed, PINV had been identified with a "delayed CNV resolution" (Mantanus et al., 1981; Roth et al., 1977) or "'prolongated duration' CNV" (Timsit-Berthier et al., 1987). Furthermore, CNV/PINV complexes were called "CNV resolution modes" (Timsit-Berthier et al., 1973), and the PINV had been considered to be part of the CNV (Mantanus et al., 1981).

If PINV is a prolongation of the CNV -and not an independent component- pre- and post-IS negativity should be positively related. Some results indicate indeed that this may be the case. PINV and CNV correlated significantly in the studies of Mantanus et al. (1981), Rockstroh et al. (1994) and Roth et al. (1979). In depressive patients those subjects who had relatively large CNV amplitudes also had relatively large PINV amplitudes (Timsit-Berthier et al., 1987). Wagner et al. (1995) found both CNV and PINV reductions due to probe stimuli immediately preceding the WS.

Other results suggest that CNV and PINV have not much variance in common. PINV amplitude and duration were not related to tCNV peak amplitudes (Bachneff & Engelsmann, 1980), and CNV and PINV proved to be separate components on the basis of PCA (Cohen et al., 1984; Van den Bosch, 1984). Furthermore, a PINV may develop also when CNV is small (Small & Small, 1971). While CNV amplitudes increased in acute patients following neuroleptic treatment, PINV amplitudes decreased (Rizzo et al., 1983). Also, patients with Schneiderian first-rank symptoms when compared to patients without these symptoms exhibited both a diminished CNV and an augmented PINV (McCallum & Abraham, 1973).

These altogether puzzling results, obtained for rather different aspects of variation (groups, intraindividual-experimental, etc.), indicate that among the multitude of factors affecting CNV and PINV some, but not all, are common to the two components.

Since PINV is regularly preceded by the tCNV, and since tCNV and BP presumably are not identical, it may be interesting to know whether PINV occurs also after self-paced instead of warned reactions. Evidence concerning the development of a post-movement negativity is, however, conflicting.

Dubrovsky & Dongier (1976) report for the sample of Dongier et al. (1973, cited after Dubrovsky & Dongier, 1976) a *N*egativity *P*ersisting after *R*eadiness *P*otential (NPRP) but not the PINV "to be a significant feature in psychotics in evolution or remission" (p. 152). Within the sample of Dubrovsky & Dongier (1976) 17 out of 19 acute psychotics and 7 out of 18 chronic psychotics showed a PINV, but only 12 out 19, respectively 1 out of 18 patients exhibited a NPRP. Furthermore, Singh et al. (1992) found negative "prolongation" of the RP in schizophrenic patients. This held also for the patient group of the Chiarenza et al. (1985) study. In this investigation, however, a special experimental RP task was provided, which will be discussed in *chapters 6* and *9*.

While experimental conditions that induce a PINV in healthy subjects have been established and allow for an operational definition of PINV in this group, no experimental variation has been undertaken to uncover cognitive determinants of the PINV in schizophrenia. Since multiple determinants are conceivable (or established) for any ERP component, determinants of PINV in controls must not necessarily underlie PINV in schizophrenia. As a matter of fact, only speculations can be given as to the processes responsible for PINV in schizophrenia.

Dongier (1969) was, to my knowledge, the first who speculated about determinants of PINV in schizophrenic patients. He suggested to call the PINV "Conflicting-" or "C-Wave", conforming with the CNV terminology of that time ("E-" and "O-waves"). Dongier (1969) suspected PINV to reflect an interpersonal conflict due to subjective ambivalence and difficulties in fitting the response into a symbolic scheme.

Callaway suspected that "the psychotic patient may be uncertain about the results of any act he undertakes" (Callaway, 1975, p. 120; cited after Cohen, 1990). This idea was further developed by Cohen (1990) who argued: "It seems as if the patients fail to appreciate having completed their task; instead their own response seems to trigger processes similar to those known from normals when controllability of aversive stimulation is at issue." Cohen (1990) therefore hypothesized that PINV in schizophrenic patients reflects "response uncertainty" and is related to "abnormalities in the generation and monitoring of self-initiated actions", as theoretically elaborated in the neuropsychological theory of Frith (1992, see *chapter 2*).

3.6 Does the Prefrontal Cortex Contribute to the Occurence of the PINV ?

Since PINV is currently operationalized by ERP recordings with only very few recording sites, at present there is no physiological evidence that gives any hints concerning the brain structures involved in PINV generation. Nonetheless, some speculations on this topic may be given which all rest upon the assumption that the situation (or experimental task) under which an ERP can be observed reveals cues to the cognitive and brain processes involved in the *functional* (not necessarily the bio-electrical) generation of the ERP. Such tasks can, in principle, be described in different terms, for example the cognitive impact of the respective task, but also its underlying brain structures, if known. These cognitive and neurophysiological descriptions of the ERP eliciting situation then provide defining features of the ERP itself.

Concerning the areas of the brain possibly involved in the functional PINV generation, two groups of well-replicated results shall be considered. Firstly, PINV in schizophrenia under standard CNV conditions; secondly, PINV in healthy subjects after loss of control; additionally, the PINV in patients with frontal lobe lesions (Rockstroh et al., 1982, cited after Birbaumer et al. 1986).

As has been outlined above, the standard CNV task can be considered as an elementary version of the delay tasks (Fuster, 1989). The delay task performance and the development of a CNV critically depend upon the integrity of the prefrontal cortex (see *paragraphs 3.2* and *5.1*), which has been shown to be dysfunctional in schizophrenic patients (see *chapter* 2). The rCBF studies of Weinberger and collegues (e.g. Weinberger et al. 1986; see also *chapter 2*) have demonstrated that neurophysiological anomalies such as "hypofrontality" may be observable in schizophrenic patients only in task situations which in healthy subjects activate the dysfunctional brain region, e.g. the dorsolateral PFC. Berman (1987) coined the term "cortical stress test" for such tasks. Following this reasoning PINV, as a neurophysiological anomaly, may be related to the "stressing" of prefrontal cortex functions by (elements of) the CNV task.

Concerning PINV after loss of control in healthy subjects, the assumption of a PFC involvement is also plausible. Loss of control over an aversive stimulus is certainly associated with subjective stress. In animal experiments, increased dopamine turnover has been found *both* in mesocortical and mesolimbic areas after exposure to electric footshocks; in contrast, re-exposure *without* footshocks to the environment where footshocks had been experienced before increased dopamine turnover *only* in the anteromedial frontal cortex (Herman et al., 1982). These results suggest that among the stress-responsive brain systems mainly the fronto-cortical dopamine system may be sensitive to anticipated stress even in the absence of the visceral stressor itself. These results somewhat parallel the "loss of control" experiments. Here, too, PINV could not entirely be explained by the visceral stressor itself but was associated with the experience of uncontrollability (Kathmann et al., 1990; Rockstroh et al., 1979).

Timsit-Berthier et al. (1987) induced a PINV in healthy subjects by nitrous oxide inhalation. A PET study of Niethammer et al. (1982; cited after Timsit-Berthier et al., 1987) revealed that inhaled nitrous oxide was detectable 6 minutes after inhalation only in the frontal cortex, but not in other cortical areas. In addition to these selective frontal effects, unspecific effects such as concentration difficulties, altered temporal experience, distorted proprioception, and pseudo-hallucinations were present in the experimental subjects.

Further evidence for a linkage between PINV and the PFC came from studies with patients suffering from bilateral frontal lobe lesions (Rockstroh et al., 1982, cited after Birbaumer et al. 1986). These patients exhibited a PINV even under conditions of control over the aversive IS.

To summarize, it is conceivable that the occurence of a PINV is due to the "stressing" of PFC functions. Concerning the two groups of replicated results on PINV, this hypothesis includes the "stressing" of putatively deficient PFC functions of schizophrenic patients by the CNV task, as well as the "stressing" of the anteromedial stress-responsive dopaminergic system by loss of control and related experiences in healthy subjects.

In *chapters 5* and *6* experimental results pertaining to these two groups of replicated PINV findings will be presented.

3.7 P300

Although the P300 component is not of prime importance to the experiments presented in this dissertation, it might contribute to an understanding of the experimental variations of the PINV, since the P300 is associated with processes in the flow of information intake, encoding, and evaluation. Thus, if reduced CNV or PINV were a consequence of deficits in such processes, their counterpart, the P300, should be atypical as well[18].

A review of the abundant literature on P300 is beyond the scope of this chapter. None the less, a few theoretical models of P300 will be sketched, in particular the Triarchic Model by Johnson, Jr. (1986), the Context Closure Model of Verleger (1988), and the Context Updating Model of Donchin & Coles (1988).

[18] Since the term P300 is generally associated with the "Oddball" task, one may refrain from calling the positive peaks developing during a CNV task P300. These positivities will be called Late Positive Complex (LPC) in *chapters 5 to 8*. Nonetheless, the present discussion is relevant, since the P300 constitutes a major proportion of the LPC (cf. Sutton & Ruchkin, 1984) .

3.8 Models of the P300

The Triarchic Model of the P300 (Johnson, Jr., 1986) describes 3 determinants of this component: (1) subjective probability, (2) stimulus meaning, and (3) information transmission.

The term *subjective probability* refers to the relative frequency and the position of a given event within the sequential structure of events, determining subjective expectations about its occurence. The larger the reduction of uncertainty provided by a given event, the larger the amplitude of the P300, according to Johnson, Jr. (1986).

The *stimulus meaning* of an event is related to the demands required by the task and is determined by the task complexity (e.g. amount of information delivered by a stimlus), the stimulus complexity (e.g. words versus auditory stimuli), and the stimulus value (e.g. incentives). The larger the stimulus meaning, the larger the P300 amplitude.

The third determinant of the P300 amplitude within the model of Johnson, Jr. (1986) is the *information transmission*, which means the proportion of information delivered by the stimulus which is received by the subject. Information transmission is influenced by the allocation of attention and by equivocation. *Equivocation* is the amount of loss of information during the stimulus presentation due to an *aposteriori* uncertainty about having correctly perceived a given event. According to the Triarchic Model information transmission is thought to modulate the amplitude contributions of subjective probability and stimulus meaning determinants.

P300 latency reflects the amount of time needed in order to evaluate and categorize a given event, according to Johnson, Jr. (1986). Inverse relations between event probability and information transmission on the one hand and P300 latency on the other hand, as well as a proportional association between stimulus complexity and P300 latency have been postulated by Johnson, Jr, (1986).

Another interpretation of the P300 is given by Verleger (1988), who assumes the P300 to reflect "context closure" at the end of a perceptual epoch. According to this model, P300 can occur only within structured and repetitive situations which allow for the integration of stimuli into meaningful contexts. Stimuli which evoke a P300 are expected stimuli which "close" the context. According to Verleger (1988) the amplitude of the P300 depends upon the "closing" characteristics of an expected event (yes or no), the amount of effort investigated in the to-be-closed perceptual epoch (positively correlated), the amount of processing ressources required by the stimulus (negatively correlated), and the number of currently "open" epochs (positively correlated). Verleger (1988) assumes that the P300 is generated in the parietal cortex and reflects the inhibition of these tertiary zones of perceptual integration and perceptual control.

A third model of the P300 is the "Context Updating" hypothesis of Donchin & Coles (1988) who assume that the P300 reflects strategic information processing serving the planning and control of behavior. P300, which is related to stimulus evaluation but not to response execution according to Donchin & Coles (1988), reflects those control processes which are related to the maintainance of a correct representation of the environment, e.g. a scheme. Representing the environment means availability/accessibility of those parts of the total of representations which pertain to the current environment. Donchin & Coles (1988) assume that the amount of change of the model of the environment is reflected by the amplitude of the P300. This process is called "Context Updating". One part of the empirical evidence critical to the "Context Updating" model is the sensitivity of the P300 to the duration of the interstimulus interval (ISI), which Donchin & Coles (1988) assume to reflect an inverse relation between the P300 amplitude and the decay of the representation of the previous trial.

Although these 3 models of the P300 focus rather different aspects of the same phenomenon, they bear some conceptual overlap. For example, both Verleger (1988) and Donchin & Coles (1988) underline the importance of expectancy for the occurence of a P300. This aspect has been taken up by Birbaumer & Elbert (1988), who argue that the slow negative potential shifts, i.e. the CNV, preceding expected events must be taken into account. A large CNV should be followed by a large P300, according to Birbaumer & Elbert (1988).

With this theoretical background results on P300 in schizophrenia will be reviewed briefly. It should yet be mentioned in advance that only a few of the total of the respective studies combined models of information processing in schizophrenia with models of the cognitive processes underlying P300 generation.

3.9 P300 in Schizophrenic Patients

With schizophrenic patients, mainly the P300 following auditory stimuli has been investigated. A large number of studies has reported reduced P300 amplitudes in the auditory "Oddball" paradigm (e.g. Barrett et al. 1986; Blackwood et al., 1987; Duncan et al., 1987 a,b; Faux et al., 1993; Grillon et al., 1990; Holinger et al., 1992; McCarley et al., 1993; Michie et al., 1990; Levit et al., 1973; Morstyn et al., 1983; Pfefferbaum et al., 1984, 1989; Romani et al., 1987; Roth et al., 1972, 1979, 1980) but also with other tasks (Verleger & Cohen, 1978; Saitoh et al., 1984; Shagass et al., 1977).

The deficit of schizophrenic patients in P300 generation is apparently independent of task parameters such as motor reaction (e.g. Pfefferbaum et al., 1989; Roth et al., 1980) or counting (e.g. Holinger et al., 1992; Morstyn et al., 1983).

Only a few studies failed to replicate smaller P300 amplitudes in schizophrenic patients compared to healthy control subjects (Krieger et al., 1989; Roth et al., 1979). Therefore, auditory P300 reductions appear well established.

Although a number of studies has reported P300 amplitude reductions following visual stimuli (Brecher & Begleiter, 1983; Brecher et al., 1987; Koga &

Narita, 1987; Pass et al., 1980) or reduced N1-P3 peak-to-peak differences (Levit et al., 1973; Verleger & Cohen, 1978), only trends towards group differences (Pfefferbaum et al., 1984, 1989) as well as nonsignificant results (Wagner et al., 1989) have been reported.

Following Duncan et al. (1987) different amounts of state-dependency of auditory versus visual P300 might account for the divergency of results. Duncan et al. (1987) observed in neuroleptically responsive, but not in non-responsive, schizophrenic patients visual P300 amplitude increases and peak latency reductions during the course of clinical improvement. While visual P300 amplitude increases were correlated by r=-.88 with clinical improvement, no such correlation was found for the auditory P300 (r=.13). Unfortunately, these results can only be taken as preliminary, since only 7 patients participated in the study.

In contrast to the P300 amplitude, the P300 latency did not reveal consistent differences between schizophrenic patients and healthy control subjects. This holds even for the auditory modality, for which both significant (Blackwood et al., 1987; Duncan et al. 1987; Faux et al., 1993; Romani et al., 1987; Roth et al., 1979) and nonsignificant (Faux et al. 1988; Levit et al., 1973; McCarley et al., 1993; Pfefferbaum et al., 1989; Roth et al., 1980) results have been reported. Duncan et al. (1987) reported significantly increased P300 peak latencies only under the condition of minimum relative frequency of the target stimulus (90:10 versus 30:70 or 50:50). Visual P300 peak latency has been investigated less than its auditory counterpart, but yielded likewise inconsistent results (Levit et al., 1973; Pfefferbaum et al., 1984; Wagner et al., 1989).

In general, group differences between schizophrenic patients and healthy controls appear larger following target than following standard stimuli, both concerning the auditory (Barrett et al., 1986; Duncan et al., 1987; Roth et al., 1980; Saitoh et al., 1984) and the visual (Pass et al., 1980) modality. In addition, distraction effects on P300 were larger in schizophrenic patients than controls (Grillon et al., 1990).

An intensified speed instruction or an increased discrimination of visual stimuli resulted in a less than normal increase in P300 amplitudes in schizophrenic patients (Brecher et al., 1987).

The critical question of possible influences of the neuroleptic medication has been investigated using several approaches. No P300 differences have been found repeatedly between medicated and non-medicated schizophrenic patients (Pass et al. 1980; Saitoh et al., 1984; Shagass et al., 1986). Furthermore, no intraindividual effects of neuroleptics could be observed after 1-4 weeks and 6-24 months (Blackwood et al., 1987), and reduced P300 amplitudes were found in neuroleptic "naive" patients and patients free of medication for about 6 months (Connolly et al. 1985; Kameyama et al., 1987; Michie et al., 1990; Ogura et al., 1987).

If neuroleptics have any effect on P300 amplitudes, it appears to be an increasing one (Duncan et al., 1987; Saitoh et al., 1984).

Since multiple generators of the P300 have been assumed (Johnson, 1993), topographical studies may be especially important both concerning the differentiation between schizophrenic patients and controls and the differentiation between the auditory and the visual modality. However, studies on P300 topography in schizophrenic patients are rather heterogeneous concerning results and methods. A common feature of all studies cited in the following paragraphs is the use of a linked ears or mastoids reference.

Reduced parieto-frontal auditory P300 gradients in schizophrenic patients were reported by Grillon et al. (1990), while auditory and visual P300 reductions were especially pronounced frontally in the study of Pfefferbaum et al. (1984). Also, Raine et al. (1990) mention reduced P300 amplitudes only for the F3 and F4 leads.

More consistent results have been reported concerning maximum P300 amplitude reductions at leads located above the left-temporal region following auditory stimuli (Faux et al., 1988, 1993; Morstyn et al., 1983) or linguistic stimuli of different sensory modalities (Koga & Narita, 1987).

Other studies found no topographical differences between schizophrenic patients and healthy subjects (Pfefferbaum et al., 1989). As to other studies not reporting topographical differences, it is not clear whether such differences had been investigated at all (Brecher et al., 1987; Shagass et al., 1977).

Altogether, left-temporal differences have more consistently been observed than other topographical results. The assumption that structural temporo-limbic anomalies in schizophrenia cause reduced auditory P300 amplitudes has seemingly been supported by the results of McCarley et al. (1993). In this study significant positive correlations ($r=.55$) were found between P300 amplitudes at T3 and left superior temporal gyrus volumes. In this study, however, *even larger* correlations were found between several right-hemisphere P300 amplitudes and left superior *frontal* gyrus reductions ($.66<r<.69$) which correlated significantly also with the right-hemisphere MRI results ($.54<r<.58$)[19]. It should also be noted that for biophysical reasons the hippocampus cannot play a major role in the P300 generation (Braun et al., 1990).

Clinical correlates of the P300 reduction in schizophrenic patients are currently underinvestigated. Correlations with thought disorders have repeatedly been reported (Morstyn et al., 1983; Romani et al., 1987; Roth et al., 1979, 1980; Wagner et al., 1989).

Although in the present experiments no "oddball" task was provided, the P300 results surveyed so far are pertinent since the P300 can be considered a main part of the *L*ate *P*ositive *C*omplex (LPC; Sutton & Ruchkin, 1984) evoked by the WS and IS.

---

[19] The authors discarded the latter group of results from the further discussion by a disproportionately rigorous statistical argument.

Chapter 4 The Basic Experimental Design and Methodological Communalities of the Reported Experiments

4.1 The Basic Experimental Procedures

This dissertation comprises 4 experiments, A to D, which are all based on a single basic experimental design with one or more modifications of its features. These modifications will be described in this chapter. The theoretical background will *not* be discussed here, but in the respective experimental chapters (*chapters 5 to 8*).

The basic design is illustrated in **figure 4.1**. As Warning Stimulus (WS) two horizontally arranged green diamonds of clearly distinguishable sizes were presented on a PC monitor. The lateral arrangements (small diamond on the left, right diamond on the right, or the reverse) varied pseudo-randomly from trial to trial, equally distributed between both arrangements. The geometrical centre of both WS diamonds lay on the horizontal central axis of the monitor and on a vertical axis. The diamonds were equally distant from the vertical central axis of the monitor. The small diamond covered 6 x 3 cm, the large diamond 12 x 6 cm. Each of the two WS combinations was presented for 100 ms. With a Stimulus Onset Asynchrony (SOA) of 4.0 s a single green diamond was presented at the centre of the monitor for 100 ms as Imperative Stimulus. For one-third of the trials the IS diamond was identical with the small WS diamond and for another third it was identical with the large WS diamond. For the remaining one-third of the trials the IS diamond was of intermediate size. This was the size of the small WS diamond plus 40, 50 or 60% of the difference between the two WS diamonds. Subjects were instructed to attach the IS diamond to one of the WS diamonds by pressing the response button of the left or the right hand.

Modifications of this basic design included:

1. Variability/ stability of the arrangement of the WS diamonds (experiment A)

Within the two sessions of experiment A the variability of the arrangement of the WS diamonds was changed randomly. While the horizontal arrangement varied from trial to trial during one session, it remained constant (e.g. small diamond always left, large diamond always right) during the other session (counterbalanced).

2. Go/Nogo stimuli (experiment B)

In experiment B a third additional type of IS was included. During 23.5% of the trials a red ellipse was presented in the middle of the screen for 100 ms. (The remaining IS types occurred during 23.5% of the trials (small diamond, large diamond) or during 29% of the trials (intermediate size diamond)).

3. Delivery of acoustic "probe stimuli" (experiments A and B)
In addition to the visual WS and IS acoustic "probe" stimuli were presented in experiments A and B via headphones. In experiment A a probe click was presented continually every second, starting with a delay of 0.5 s relative to the WS onset. In experiment B two probe tones per trial were presented during the pretrial-baseline, iCNV-, tCNV-, or PINV-segment. The 6 possible pairwise combinations occured with equal probability. Details of the probe stimulus presentation will be given in *chapter 7*. In both experiments probe stimuli were task-irrelevant (avoiding a dual task situation) and were introduced as such to the subjects.

In addition, the experiments differed in the number of trials. While *each* of the four experiments began with 20 practice trials of the basic design type (varying WS arrangements, no intermediate size IS, no probe stimuli), the number of trials for the experiments A to D varied. Experiment A comprised two experimental sessions with 148 trials each. The experiments B, C, and D had 204, 205, and 205 trials, respectively.

4.2 The Laboratories

Experiments A to C were conducted in the psychophysiological laboratory of the University of Konstanz in the local state hospital Reichenau (PLK), which is situated adjacent to the research ward of the university. The laboratory comprises a separate subject cabin and 3 control rooms, which are equipped with six interconnected 386 PCs, including one server, during the conduction period of experiments A to C. Subjects were tested in an electrically shielded sound-attenuated 3 by 2 m cabin. The windows of the cabin were closed by venetian blinds. A PC monitor was situated at the back and a comfortable armchair in the front half of the cabin. The EEG connection box, including preamplifiers, was located behind the armchair. The EEG amplifier (MES, Munich) was located outside the cabin.

Experimental stimuli were controlled by the hardware/software package NEUROSTIM (NEUROSCAN), which was installed on a 386 PC. Two VGA monitors were connected with the stimulation PC, one in the subject cabin, one in the experimenter room.

On a second 386 PC in the experimenter room the recording software was installed. The program DCMES (MES, Munich) displayed the biosignals on the monitor and stored them for off-line analyses. Software as well as hardware parameters such as digitization rate, upper and lower cut-off frequencies, amplifier gain, etc. could be separately defined within a DCMES initialization file for each experiment.

Two response pads with one response button each were provided in order to register the subject's responses. These pads were approx. 10 x 5 x 4 cm. Subjects could easily grip a response pad with one hand and use the index finger for responding. When a response button was pressed, a signal was sent to the recording system. Here, one of the two bits of the event marker channel reserved for responses was set. The DCMES event channel received additional input from NEUROSTIM (see below).

Experiment D was conducted in the laboratory of the Department of Psychology at the University of Konstanz. All software and hardware components were identical in the two laboratories.

4.3 The Recruitment of Experimental Subjects

4.3.1 Schizophrenic Patients

Schizophrenic patients were asked to participate only if the responsible clinical psychologist or psychiatrist deemed them capable of participating in the experiment. Although it may be advisable to select only male subjects for the participation in psychophysiological experiments in order to keep within-subject variance small, generalizability problems with this method are obvious. As Cohen (1990) pointed out, much of the generalizability of research on P300 in schizophrenia is questionable since most of the P300 research has been carried out with males. For this reason, a balanced proportion of male and female subjects was provided for the experiments with schizophrenic patients[20].

4.3.2 Healthy Control Subjects

Healthy control subjects were recruited always with a certain delay relative to the patient sample in order to allow matching for sex, age, and educational grade (given the lower eductional grade of many schizophrenic patients). Control subjects were recruited, e.g., from the university library, from shops, etc.

4.4 The Assessment of the Psychopathological State in Schizophrenic Patients

Assessments of current psychopathology and medication were done for descriptive purposes only. A differential or correlative investigation of possible

[20] Even if male and female subjects participate, EEG research is certainly still based on a highly selective subject sampling. For many disease-related reasons (e.g. motor disturbances, impaired comprehension of the instruction, paranoid-hallucinatory apprehension of the testing situation (e.g. thought insertion or mind reading by means of the EEG), and motivation) schizophrenic patients will refuse participating in an EEG experiment. Some of those who participate will be excluded afterwards because they produced, for other more or less disease-related reasons, too many artifacts. The resulting sampling bias is certainly not confined to EEG research and should at least be taken into consideration when inferential statistical results are interpreted.

relations between these variables and the ERP, although interesting from a theroretical point of view, does not appear fruitful within the present experimental approach. Firstly, any correlational analyses would require sample sizes well beyond the possibilities of the time-consuming ERP research. Secondly, a correlational approach would only benefit from sample compositions which are more heterogeneous than the available. Furthermore, if such analyses are undertaken in order to explore medicational *effects* on ERP measures, the medicational treatment must be randomly assigned. Otherwise, the possibility that a common factor (e.g. the presence of hallucinations), accounting both for the medicational dose and the magnitude of the ERP variable, is responsible for a given correlation between the dependent variables cannot be ruled out. Random (drug) treatment assignment is in most instances certainly not possible, and medicational variance therefore includes uncontrolled (presumably) clinical variance.

For these reasons, psychopathological assessment was confined to the Brief Psychiatric Rating Scale (BPRS, Overall & Gorham, 1962; Lukoff et al., 1986). The BPRS includes a variety of psychopathological aspects and its total score can be considered an overall index of the degree of psychopathology.

In addition to the BPRS the medication on the day of testing was expressed as chlorpromazine equivalents (Jahn & Mussgay, 1989).

4.5 The Testing Procedure

4.5.1 Having the Subjects Become Acquainted With the Laboratory, Electrode Application, and Some Questionnaire Data

After having recruited a subject for an experiment (which included a brief description of the experiment and its purpose) a date (or two, in case of experiment A) was fixed for the experimental session. All patients received a note with their appointment date(s).

Each session began with a short introduction to the equipment of the experimenter room and the subject cabin. All subjects were told that the experimenter would be present in the experimenter room throughout the session. Then subjects were led into the cabin and were seated in the armchair. Sitting in front of the subjects, the experimenter explained the investigation. In particular, the manner in which electrodes would be affixed (determination of the electrode positions, cleansing of the skin, etc.) and the global scheme of task and resting blocks was explained. Each subject was encouraged to "complain" or to ask questions throughout the session, if desired. Furthermore, all subjects were told that they had the opportunity to terminate the session at any time if necessary. Especially during this initial talk subjects, and in particular schizophrenic patients, were given the opportunity to express reservations or anxiety concerning the EEG measurement.

Following this, all subjects were asked to sign an agreement concerning their participation and the statistical analysis of their biosignals. Only one schizophrenic patient refused to sign, leading to the termination of the session.

EEG positions were measured and marked on the head with a wax crayon. Since electrode application procedures differed between the experiments A and B on the one hand and experiments C and D on the other hand, both procedures will be described separately.

During experiments A and B the measurement of the electrode positions was done by hand. First, Cz was determined as the centre between nasion and inion and the two preauricular points, as described in Lutzenberger et al. (1985). After that the remaining positions were determined according to the international 10-20-System. For experiment A positions F3, F4, F7, F8, C3, C4, T3, T4, P3, and P4 and for experiment B the positions F3, Fz, F4, C3, C4, P3, Pz, P4, and FCz as common recording reference were marked. Cz was the common recording reference in experiment A. In both experiments the mastoid leads (A1, A2) were applied. In addition, in experiment A and B the EOG was recorded from the outer canthi (horizontal EOG) and at positions 1-2 cm above or below the left eye (vertical EOG). In all experiments a ground electrode was placed at the forehead.

Since experiments C and D were multichannel recordings with 32 electrodes[21] (plus recording reference), during these experiments most of the electrode positions were determined by means of an "empty" electrode cap (Electrocap). These caps do not have integrated electrodes, but 5-6 mm large wholes at the appropriate positions. Therefore, they may be used as a stencil for marking the electrode positions. The Cz location was measured as described above and the cap was adjusted to ensure the symmetrical placing of positions Fp1 and Fp2. After that the electrode positions were marked by a wax crayon and the cap was removed. This procedure took 2-3 minutes.

The following steps of affixing the electrodes were identical for all 4 experiments.

A cotton stick and an abrasive paste (OMNIPREP) were used for cleansing the skin, and the electrolyte (GRASS EC2 electrode paste) was rubbed in the skin. Then Ag-AgCl sintered electrodes (ZAK) were affixed with additional electrode paste and stabilized with LEUKOSILK adhesive tape. This procedure initially took about 30 minutes for the experiments A and B (18, respectively 16 electrodes) and 45 minutes for the experiments C and D (34 electrodes). After a few sessions 15-20 minutes for experiments A and B, and less than 30 minutes for experiments C and D sufficed. In all cases electrode impedances were kept below 5 kΩ.

Finally, the experimenter checked the spontaneous EEG. Electrode drifts at the beginning of an EEG session may sometimes occur due to electrochemical changes and are of a transient nature. If drifts persisted for more than 5 to 10 minutes the electrodes were replaced or reattached.

[21] See paragraph 8.8 for electrode positions.

Prior to the task information about age, education, smoking habits, and handedness were gathered.

4.5.2 Recording of Defined Eye Movements of Large Amplitude

While in experiment A only eye blinks were corrected during off-line analyses, more elaborate correction methods for eye blinks, horizontal and vertical eye movements were applied in the experiments B, C, and D. For this purpose, prior to the presentation of the experimental tasks all subjects had to execute defined and large-amplitude eye movements and "normal" eye blinks. The further processing of these eye movements will be described in the eye artifact correction sections. By means of a NEUROSTIM program, a red arrow was presented in the middle of the screen which pointed either to the right, to the left, upwards, or downwards. At appropriate positions on the cabin wall green spots were affixed as target points for the execution of the horizontal and vertical saccades. Each 1.5 s the red arrow disappeared from the screen for 200 ms. Subjects were instructed to execute a fast eye movement to the green spot to which the arrow had pointed and back again to the arrow, briefly after each re-appearance of the arrow. In this way 20 large-amplitude saccades subtending an angle of about 30 degrees were executed in each of the 4 directions. Finally, 20 "natural" eye blinks were executed, timed by the regular appearance and disappearance of a schematically drawn eye presented on the screen. Subjects were instructed not to move their head and to try to synchronize eye movements as accurately as possible. This procedure was regularly finished within 5 minutes.

4.5.3 The Experimental Instruction

The experimental instruction is certainly one of the crucial components of each psychophysiological investigation since it has an impact on the behavior of the subject and may be a source of a difficult-to-control experimenter bias. Therefore, a standardized deliberate presentation of the instruction, *identical* for all subjects, is generally necessary. A written instruction may fulfill these requirements.

However, in case of experiments with schizophrenic patients a written instruction appears for several reasons neither possible, nor advisable, and individual oral presentations appear superior, especially if difficult tasks are provided. "Not possible" since many schizophrenic patients have difficulties in understanding the instruction and ask questions after reading the instruction which must be answered individually—and therefore, in a non-standardized manner. "Not advisable" since difficulties in understanding the instruction or even misunderstandings may not necessarily be expressed by the patients. Resulting "false habits" may prove extremely difficult to correct.

Therefore, an oral presentation of the instruction must be favored. Thus, difficulties in understanding the instruction may be detected and corrected readily.

During the experiments A, C, and D, which the author conducted, the experimental instructions were presented orally. During experiment B, which was conducted by a diploma student, the instruction was presented in a written form and the subjects had the opportunity to ask questions afterwards.

The instruction for experiments A, B, C, and D read as follows:
"You will see two pictures on the screen again and again, a first one and a second one. The first picture is composed of two diamonds. Diamonds are geometrical shapes. One diamond is small and the other diamond is large. Sometimes the small diamond is on the left, sometimes it is on the right side of the screen. The large diamond is sometimes on the right and sometimes on the left side of the screen. Hence, both diamonds occur simultaneously but change their positions from trial to trial. - This is the first image. - The second image, which appears some seconds after the first, is composed of a single diamond in the middle of the screen. This single diamond is identical either with the large diamond or with the small diamond of the first picture. Hence, it is identical either with the left or with the right diamond of the first picture. - It is your task to match the second, single diamond with the identical diamond of the first picture as soon as you see it. You can do this by pressing a reponse button with the left or with the right hand as fast and accurately as possible."

Only during session A of experiment A instead of the sentences 5 and 6 the following sentence was spoken: *"The small diamond is always on the left, the large diamond always on the right of the screen."* Sentence 7 was skipped.

After having presented the instruction the experimenter ascertained that the subject had understood the instruction by verbally presenting one or two stimulus combinations and asking the subject for his/her response. With some schizophrenic patients it was necessary to exercise the task with the aid of graphical representations of the task structure.

After that all subjects were asked not move head and eyes and to fixate a small fixation cross in the middle of the screen; furthermore, to respond as fast and accurately as possible.

Finally, the experimenter left the cabin and started the program with the first part of the experiment.

This part of the experiment lasted generally no more than about 5 minutes, with some patients 7-8 minutes.

4.5.4 Practice Trials

All experiments began with the presentation of 20 practice trials without IS diamonds of the intermediate category and without probe stimuli. Furthermore, the ellipse ISs of experiment B were also not presented during these trials.
After completion of the practice trials subjects had the opportunity to ask questions and the experimenter gave detailed feedback about the subject's artifact-relevant and response behavior. Most subjects, including patients, were able to reduce eye movements and to respond faster after each feedback.

With a few patients, a repetition of instruction and practice trials was necessary when responses during the practice trials were wrong. With one patient even this measure proved unsuccessful and the recording was terminated. For the execution of this part of the experiment about 5 minutes were needed.

4.5.5 Experimental Trials

The total of experimental trials of all experiments, comprising 148 to 205 trials, was subdivided into 3 parts in order to provided sufficient refreshing pauses. If desired subjects received some cakes and mineral water or orange juice. Consumption of coffee or cigarettes was not allowed. During the breaks each subject received a detailed feedback about his past artifact-relevant and response behavior. At the end of each break all subjects were reminded not to move head or eyes and to response as fastly and accurately as possible.

The duration of experimental trials was 30 minutes in experiment A and 41 minutes in experiments B, C, and D, plus 2 breaks of at most 5 minutes.

4.5.6 Final Steps

In experiments A and B after finishing the experimental task electrodes were removed immediately. In experiments C and D, the electrode cables were unplugged and the subject was led into the experimenter room. Here, a Digitizer (Polhemus ISOTRAK) for measurement of the exact 3-dimensional electrode locations was situated. For this purpose the subject had to sit down on a chair and a device on which the source of an electromagnetic field was mounted was brought into a nearby and stable position relative to the subject's head. In a fixed order guided by a computer program, the centres of the upper sides of all electrodes were touched with an electromagnetic sensor. Pressing a mouse button made the computer program record the sensor position. After this procedure, which needed about 3-4 minutes, each subject was given the opportunity to remove remnants of the electrode paste from face and hair. Finally, all subjects received their fee and the session was terminated.

4.6 Presentation of Experimental Stimuli and Physiological Recordings

4.6.1 Amplifier Settings

In **Table 4.1** the amplifier settings used throughout the 4 experiments are presented.

Table 4.1 Amplifier Settings in Experiments A-D

Sampling Frequency	100 Hz
Upper cut-off frequency	30 Hz
Lower cut-off frequency	DC
6 dB/octave; 5 bins/µV	

4.7 Data Reduction and Statistical Analysis

The following paragraphs will provided a synopsis of the data reduction and statistical analysis procedures used in experiments A to D. Although for comparison purposes these procedures were kept as similar as possible, experimental particularities as well as software progress sometimes led to different procedures.

As already mentioned, different procedures were applied for eye artifact treatment.

Parametrization intervals of the PINV could not always be held constant between experiments since in some experiements (A and B) intervals containing acoustical probe-evoked potentials had to be excluded. Since in all experiments at least two successive intervals were chosen, and since most experimental effects were identical for these intervals, the comparability problem resultig from the use of different intervals is probably minor.

Finally, while for experiment A a linked-mastoid reference was chosen, increasing awareness of topographical issues shifted preferences to the average reference.

4.7.1 Correction of DC Shifts

The initial data processing step was always the correction for DC-shifts with the program DCAUS (MES, Munich). For each channel a fourth-order polynomial was fitted to the DC levels during a 1 second pretrial "baseline". Then all data points were subtracted from this fit (cf. Bötzel et al., 1993).

4.7.2 Further Data Processing Steps With the KN-Programs

All further steps in data processing preceding the statistical analyses will at first be presented in a tabular manner in **table 4.2.**

Data Transformation and Handling of Event Marker DCTORAW transforms files from DCMES data format into the Konstanz data format, creating a new data file containing the sequence of single trials. These trials are defined with respect to one or more event marker in the DCMES file in the runstring of the DCTORAW statement. In this runstring other parameters have to be specified as well, for example the subject number and the numerical resolution of the signal in bins/µV.

In all 4 experiments trials of 9 s duration were defined with respect to the IS marker. A pretrial-baseline of 1 s, 4 s Inter-Stimulus Interval (ISI), and 4 s PINV interval were provided.

BASE performs an initial baseline subtraction. For all experiments reported here, the first 10 data points (100 ms) were chosen for this baseline.

The program CHREIZ identifies experimental conditions by reading the configuration of the event marker of each trial. A condition number is written into the trial header of each trial. Different versions of CHREIZ were used in the experiments since different marker combinations were present in the different experiments. In experiments A, C, and D the single trials were categorized depending on the combination of WS-, IS-, and response markers. In experiment B, an additional categorization of the trials according to the above criteria *plus* a categorization according to the combination of acoustic probes presented during each trial was provided. Furthermore, all versions of CHREIZ computed mean reaction times for each differentiated condition.

AVEREF computes an average reference (the data points of each channel are subtracted from the mean of the data points of all channels), thereby creating a new data file. With the average reference an additional channel is created, and the recording reference (Cz in experiment A, Fcz and AFz in experiments B-D) is added to the other channels.

REREF references the average referenced data to a new reference which can be chosen arbitrarily. For all experiments the average mastoid reference was chosen for ERP parametrization. This reference was also used for the grand average curves in experiments A and B, but maps for the multi-channel data of experiments C and D used the average reference.

Table 4.2 Sequence of Applied KN-Programs

	Experiment A	B	C	D	Function
Data Transformation and Event Marker-Handling					
DCTORAW	+	+	*	*	Transformation from DCMES to KN format
BASE	+	+	*	*	Baseline Subtraction
CHREIZ	#	#	#	#	Condition-dependent categorization of single trials
AVEREF	+	+	*	*	Transformation from the recording to the average reference
REREF	+	-	-	-	Transformation from a given reference to a new reference (e.g. linked mastoids)[22]
Control of artifacts					
ARTIFACT	+	-	-	-	Detection of artifacts within single trials
BLINKREG	+	-	-	-	Computation of the blink transmission coefficients
REGREYE	-	+	-	-	Computation of the transmission coefficients for predefined eye movements (e.g. saccades)
CORREYE	-	+	-	-	eye activity correction
DOPCAKN	-	-	+	+	PCA computation of eye movment topography
3EOGCE	-	-	+	+	eye artifact correction (MSEC algorithm)
Data Inspection					
LOOK	+	+	*	*	Single-trial inspection
Averaging and Filtering					
AVERAGE	+'	+	*	*	Averaging (condition-wise)
CONDMEAN	+	+	*	*	Averaging (condition-pooling)
FILTER	+	+	*	*	High-pass and low-pass filtering
Interpolation					
ELINTERP	-	-	+	+	Interpolation of topographical data (e.g. from measured to standard electrode positions)
Parametrization					
AREARNGE	+	-	-	-	Computation of mean amplitudes of two non-adjacent intervals
AREANOTX	-	+	+	+	Computation of mean amplitudes
MINMLATS	+	+	-	-	Computation of peak amplitudes and latencies

+ = 286 PC version of a program; * = 386 PC version of a program; # =experiment-specific version of a program; - = program not applied; +' application includes eye activity correction in that case

Control of Artifacts The different methods used for artifact detection and/or correction will be described in detail in *paragraph 4.7.3*. They will be mentioned

[22] Only average reference data can be re-referenced.

here only briefly. In experiments A and B conventional regression analytic methods were used for the correction of EOG artifacts. In experiment A blinks and in experiment B blinks, vertical, and horizontal saccades were corrected. In experiments C and D the *M*ultiple *S*ource *E*ye *C*orrection (MSEC) method by Berg & Scherg (1994) was used in order to correct blinks, horizontal, and vertical eye movements.

Visual Inspection of the EEG data After the eye activity correction all single trials were visually inspected in order to evaluate the quality of the correction in a given trial. This inspection was done within the LOOK program. Single trials still containing eye movements, drifts or other artifacts were excluded from further analysis.

Averaging and Filtering The artifact-corrected trials were averaged with the program AVERAGE. Condition-specific averaging depended on the experiment, as described below. All negative potential shifts were lowpass-filtered with 7 Hz.

Experiment A Initially, each single-subject average differentiated 9 conditions for session A (1 WS-combination *by* 3 IS types *by* 3 response types (left hand, right hand, and no response)) and 18 conditions for session B (2 WS-combinations *by* 3 IS types *by* 3 response types each). These conditions were merged into 2 final conditions in the following manner. Firstly, all trials with false or missing responses (determined by CHREIZ) were skipped; secondly all conditions which differed exclusively in the responding hand were pooled. Thirdly, those conditions were pooled which differed only in the arrangement of the WS diamonds. The result was 2 conditions, clear versus ambiguous, for sessions A and B, respectively, which did not differentiate WS combinations (only for session B) and responding hand (for sessions A and B).

Experiment B Initially, each single-subject average differentiated 24 conditions (2 WS-combinations *by* 4 IS types *by* 3 response types (left hand, right hand, and no response)). These 24 conditions were then merged into 3 final conditions in the following manner. Firstly, all trials with false, missing or too many responses (determined by CHREIZ) were skipped; secondly all conditions which differed exclusively in the responding hand were pooled. Thirdly, those conditions were pooled which differed only in the arrangement of the WS diamonds. The result was 3 conditions which did not differentiate WS combinations and responding hands, but distinguished the IS type (clear, ambiguous, and nogo).

Experiments C and D Initially, each single-subject average differentiated 18 conditions (2 WS-combinations *by* 3 IS types *by* 3 response types). These 18 conditions were merged into 2 final conditions in the following manner. Firstly, all trials with false or missing responses (determined by CHREIZ) were skipped; secondly all conditions which differed exclusively in the responding hand were pooled. Thirdly, those conditions were pooled which differed only in the arrangement of the WS diamonds. The result was 2 conditions, clear versus ambiguous, which did not differentiate WS combinations and response hand.

Table 4.3 ***Parametrization Intervals (in seconds) for the Components Analyzed in Experiments A-D+***

Negative Potential Shifts

	iCNV*	tCNV*	PINV**
Experiment A	1.0-1.5	3.0-4.0	1.0-2.0; 2.0-3.0
Experiment B	-	3.0-3.5	1.0-1.5; 2.0-2.5; 2.5-3.0
Experiment C	-	3.0-3.5	1.0-1.5; 1.5-2.0; 2.0-2.5; 2.5-3.0
Experiment D	-	3.0-3.5	1.0-1.5; 1.5-2.0; 2.0-2.5; 2.5-3.0

* = relativ to WS onset; ** = relative to IS onset

Positive Potential Shifts

	WS-LPC amplitude	IS-LPC amplitude
Experiment A***	0.26-0.30	0.36-0.51
Experiment B****	-	-

* = relativ to WS onset; ** = relative to IS onset
*** = see *paragraph 5.8* for a detailed description;
**** = see *paragraph 6.8* for a detailed descritpion
+=The parametrization of probe EP is described in *paragraph 7.5*

Parametrization The dependent variables of the slow negative waves to be statistically analyzed were computed as mean amplitude measures (Donchin & Hefley, 1978) for intervals described in **table 4.3.** No PINV interval overlaps with the Late Positive Complex (LPC; see Elbert et al., 1982) or with reaction times (see *chapters 5*, *6*, and *8*). Parametrization was done with the program AREANOTX.

In addition to these slow negative potential components the LPC following the WS and/or the IS was analyzed in experiments A and B. Peak or mean amplitude measures (which yielded almost identical results) were determined according to condition and group specific peak latencies of the respective LPC over parietal channels of the grand averages, which were determined by the "cursor-"option of the program LOOK. In case of peak amplitudes and latencies determined by the MINMLATS program, windows for peak detection were set -200 ms to +200 ms relative to the parietal peak latency. Within these windows channel specific peak amplitudes and peak latencies were determined. In the case of mean amplitude parameters (which were computed mainly for comparison purposes with the AREANOTX program) parametrization intervals were determined by the parietal channel latency (-200 ms to + 200 ms). The tCNV was used as the baseline for the PINV parametrization in the "experimental" parts (*chapters 5* and *6*). ANOVA results for the 1.0-pretrial and/or IS-LPC-peak baseline can be found in **appendix A.**

4.7.3 Correction of Artifacts

A fundamental methodological problem of psychophysiological measurements is the treatment of artifacts. The strategy of out-selecting artifact contaminated trials might be practicable in studies with healthy subjects only. For at least two reasons it is not applicable to the investigation of schizophrenic patients. Firstly, due to the high prevalence of artifacts a very large number of trials would be needed in order to obtain a sufficiently high signal to noise-ratio. Secondly, the occurrence of some artifacts such as blinks or saccades may be expected to be related to the current state of a patient. Excluding the respective trials would amount to a selection bias.

For these reasons artifact *correction* is indispensible in studying ERPs in schizophrenic patients. Depending on method-related "pre-requisites" (such as the number of EEG channels) and the availability of correction programs different methods were applied for the experiments decribed here. These programs attempted to correct for EOG artifacts, while other types of artifacts could not be corrected.

Experiment A The eye activity correction in experiment A was done by the combined ARTIFACT, BLINKREG, and AVERAGE programs. The regression analytic approach described by Berg (1986) was applied. At first the single trials were categorized according to the occurrence of blinks, complex signals (due to, e.g., double blinks or eyelid flutter), DC shifts beyond the amplifier range, and drifts. For those trials identified as blink contaminated only, the BLINKREG program computed the EOG-EEG transmission coefficients of each channel. These coefficients were finally applied by the eye correction option of the AVERAGE program.

Experiment B In this experiment the eye activity correction method described by Elbert et al. (1985) was accomplished by the programs REGREYE and CORREYE.

Three EOG channels (horizontal, vertical, radial) are constructed from the signals recorded near the eyes. A multiple regression between these 3 channels and each EEG channel allows the correction of horizontal and vertical eye movement and eye blink artifacts. Whereas the horizontal EOG records the horizontal movements, the combination of vertical and radial EOG serves to differentiate vertical eye movements and blinks. At first, regression weights for the transmission of the EOG activity into the EEG channels are determined by the program REGREYE. These transmission coefficients are then applied by the

CORREYE program, which corrects for the pre-defined eye movements, thereby creating a new data file which contains the corrected bio-signals plus 3 separate channels containing the averaged EOG activity.

Experiments C and D Eye artifact correction in experiments C and D was accomplished by the MSEC method by Berg & Scherg (1994). The MSEC method tries to overcome a critical flaw in traditional eye activity correction methods which use regression analytic (or similar) defined amplitude relationships between EOG and EEG channels (transmission coefficients). With these approaches the estimated eye activity contains a proportion of EEG activity. Subtracting the estimated eye activity from the EEG by using regression coefficients subtracts EEG from EEG activity, especially at positions near the eyes (Berg & Scherg, 1993). In contrast to traditional EOG correction methods, with the MSEC method eye activity is described by empirically determined components. The components are treated like dipole sources which model the EEG activity. While both eye and cortical activity are assumed to be recorded by *all* applied channels, albeit to a different degree, the topography of eye versus cortical activity, as described by source vectors, differs. In the case of experiments C and D the eye source vectors were determined empirically by applying a Principle Component Analysis (PCA) on covariances of all time points of the calibration data (see *paragraph 4.5.2*). The resulting source vector defines the topography of the eye activity, the source waveform describes the magnitude of the source over time. In order to correct the EEG for eye movement artifacts, the source wave rather than the EOG signal is subtracted from each EEG channels in proportions which are defined by the respective source component. Three types of eye movement were corrected: vertical and horizontal eye movements, and blinks. In addition, the source waveform identifies the eye activity. By averaging these waveforms along with the EEG, an estimate of the time of locked eye activity is obtained.

4.7.4 Statistical Procedures

The inferential statistical analysis of ERP investigations in clinical settings, especially when schizophrenic patients are involved, is associated with a number of unique methodological problems.

The following features of such studies are pertinent to these problems.
1. In general, non-representative patient and control samples are investigated. 2. In general, heterogeneous patient samples are investigated. 3. Research questions have frequently a mixed confirmatory-exploratory character. 4. Samples with small or very small sizes (relative to the number of dependent variables) are investigated. 5. Repeated measures designs are used.

Although the related problems cannot be solved within this dissertation, they should be discussed at least, in order to provide an interpretational “frame“ for the statistical analyses using variance-analytical methods.

The Analysis of Variance (ANOVA) is an inferential statistical procedure which is used to infer from a sample to a population (Bortz, 1993). This inferential conclusion is possible only if the sample is globally or specifically representative, which means that a random or a stratified sample must be drawn (Bortz, 1993). Since schizophrenic samples participating in ERP studies are in most instances neither stratified nor random, but ad hoc samples selected by a number of clinically relevant criteria (cooperativeness, general psychopathological state, specific aspects of the clinical state such as absence of artifact producing symptoms (excessive blinks, dyskinesias, etc.) or paranoid-hallucinatory symptoms apt to include the procedure of recording the EEG ("mind reading", etc.)), the nature of the target population of statistical inferences remains unknown.

A second problem pertinent to planning and interpretation of ANOVA of studies with schizophrenic patients results from the mixture of confirmatory and exploratory research questions which are frequently present in ERP research with schizophrenic patients[23]. Both groups of research questions are frequently analyzed simultaneously (e.g. within one and the same ANOVA), although a successive treatment has been suggested (Muller et al., 1983). If orthogonal ANOVA designs are realized, the successive/simultaneous-differentiation is no problem. In any event, this differentiation should be made explicit.

The remaining problems are not specific to clinical ERP studies and result in the frequently discussed question whether multivariate or univariate statistical designs should be used. The choice of a univariate or multivariate design is governed by considerations of the validity of the nominal alpha-level and the level of the statistical power. The validity of these decision criteria depends largely on the validity of some of the statistical assumptions underlying a given statistical procedure, such as ANOVA, repeated measures ANOVA, in the case of most ERP studies, or MANOVA, the multivariate extension of the ANOVA.

Both ANOVA and MANOVA assume uni- or multivariate normality of the observations, and both are rather robust concerning violations of this assumption (cf. Stevens, 1986, p. 199, p. 205).

While ANOVA assumes the equality of the population variances (variance homogeneity), MANOVA assumes the equality of the population covariance matrices of the p dependent variables. Again, Monte Carlo studies revealed the robustness of both procedures concerning violations of these assumptions, if samples of equal or about equal (less than 1.5 : 1.0) sizes were compared (cf. Stevens, 1986, p. 199, p. 205).

Both procedures assume independence of the observations, which can be assessed by the intraclass correlations (Stevens, 1986, p. 202) and can be expected to be fulfilled when subjects are tested separately. Violations of the independence assumption result in extremely biased F-tests.

---

[23] While variations of a the experimental situation or task are certainly always to some degree confirmatory, most research questions pertaining to the evolving field of topographical characteristics of ERP generation in schizophrenia are at present still exploratory.

In contrast to MANOVA, repeated measures ANOVA assumes sphericity of the variance-covariance matrix, or, in the case of simultaneously studying repeated mesures and group factors, multisample sphericity (Keselman & Keselman, 1988). This is a critical assumption which has been ignored frequently and will be discussed now. The sphericity assumption is pertinent to the *error term estimate* in the case of intercorrelated dependent variables. The error term of the dependent t-statistic is the *pooled* variance estimate, $s^2_1+s^2_2-2s_{12}$. In the case of repeated measures F-tests several of these paired comparisons are done. This suggests the pooling of the error terms of each of the paired comparisons as a *general* error term. Evidently, this is meaningful only if all possible error terms estimate the same quantity, which means that all error terms are of equal size. The assumption of equal error terms is the sphericity assumption. This assumption is fulfilled if the variance-covariance matrix of the dependent variables is spherical, which means that the variances of all *contrasts* of the repeated measures factor are equal. Therefore, each repeated measures factor has its own sphericity assumption. If T is the number of repeated measures factors, the number of sphericity assumptions is equal to (2^{T-1}). If several groups are investigated the additional assumption of the identity of the variance-covariance matrix of the contrasts in the respective groups must hold.

In the case of violations of the sphericity assumptions, true alpha values are different from nominal alpha values. Both conservative and liberal F-tests are possible (cf. Vasey & Thayer, 1987). For this reason, an epsilon adjustment of the F-values has been suggested by Box (1954). If sphericity is violated the F-ratio is approximatively F-distributed with $\varepsilon(k-1)$ nominator and $\varepsilon(k-1)(n-1)$ denominator degrees of freedom (df; k=number of repeated measures, n=numer of subjects). ε has values between 1 and $1/(k-1)$. Since ε depends on unknown population variance-covariance matrices, it can only be estimated (Keselman & Rogan, 1980). The ε estimate by Geisser & Greenhouse (1958), which assumes maximal non-sphericity and sets ε to $1/(k-1)$, is too conservative. Other ε estimates were provided by Greenhouse & Geisser (1959) and by Huynh & Feldt (1970). Greenhouse-Geisser adjustment is superior in the case of moderate to heavy non-sphericity, Huynh-Feldt is superior in the case of light non-sphericity (cf. Vasey & Thayer, 1987).

Due to the df reduction, univariate repeated measures ANOVA looses statistical power in the case of non-sphericity. In ERP research non-sphericity is almost certain, especially in the case of multi-channel recordings. Jennings (1987) reported for an electrode factor with 3 to 7 levels a Greenhouse-Geisser ε of 0.60-0.30.

The reduction of statistical power suggests the search for alternative statistical procedures. The multivariate ANOVA extension, MANOVA, is recommended since it uses inherently *specific* error terms, thereby avoiding the sphericity problems of the general error term (O'Brien & Kaiser, 1985). This great advantage of the MANOVA is counteracted by the requirement of relatively large samples, in any case more subjects than dependent variables (cf. Muller et al., 1983). This may be a handicap, especially in studies with psychiatric patients.

Table 4.4 Number of Subjects and Number of Dependent Variables in Experiments A-D

	A	B	C	D
n	38 (2x19)	24 (11+13)	23 (11+12)	38 (20-18)
k	40	18	18	18

k=number of dependent variables, n=number of subjects

Unfortunately, there are only few systematic comparisons of the ANOVA versus the MANOVA power. Davidson (1972) compared the ε-adjusted ANOVA with the MANOVA in two situations: (1) $n=k+6$, (2) $n>k+20$ (k=number of repeated measures, n=number of subjects) and recommended the ANOVA in case (1) and the MANOVA in case (2). Vasey & Thayer (1987) recommended the ε-adjusted ANOVA if n is approximately equal to k, while Jennings (1987) suggested that the results of both procedures should be compared. For the experiments reported here, k and n are listed in **table 4.4**.

While experiments A, B, and C should be analyzed by the adjusted ANOVA, experiment D *should* be analyzed by the MANOVA. For comparison purposes experiment D will be analyzed by ANOVA, too, and multivariate results of experiment A will be reported in the **appendix** A for comparison purposes. Unbiased F-tests and the best possible power should be provided by this choice.

The statistical designs reported in *chapters 5 to 8* have a mixed confirmatory-exploratory character, since there are theoretically deduced hypotheses for *some* of the effects, while the remaining effects of the "full factorial design", especially those including "topographical" factors, are analyzed for exploratory purposes. This differentiation of effects has been called the "leapfrog"-strategy (Muller et al., 1983) and is legitimate in principle. Since, however, many significance tests are performed, the type-1 error probability of which is $p=1-(1-\alpha)^k$ in case of independent test ($p<k$ times α in case of dependent tests, Muller et al., 1983), controlling for the overall type-1 probability is necessary. This control is regularly done by Bonferroni approaches or the Bonferroni-equivalent sequential approach of Holm (1979).

Another comment should be given as to the result description. Within the full factorial ANOVA design, main effects and interaction effects of different order are tested and may become significant. Each significant interaction effect renders the included main effects and the included lower order interaction effects uninterpretable in a strict sense, since statistical interaction of (e.g.) two ANOVA factors means that the effect of factor A depends upon a given level of factor B, and vice versa. For this reason it is in principle warranted to skip the description of a given effect if it is included in a significant interaction. This procedure will none the less *not* be followed in the presentation of the results of experiments A to D for the following. Even triple interactions may be difficult to grasp in the case of multiple factor levels (e.g. 2 x 2 x 5 or 3 x 2 x 3 cells). It is considered more convenient, especially to the reader, if such interactions are successively "built up" out of the underlying main effects or interaction effects of lower order. In the case of ordinal interactions (e.g. level m of factor A achieves larger values than level n of factor A *at all levels of factor B*) one basic interpretation of lower order interactions or main effects (e.g. level 1 of factor A *increases* values over level 2 of factor A) must not be restricted. In the case of disordinal interactions (e.g. *only* at level x of factor B does level m of factor A have lower values than level n of factor A), the restrictions can be made explicit. For this reason, a sequential presentation of results, beginning with the simpler effects and ending with the complex effects, will be chosen.

Another issue pertaining to the statistical tests is the normalization of the data. At least two reasons for normalizing ERP data before applying ANOVA (or related) procedures have been suggested. Firstly, the biasing effects of outliers can be controlled. Secondly, effects of changes in the generator structure of the ERP can be controlled for effects of amplitude changes, which are —ambiguously— not distinguished within a condition by electrode interaction of the ANOVA (McCarthy & Wood, 1985). Such data changes are, in my opinion, anything but unproblematic. Take for example the minimum-maximum normalization suggested by McCarthy & Wood (1985). After applying this algorithm for each subject and condition the amplitude range of each electrode channel is restricted to the range from 0 to 1. As a consequence, the *means* (over all channels) of the different experimental subjects will probably vary within a restricted *sub*range of this 0-1-range. Hence, a reduction of the variability of the ANOVA subject factor is artificially imposed upon the data. This is not only a problem for the testing of the group factor (within which the subject factor is nested and the testing of which is renounced by the normalization procedure in advance); this is also a problem for *each* within-subject factor and interaction in split-plot designs, since the (artificially diminished) subject factor is part of their error terms (Nagl, 1992).

As for the controlling of gross outliers, the dependent variable distribution should be considered. In the case of heterogeneous groups, as has to be expected when studying schizophrenic patients, the outlier definition may be somewhat arbitrary. If no gross outliers are included in the sample, the amount of heterogeneity within the group(s) is controlled for in the ANOVA (error term).

Although the differentiation of generator structure versus amplitude changes, possibly associated with a significant condition by electrode interaction, is an important task in ERP research, one may question, firstly, whether this should be done at all when using only few electrodes, and secondly, whether this is really accomplished by data normalization. An argument frequently put forward against source localization methods is that each potential distribution at the scalp has, in principle, an infinite number of possible underlying current density distributions within the brain (e.g. Lütkenhöner, 1992). As a consequence it should be possible to "simulate" each change in the *strength* of a generator structure (i.e. an amplitude effect) by an infinite number of brain current densitiy distributions. In such a case the data normalization would erroneously suggest an amplitude effect if a given condition by electrode interaction became insignificant after normalization. And what about a possible overlap of changes in both amplitude and generator structure, if, for example, source A in condition 1 increases in strength under condition 2 *and* is supplemented by a source B? Following the data normalization one would conclude that only the second change had taken place, again erroneously. This list of ambiguities associated with the normalization of ERP data may be continued.

The conclusion to be drawn from these considerations is that inferences (not: explicit speculations) about condition specific ERP generators or strengths should be restricted to the application of source localization methods (e.g. Scherg (1992), Scherg & Berg (1991), Scherg & Ebersole, 1993; Scherg & Picton (1991)) on multichannel recordings. When using fewer electrodes, only cautious conclusions should be drawn from the results of normalized data. These results will be reported in the footnotes of the result tables.

Chapter 5 Cognitive Determinants of the PINV

Summary of Chapter 5 A common assumption both of the neuropsychological theory of Schizophrenia by Frith (1987, 1992, 1994; Frith & Done, 1988) and the neurodevelopmental theory of Schizophrenia by Weinberger (1991, 1994) is an impaired communication between the prefrontal cortex and the temporo-limbic system. This deficit should result in impaired "information networking" (Goldman-Rakic, 1994) in (task) situations which stress prefrontal cortex functions. According to the theory of the prefrontal cortex by Fuster (1989) delay tasks, an elementary variant of which the standard two-stimulus paradigm is, are subserved by the prefrontal cortex in conjunction with other brain structures, most notably the temporo-limbic system. With the working hypothesis that the occurrence of the PINV in schizophrenic patients is related to the "stressing" of prefrontal cortex functions, the task demands on prefrontal provisional memory functions were experimentally varied within a modified visual delayed matching-to-sample task. In order to induce a PINV in healthy subjects, a proportion of the to-be-matched imperative stimuli (IS) was ambiguous with respect to this task.

While schizophrenic patients exhibited no overall signs of diminished iCNV, tCNV, WS-LPC, and IS-LPC amplitudes, they showed significantly increased amplitudes of the fronto-central PINV, as compared to healthy control subjects. PINV was larger over the left than over the right hemisphere in patients, while controls exhibited the reverse asymmetry. Ambiguous IS increased PINV amplitudes in both groups to a similar extent, when compared to IS that could be matched clearly. High compared to low provisional memory load augmented PINV amplitudes only in schizophrenic patients. These results suggest stimulus ambiguity to be a general determinant in PINV generation and the stressing of prefrontal cortex functions to be a determinant of the patients' PINV. The different lateralization of the PINV in patients and controls stimulates further topographical studies.

5.1 The Theoretical Background of Experiment A

This experiment was aimed at uncovering cognitive determinants of the PINV in schizophrenic patients and healthy subjects. Two different lines of reasoning are merged in this experiment. They will be described at first.

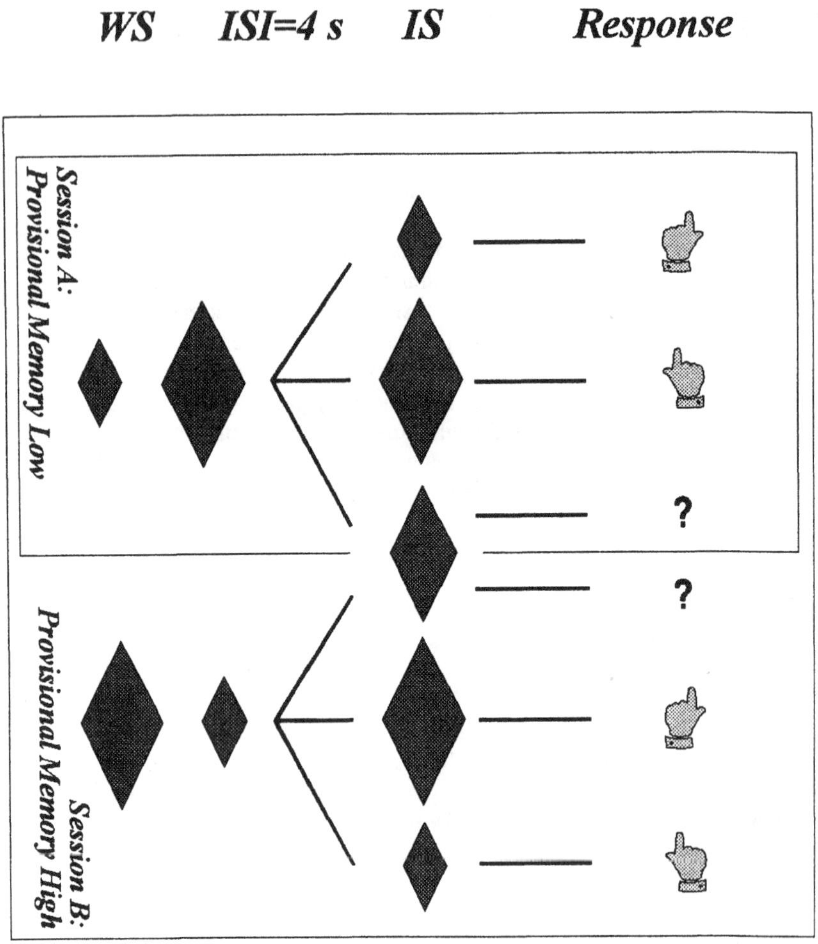
WS
ISI=4 s
IS
Response
Session A:
Provisional Memory Low
?
?
Session B:
Provisional Memory High

Figure 5.1

In *chapter 2* evidence suggesting reduced prefrontal metabolism under conditions which activate this region in healthy subjects and of structural abnormalities in temporo-limbic regions in schizophrenia was discussed.

Weinberger (1991) assumes that the prefrontal hypometabolism, occurring mainly under conditions which activate the PFC in healthy subjects, is secondary to a deteriorated "information networking" with damaged temporo-limbic structures. Although it is questionable that deficient prefrontal-temporo-limbic communication alone can explain schizophrenic psychopathology, at least some positive symptoms of schizophrenia which are considered to reflect disorders of self-monitoring, are assumed to result from deficient communication between the prefrontal and the temporal cortices (Frith, 1992, Frith & Done, 1988). Supportive physiological evidence for this deficit exists for the execution of some tasks and has been reviewed in *chapter 2.2*. These tasks have in common that they are assumed to tap PFC functions and that task performance must be guided by representations, a function supported by working memory processes.

The term "working memory" as used by psycho- and neurophysiologists refers to functions which are assumed to be subserved by the PFC. "(...), the prefrontal cortex directs behaviour through the access, maintainance and updating of "representations" (memories) over time when the stimuli are no longer present." (Fleming et al., 1994). Working memory functions have most thoroughly been investigated with delay tasks, which are among the best-established tests of PFC functions. A typical kind of delay task provides the presentation of two food wells, one of which contains food, to an experimental animal. After this initial presentation the two food wells are covered, for instance, with opaque plaques. Then the whole apparatus is hidden for a certain period of time, the delay. Finally, the animal is shown the apparatus again and is allowed to uncover just one of the wells in order to get the food. The crucial feature of delay tasks is that the choice of the correct well cannot be guided by external cues (both covered wells are identical in appearance), but must be guided by remembering where the food was hidden, i.e. by a cross-temporal representation of the location of the food.

The task of the standard two-stimulus paradigm is considered an elementary form of the delay task (e.g. Fuster, 1989), and each delay task is capable of evoking a CNV. Several subfunctions are assumed to be required by a such a task: orienting and contingency formation, motor preparation, stimulus anticipation, and response execution.

Since PINV occurs after the execution of a delay task, this type of task shall now be considered. It should be mentioned in advance that thereby a description of the basic experimental design of this dissertation is given.

Stamm & Rosen (1969, 1975; cited after Fuster, 1989) observed a prefrontal cortical surface-negative potential in monkeys performing a delay task, which they identified as the CNV. This negativity was associated with anticipatory functions and was assumed to be related to "the formation of short-term memories or to the programming of appropriate responses" (Fuster, 1989, p.

96). The investigations of Caspers have confirmed that negative cortical potential shifts reflect increasing unit activation (Caspers et al., 1984, see also *paragraph 7.1, 7.2, and 8.2*).

As Fuster (1989) argues, prefrontal cell discharge rate is strongly related to the main events in delay tasks: presentation of the cue, delay, presentation of the sample, and response.

There are cells in the DLPFC which increase their firing with the *onset of the cue* (comparable to the WS in the two-stimulus paradigm), with an early, unspecific and attention related component after about 150 ms, and a second component which is exhibited by fewer units and related to trial-specific and task-relevant features of the cue (WS).

Even *during delays* of 1 minute or longer, sustained discharge changes can be observed in PFC units (situated mainly in the dorsolateral PFC), which resumes pretrial rates after the end of the trial. To a lesser extent such delay-spanning cell activity has also been found in the dorsolateral thalamic nucleus (Fuster & Alexander, 1973) and the inferotemporal cortex (Fuster et al., 1981; Fuster & Jervey, 1982).

Activity of PFC cells only becomes apparent with the learning of the task, and the delay activity of these cells has been found to be related to task performance.

Two types of gradual firing changes during the delay, deceleration and acceleration, are prevalent. While the former appears to subserve *short-term memory functions*, the other has been suggested to reflect *preparatory set* (see below; Fuster, 1989, p. 114).

The critical condition for the occurrence of the delay-bridging activity is the existence of a cross-temporal contingency between cue and response. If the cue is experimentally deprived of its "significance as the indicator of a particular choice and motor response" (Fuster, 1989, p. 109) firing changes during the delay disappear. Since the discharge of some PFC units is sensitive to response relevant features of the cue, and since this cue-specific discharge is sustained along the delay period, these units were considered to partake in the retention of the cue representation, hence in *short-term memory functions* of the DLPFC (Fuster, 1989, p. 110).

During the *motor* response, most of the PFC units′ change in firing rate is an increase. Also under nogo-conditions some PFC units are active during non-responding (Watanabe, 1986). Most characteristically, however, PFC units change their firing rate earlier than premotor or motor cortical cells and sometimes several seconds prior to the response, independent of the type of the response that will follow. Due to this firing pattern these cells appear involved in *preparatory set.*

Both terms, preparatory set and short-term memory have a unique meaning within the context of the theory of the PFC by Fuster (1989) and will be described below.

The aforementioned studies on unit activity in prefrontal cortices of experimental animals during delay tasks are instructive, but restricted in their generalizability to human cortical activities during delay tasks and its measurement by scalp EEG. Notwithstanding the observed association between negative potential shifts and increased unit activation (Caspers et al., 1984), in *chapter 8* it will be argued that action potentials do not contribute much to the surface potentials recorded with the EEG. As a consequence more direct evidence of prefrontal involvement in delay tasks in man is required. This evidence will be presented some paragraphs later.

Delayed response, delayed alternation, and delayed matching-to-sample tasks are different types of delay tasks. The successful execution of these delay tasks amounts to the involvement of the cardinal function of the PFC, according to the theory of the PFC by Fuster (1989), the *"mediation of cross-temporal contingencies"* (Fuster, 1989, p. 159). The crucial feature of each type of delay task is the mediation of cross-temporal contingencies *on a trial-by-trial base.* That means, that although in *each* trial of a delay task the task components (cue, sample, response) are the same, their behaviorally relevant *contingency* must be formed anew. This feature distinguishes delay tasks from, for instance, conditional association tasks, where temporally stable associations must be built.

According to Fuster (1989) *three prefrontal subfunctions* are involved in the formation of cross-temporal contingencies. The first is called *short-term or provisional memory*. During delay tasks it is necessary to refer any event of the behavioral sequence to both preceding events and the scheme of action (the task structure). Both, scheme of action and preceding events must therefore be represented and retained temporally *in order to* make this future reference possible. This temporary retention of action-relevant or action-oriented information amounts to the short-term or provisional memory function, the crucial feature of which is its *operational character*, but not its content or duration.

It should be noted that some dissension exists concerning the labeling of this PFC function. While Fuster (1989) offers two terms for the same function, short-term memory and provisional memory, he mentions that terms like working memory and operant memory have also been used. Goldman-Rakic (1987) used the term "representational memory" and applies the term working memory, too (Goldman-Rakic, 1991, 1994). Goldman-Rakic (1991) explained the term "working memory" by referring to the "ability to bring to mind events without direct stimulation".

I believe that neither short-term memory nor working memory should be used for this specific memory function, since both terms have a predefined meaning in cognitive psychology that is not entirely congruent with the use in the neurophysiological disciplines[24]. The term representational memory is clearly

[24] For instance, the "articulatory rehearsal mechnism" is one of two subsystems of the "articulatory loop" which is one of the two modality-specific short term stores in the working memory model of Baddeley (1990, cited after Fleming et al., 1994). But articulatory rehearsal

superior, because it underlines that with the assistance of this memory function behavior is guided through internal representations and not external stimulations, and this is certainly a characteristic feature of prospective or goal-oriented behavior. Nonetheless, as Fuster (1989) argues, each memory is representational in nature and the use of this term would deny the uniqueness of this prefrontal function.

For these reasons I will use the term *provisional memory*, since this term connotes the operant, time-bridging, and goal-oriented features of the type of memory that Fuster (1989) theoretically elaborated.

While past experiences and the scheme of action are retained in provisional memory in order to cope with future events, the second prefrontal function is to prepare future actions on the basis of past experiences. Fuster (1989) calls this prospective function *anticipatory set* or *preparation*. Again, other terms have been used for related or similar functions: Luria´s "Presynthesis", Piaget´s "Schema", and Neisser´s "Anticipatory Schema" (cf. Fuster, 1989). *Based* on these schemes, which may be the current content of provisional memory, anticipatory set emerges and preparation "for a range of possible contingencies" (Fuster, 1989, p. 164) becomes possible.

Concerning temporal integration, provisional memory and anticipatory set have complementary roles. Based on evidence with experimental animals Fuster (1989) assumes both functions to be subserved by the dorsolateral prefrontal cortex (DLPFC).

The third PFC function, also required during delay tasks, is called *control of interference* and is assumed to be mediated by the ventral PFC (orbito-frontal cortex). During the delay external and internal interfering influences must be averted. Under the latter category mnemonic traces of previous behavioral structures similar to the present or of contingencies of previous trials can be subsumed.

One of the most important features of the PFC —as a higher sensory-motor association cortex— is the functional interrelatedness with other brain structures. Such functional interrelatedness has been experimentally established by cortical cooling methods (inducting of hypothermia) for the PFC and the mediodorsal thalamic nucleus (cf. Fuster, 1989, p. 117). The mediodorsal thalamic nucleus presumably relays information about drives and the internal milieu to the PFC (Fuster, 1989, p. 184). Lesions of this nucleus were associated with impaired spatial delayed alternation performance (Isseroff et al., 1982). Enhanced local cerebral glucose utilization (LCGU) was observed in anterior and mediodorsal thalamic nuclei during delay tasks (Friedman et al., 1990). Furthermore, intensified firing during the delay has been found in the head of the caudate nucleus (Hikosaka et al., 1989)

Another group of results which is of particular importance in the present context, suggests complex functional relationships between the PFC and the

---

will hardly take place in animals. "Short-term memory", as defined in Cognitive Psychology, has not the operant connotation intended by Fuster (1989).

inferior temporal cortex during visual delay tasks. Cooling of one of these areas resulted in marked excitatory and inhibitory changes in neuronal activity of the other (Fuster, 1989, p. 119), and the cooling of each of these sites was associated with impaired delayed matching-to-sample performance, in contrast to the cooling of the parietal cortex (Fuster, 1989, p. 119).

Since some results indicate that more inferotemporal than prefrontal cortex units react differentially to the cues of a delay task, cue representation even during the delay is more likely to be accomplished by the inferotemporal cortex (Fuster, 1989). Assistant PFC unit activity, on the other hand, is assumed to be necessary in order to have inferotemporal units retain these representations over the delay period for subsequent availability.

Another brain region involved in delay task execution is the hippocampus, since hippocampal lesions in experimental animals resulted in impaired delay task performance (cf. Fuster, 1989; Squire & Zola-Morgan, 1991). Only lesions of the hippocampus (plus amygdala, whose lesion alone has no effect) itself, but not lesions of its white matter, were associated with impairments in delayed nonmatching-to-sample tasks (cf. Squire & Zola-Morgan, 1991). With hippocampal lesions delay task deficits may emerge dependent upon the length of delay, with longer rather than shorter delays resulting in delay task impairments (Zola-Morgan & Squire, 1985). Concerning PFC lesions, the issue of the delay length is not clear. While Diamond & Goldman-Rakic (1989) observed *no* deficits in delayed response or Piaget's "A-non B-Tasks" with zero delays, Winokur (1992) found deficits in successive paired comparison tests with a go/nogo response differentiation after PFC lesions even with a zero delay.

Hippocampal involvement in delay task execution has also been emphasized by results of Friedman & Goldman-Rakic (1988). In monkeys they found increased LCGU during several delay task types in the CA1 and CA3 regions of the hippocampus and in the dentate gyrus, but no significant changes in the amygdala when compared with control tasks. These results indicate cooperation of hippocampus and PFC during "working memory" (Friedman & Goldman-Rakic, 1988) tasks of the delay task type. This cooperation can be assumed to be anatomically enabled by the entorhinal and subicular cortices (see *chapter 2*).

The aforementioned results from studies with experimental animals (including primates), are supplemented by comparable results obtained from human subjects. Brain imaging studies measuring rCBF yielded evidence for increased prefrontal metabolism during auditory delayed matching-to-sample tasks (Mazziota et al., 1982; Roland et al., 1981; Roland, 1982). Furthermore, with a computerized version of the Delayed Response Alternation task, a task combining elements of the delayed response and the delayed alternation task, metabolic activities in the dorsolateral PFC similar to those reported for the Wisconsin Card Sorting Test were found (Gold et al., 1991, cited after Fleming et al., 1994). Berman et al. (1991) reported a bilateral augmentation of rCBF in the dorsolateral PFC not only during the Wisconsin Card Sorting Test but also during a Delayed Alternation task. Finally, Freedman et al. (1986) reported significantly more errors in Delayed Alternation and Delayed Response tasks of patients with bilateral frontal lesions, compared to alcoholic subjects. These deficits were not correlated with memory functions such as those assessed by the Wechsler Memory Scale including paired associate learning, which are associated with the left temporal lobe (Kolb & Wishaw, 1990, p. 455), but correlated significantly with performance in the Wisconsin Card Sorting Test.

Altogether these results suggest that also in man delay task performance depends critically upon PFC functions.

While, to my knowledge, there are no studies on brain metabolism during delay tasks in schizophrenic patients, impaired oculomotor delayed matching-to-sample performance has been found in this group (Park & Holzman, 1992). This deficit was not related to the sensory-motor components of the task and was moderately correlated with Smooth Pursuit Eye Movement dysfunctions (Park & Holzman, 1993), which have been suggested to derive from frontal eye field dysfunctions in schizophrenia (Levin, 1984).

Sawaguchi & Goldman-Rakic (1991) demonstrated that the local blockade of dopamine D1 receptors in the dorsolateral PFC was followed by performance deteriorations in the oculomotor Delayed Response task. Task-related activation of the PFC during the execution of prefrontal tasks was obtained by the administration of dopamine in monkeys (Sawaguchi et al., 1988) and of dopamine agonists such as amphetamine in man (Daniel et al., 1991). Dextroamphetamine administration in schizophrenic patients improved some aspects of the Wisconsin Card Sorting Performance in schizophrenic patients (Goldberg et al., 1991).

To summarize the arguments pertaining to the hypotheses on which the first experimental factor of the present experiment is based:

As outlined in chapter 2, *schizophrenia is associated with prefrontal and temporo-limbic abnormalities which may for several reasons be the basis for behavioral and neurophysiological abnormalities under conditions which require an intact communication between the two brain regions. Since behavioral, cognitive, and neurophysiological abnormalities in schizophrenia appear to occur primarily when dysfunctional brain regions are "stressed", it is important to consider under which condition slow cortical potentials such as the PINV occurs. This condition is the two-stimulus paradigm task, which is widely held to stress particularly the prefrontal cortex (see* chapter 3*). During delay tasks thalamic nuclei and temporo-limbic structures have been demonstrated to participate in the task performance. Provisional memory, anticipatory set, and interference control are the 3 aspects of the prefrontal function, the mediation of cross-temporal contingencies, which is required for delay task execution.*

Factor 1 of the present investigation attempts to realize two levels of prefrontal cortex involvement experimentally. After a presentation of the experimental design, the qualitative and/or quantitative differences between the two levels of this factor will be described in terms of the 3 prefrontal functions delineated above.

The basic design of the 4 experiments of this dissertation were in detail described in *chapter 4*. Therefore, only a brief summary will be given here.

As illustrated in ***figure 5.1***, a modified visual delayed matching-to-sample task was used. Two horizontally arranged diamonds of different size were presented as the Warning Stimulus (WS) on a computer monitor for 100 ms. With a stimulus onset asynchrony (SOA) of 4.0 s a single diamond, being identical with one of the WS diamonds, was presented as IS for 100 ms. Subjects were instructed to match the IS diamond with one of the WS diamonds by pressing a response button with the left or the right hand.

The levels of the experimental factor 1 were counterbalanced over 2 sessions scheduled on two subsequent days. During session A the small diamond was always on the left, the large diamond always on the right side. During session B this lateral arrangement of the two diamonds varied randomly from trial to trial, with the small diamond being presented on either side with 50% probability (the reverse for the large diamond).

In terms of PFC functions during delay tasks according to Fuster (1989) the two sessions differed in the following way: Firstly, *control of interference* from internal or external sources was comparable between the two sessions except for one aspect: only in session A there was no opportunity for contingencies of a given trial to interfere with contingencies of the following trial. Secondly, *anticipatory set* was expected to be present in both sessions. Thirdly, the most important difference between session A and session B was the provisional memory requirement in session B, which was absent in session A. Only in session B the positions of the two diamonds had to be retained in order to respond correctly to the IS *during each trial anew*. This difference, in particular, qualified the task of session B as a delay task, while the task of session A may be considered a choice RT task (the WS provides no information relevant to the choice of the response side; *each* small diamond of the IS may be answered by a left hand response, *each* large IS diamond by a right hand response). In the following I will refer to this experimental factor as the *provisional memory factor*.

Since healthy subjects usually do not exhibit a PINV in forewarned reaction time tasks, a closer inspection of the conditions which induce a PINV in this group is warranted. As has been reviewed in *chapter 3*, "loss of control" over an aversive IS is apt to induce a PINV in healthy subjects. Since aversive stimulation *per se* cannot account for this effect, a cognitive, presumably stressful, process must be considered. "(...) changed or ambiguous contingencies, disappointed expectations, or loss of control" (Birbaumer et al., 1986) were considered factors in evoking a PINV.

In line with this reasoning ambiguity in matching the IS with the WS diamonds was introduced as the second experimental factor. As a completely crossed experimental factor, diamonds of a middle size category as IS in one third of the trials were introduced (see ***figure 5.1***). These diamonds had sizes in-between the sizes of the small and large diamonds, respectively. This was exactly the size of the small diamond plus 40, 50 or 60% of the difference between the small and the large diamond. While small and large IS diamonds could be matched *clearly* with one of the two WS diamonds, diamonds of a middle size were *ambiguous* in this respect (see *chapter 5* for a more detailed description of the technical details). For this reason, the second experimental factor will be called *ambiguity factor*.

5.2 Hypotheses

With the specification of a full factorial variance analytical design in clinical research practice only a proportion of the statistical hypotheses tested has a research hypothesis counterpart. “Leapfrog“ strategies (Muller et al., 1983), including confirmatory and exploratory parts, prevail. If Bonferroni or related α-adjustments are included, this strategy is legitimate in the case of balanced designs. In any case the distinction should be made explicit and this will be done now. The exact definition of the ANOVA factors will be given in *paragraph 5.6*.

Confirmatory Part

CNV

Hypothesis 1a: Schizophrenic patients exhibit diminished iCNV amplitudes. *Reason:* Published results on diminished CNV in schizophrenia; literature presented in *chapter 3.3*. *Statistical form:* Group main effect.

Hypothesis 1b: The centro-frontal iCNV gradient is steeper in controls than in patients. *Reason:* Published results on relative CNV augmentations at frontal leads in schizophrenic patients, literature presented in *chapter 3.3*. *Statistical form:* Group main effect.

Hypothesis 2a: Schizophrenic patients show diminished tCNV amplitudes. *Reason:* Published results on reduced CNV amplitudes in schizophrenia, literature presented in *chapter 3.3*. *Statistical form:* Group by Electrode interaction.

Hypothesis 2b: The parieto-frontal tCNV gradient is steeper in controls than in patients. *Reason:* Published results on relative CNV augmentations at frontal leads in schizophrenic patients; literature presented in *chapter 3.3*. *Statistical form:* Group by Electrode interaction.

Hypothesis 3a,b: iCNV and tCNV are augmented with higher provisional memory load requirements. *Reason:* Association between iCNV and contingency formation, and tCNV and task complexity; literature presented in *paragraph 3.2*. *Statistical form:* Provisional Memory main effects.

PINV

Hypothesis 4: Ambiguous imperative stimuli enlarge the PINV in schizophrenic patients and induce a PINV in healthy controls. *Reason:* Hypothesized

association between ambiguous contingencies and PINV; literature presented in *paragraphs 3.5-3.7 and 5.1 Statistical form:* Ambiguity main effect.
Hypothesis 5: With higher task demands on provisional memory functions the PINV amplitudes augments in schizophrenic patients but not in healthy controls. *Reason:* Assumed association between the stressing of prefrontal functions and the PINV in schizophrenic patients; literature presented in *chapter 2, paragraph 3.6, 3.7, and 5.1. Statistical form:* Provisional Memory by Group interaction.
LPC[25]
Hypothesis 6: Schizophrenic patients exhibit diminished LPC amplitudes following the warning and the imperative stimuli. *Reason:* Published results on reduced P300 amplitudes in schizophrenic patients; literature reported in *paragraph 3.9. Statistical form:* Group main effect.
Hypothesis 7: Due to the low relative frequency or the high equivocation the LPC following the imperative stimulus is smaller under the ambiguous than under the clear condition. *Reason:* Triarchic Model (Johnson, Jr., 1986), presented in *paragraph 3.8. Statistical form:* Ambiguity main effect.
Hypothesis 8: With higher provisional memory task demands the amplitudes of the LPC following the warning stimulus becomes larger. *Reason:* Triarchic model (Johnson, Jr., 1986) presented in *paragraph 3.8. Statistical form:* Provisional Memory main effect.
RT
Hypothesis 9: Reaction Times are longer in patients than in controls. *Reason:* Results reported in the literature (e.g. King, 1991). *Statistical form:* Group main effect.
Hypothesis 10: Under the high provisional memory load condition reaction times will be augmented. *Reason:* addition stages of information processing are required under high compared to low provisional memory load conditions. *Statistical form:* Provisional Memory main effect.
Hypothesis 11: Ambiguous IS will augment reaction times. *Reason:* Resolving the "conflict" in responding due to ambiguous WS-IS and IS-response contingencies will take some time. *Statistical form:* Ambiguity main effect.

Exploratory part
Other *questions* can be put to the results of the ANOVA procedures. These questions comprise chiefly all topographical differentiation of the main or interaction effects specified in the confirmatory part. For example, does the topography of the ERP components, in particular the PINV, differ between the two groups ? If the ambiguity and provisional memory factors are effective: Are their effects restricted to certain scalp locations ? Do patients and controls differ in this condition-dependent topography ? Furthermore, if higher task load under the higher provisional memory demand condition increases the iCNV and tCNV

[25] As has been argued in *chapter 3.8* the P300 constitutes the main proportion of the LPC elicited during a forewarned reaction time task.

amplitudes: Is this enhancement more pronounced in patients than in controls, because patients engage more in a putatively deficient function ? Or will the reverse occur, since patients do generally engage less and exhibit the reduced responsiveness to varying task conditions frequently reported in the literature ? Also, if higher provisional memory load augments the warning stimulus LPC, is this effect more pronounced in patients since they must allocate relatively more attentional resources ? Likewise, is there an amplitude change of the LPC following the imperative stimulus with higher provisional memory task load ? If this effect occurs: Does it have the same strength and the same topography in patients and controls ? Furthermore, what effects - if any - will ambiguous imperative stimuli have on reaction times ? Are these effects similar in the two subject groups ? Finally, will reaction times in schizophrenic patients disproportionately enlarge under the high provisional memory load ? Will patients exhibit more response errors than controls, especially under this condition ?

5.3 The Laboratory

Experiment A was conducted in the psychophysiological laboratory of the University of Konstanz in the State Hospital Reichenau.

5.4 Recruitment of Subjects and Assessment of Psychopathological Variables in Schizophrenic Patients

Twenty-nine schizophrenic patients and 21 healthy controls were recruited for the present experiment in the way described in *paragraph 4.3*. As clinical variables the Brief Psychiatric Rating Scale (BPRS) and the medication of the day of testing were assessed (described in *paragraph 4.4*). Except for two patients, who received the final diagnosis of an affective disorder, all patients met a DSM-III-R diagnosis of a chronic schizophrenic disorder. Except for one outpatient, all were inpatients. Data of 19 out of 29 participating patients and 19 out of 21 participating healthy control subjects could be statistically analyzed. In **table 5.1** the socio-demographic features of both experimental groups are presented.

One schizophrenic patient was living outside the psychiatric hospital at the time of testing. Finally, another schizophrenic patient was recruited from a ward for long-term chronically ill patients of the Psychiatric Hospital Reichenau. For several years, this patient was also well-known to the staff of ward IIA and fulfilled DSM-III-R criteria for a schizophrenic disorder.

During the time of testing all patients were receiving neuroleptic medication (mean=185.9 mg/day CPZ equivalent; SD=85.4) and psychopathology was assessed with the BPRS (mean=41.3; range=30-61).

Table 5.1 Socio-Demographic Features of Patients and Controls

	Schizophrenic Patients	**Healthy Controls**
N	19	19
Mean Age	29.9	30.9
(SD; Range)	(5.9; 20-43)	(6.8; 25-46)
Male : Female	15 : 4	14 : 5
Smoker : Non-Smoker	11 : 8	15 : 4
Education		
< 8	1	0
8	7	8
10	5	5
12	0	2
13	5	2
>13	0	1
Right-Handedness	18	18

Most of the patients were free of florid symptoms during the time of testing. One patient reported having had visual hallucinations during the experimental sessions which, however, did not obviously deteriorate her performance when compared with the other patients. Another patient suffered from permanent auditory hallucinations and was unable to carry out the experimental task for this reason. Therefore, the first session was terminated after some minutes.

Twenty-one healthy subjects participated in the experiment. None of them reported having a neurologic or psychiatric disease or taking any medication.

5.5 The Testing Procedure

The testing procedure applied in experiment A is equivalent to the general scheme described in *paragraph 4.5*. The only difference is the testing of each subject at two successive days. After a subject had acceded to participate, an individual testing schedule comprising both experimental sessions was arranged. Prior to the first experimental session the subject was shown the laboratory and a written consent was given. After the instruction, the headset was adjusted by which the acoustical probes were delivered.

Each subject received DM 20,- for each finished experimental session and an additional bonus DM 10,- for the completion of both sessions and fast and correct responding. None of the subjects terminated the experiment by him/herself.

5.6 The Task Presentation

Figure 5.1 illustrates schematically the experimental design (see *chapter 4* for a more detailed description).

As WS a pair of horizontally arranged green diamonds was presented against a black background for 100 ms. These diamonds had clearly distinguishable sizes. During session B the lateral arrangement of the two diamonds varied randomly from trial to trial. During session A the small diamond (covering 6.0 * 3.0 cm of the screen) was always on the left, the large diamond (covering 12.0*6.0 cm of the screen) was always on the right side of the central vertical axis of the monitor. The testing sequences (A-B or B-A) were counterbalanced across subjects. Four seconds after WS onset (SOA=4.0 s) a single green diamond was presented in the middle of the monitor for 100 ms. In one third of the trials (n=48) the IS diamond was identical with the small WS diamond, in one third it was identical with the large WS diamond. During the remaining one-third of the trials the IS diamond had a size lying in-between those of the WS diamonds, but the occurrence of these middle-size diamonds was not introduced to the subjects who were instructed to match the IS diamond with the identical WS diamond by pressing the left or right hand response button. Via a headphone auditory probes (clicks of 20 ms duration and 50 dB intensity with hanning window) were delivered in regular intervals of 1.0 s, beginning 0.5 s after the WS onset. Clicks were introduced to the subjects as task-irrelevant and to be ignored.[26]

5.7 Apparatus and Physiological Recordings

The equipment of the laboratory was described in *paragraph 4.2*. Experiment A was conducted with the GENTASK version for 286 PC. The amplifier settings are listed in table 4.1. (see *paragraph 4.6.3*). EEG electrodes were affixed to the positions illustrated in **figure 5.2**. Cz was determined as the geometrical center between nasion and inion and the left and right preauricular points as described in Lutzenberger et al. (1985). Then, electrode positions F7, F3, F4, F8, T3, T4, C4, T4, P3, P4, A1, and A2 (mastoids) were determined according to the international 10-20-system as described in *paragraph 4.5*. Four EOG electrodes (*not* illustrated in **figure 5.2**) were affixed to the outer canthi and 2-3 cm above and below the left eye in order to record eye movements.
Although for the data of experiment A only a blink-correction method was applied, all detectable eye movement activity was taken into account for the selection of correct trials. During recording Cz served as common reference.

---

[26] Results of the click-evoked potentials will be described in *chapter 7*.

Figure 5.2 Electrode Montage in Experiment A

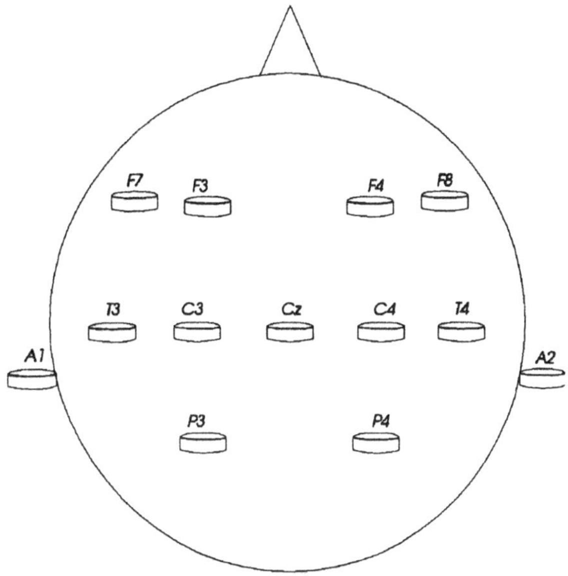

5.8 Data Reduction and Data Analysis

The EEG data amplified and sampled at 100 Hz by the DCMES system were at first corrected for DC shifts with the program DCAUS (MES, Munich; see *paragraph 4.2*). DC correction fits a fourth-order polynom to the mean DC levels of a 1.0 s pretrial baseline interval in order to compute each trial as a difference from the statistically fitted DC level.

The subsequent steps in data analysis were performed with the analysis programs described in detail in *paragraph 4.2*. The following analysis steps were performed (in that order): Transformation of data format from the DCMES to the Konstanz raw data format; categorization of each trial according to its combination of stimulus and response event marker and computation of RT; initial baseline subtraction; re-referencing of raw data on an average reference, re-referencing of average reference data on a average mastoids reference, artifact identification, computation of regression coefficients for blinks, EOG correction and condition-wise computation of single subject averages, averaging over predefined conditions (see below), and computation of mean amplitudes over predefined intervals for statistical analyses (see below).

Details of the artifact detection and EOG correction are described in *paragraph 4.7.3* (programs ARTIFACT, BLINKREG, and AVERAGE).

Table 5.2 Mean Relative Frequencies of Correct, False or Missing Responses and Artifact-Free EEG Trials (Separately for the Experimental Groups and Conditions)

(a) Relative Frequencies of correct, false or missing responses

	Session A				Session B			
Condition	clear		ambiguous		clear		ambiguous	
Response	correct	false	correct	false	correct	false	correct	false
Patients	.88	.02	.10	.09	.86	.05	.09	.08
Controls	.89	.01	.09	.07	.89	.03	.09	.08

(b) Artifact-Free EEG Trials

	Session A		Session B	
	clear	**ambiguous**	**clear**	**ambiguous**
Patients	.68	.72	.72	.75
Controls	.74	.78	.81	.85

Table 5.2 presents mean relative frequencies of correct, false or missing responses and of artifact-free EEG trials, separated for experimental groups and conditions. For the condition "clear" correct, false, or missing responses can be differentiated, while for the condition "ambiguous" only correct or missing responses can be distinguished (due to the ambiguity of the IS "correct" or "false" is not defined).

Grand averages were computed for each group, distinguishing the conditions clear and ambiguous for each session.

After these procedures, mean amplitude parameters for each subject, condition, and channel were determined for an iCNV, a tCNV, and 2 PINV intervals. For this parametrization probe-evoked potentials were considered in the following manner. By visual inspection of the grand averages of the patient and the control group, which were 3-7 Hz bandpass-filtered for this purpose, the interval superposed by the probe EP (N100-P200 complex) was defined as latency until return to the pre-probe DC level. This task was accomplished "by hand" at the PC monitor. With the LOOK program it is possible to simultaneously look at the amplitudes for a given latency of each recording channel. For all channels, intervals, and conditions the click EP returned to the pre-click baseline level within 250 ms. Therefore 250 ms segments, beginning with the probe onset, were skipped for the CNV and PINV parametrization. Mean amplitudes for the iCNV were computed over the 1.0-2.0 s interval following WS onset (excluding the 1.5-1.75 s interval). This rather late interval was selected because the visual WS evoked rather large late positive complexes (LPC). For the tCNV the interval from 3.0 to 4.0 s following WS onset (excluding the interval 3.5-3.75 s) was chosen. PINV was defined as the mean amplitude for the 1.0-2.0 s and 2.0-3.0 s intervals, respectively, following IS onset (excluding the 2.5-2.75 s and 3.5-3.75 s intervals). In addition, peak amplitude and latency values for the LPC following the WS and the IS were determined. Based on the visual inspection of the grand average data the condition and group specific latencies of the WS- and IS-LPC amplitude peaks at the parietal positions were determined with the LOOK program. The peak latency ranges were 290-300 ms and 260-300 ms for theWS-LPC peak, and 360-510 ms and 340-450 ms for the IS-LPC peak in controls and patients, respectively, depending on the task condition. Because of the differences in peak latencies, peak amplitudes and latencies were determined for each group and condition separately, defining a peak interval from -200 to +200 ms with respect to the given parietal peak latency of the grand averages. In one patient the data of the P3 electrode channel were seriously artifact contaminated for one session, while the other channels were good for both sessions. Following a recommendation of Prof. Dr. T. Elbert the P3 data were interpolated in the following way: Excluding the crucial subject the mean difference between P4 and P3 (P4-P3) in the patient group was determined and subtracted from the P4 values of the crucial patient in order to determine the P3 value.

These ERP scores were statistically analyzed with the GLM-Procedure of SAS. *Univariate* instead of multivariate analyses are reported in this chapter for reasons described in *paragraph 4.7.4*. Within a mixed-model repeated measures ANOVA the between-subject factor "GROUP" (patients versus controls) and the within-subject factors "(PROVISIONAL) MEMORY" (referring to the low versus high provisional memory requirements in sessions A and B, respectively), "AMBIGUITY" (with low, respectively high ambiguity for IS that could be matched with the WS clearly, respectively ambiguously), "HEMISPHERE" (left versus right hemisphere), and "ELECTRODE" (distinguishing the 5 electrode positions F3 (F4), F7 (F8), C3 (C4), T3 (T4), and P3 (P4) for the left (right) hemisphere) were simultaneously analyzed. In addition, reaction time values were statistically analyzed by repeated measures ANOVA, including the factors GROUP, (PROVISIONAL) MEMORY, and AMBIGUITY. Due to storage errors the RT data of 4 subjects were lost. Huynh-Feldt-corrected p-values are reported for all main and interaction effects including a within-subject factor with more than 2 levels in order to control for possible violations of the sphericity assumption of split-plot desgins (see *paragraph 4.7.4*). Planned contrasts are accomplished in order to specify the origin of a significant effect including an experimental factor with more than 2 levels. In experiment A, all levels of the ELECTRODE factor are contrasted with the (C3,C4) level. Only significant ($p \leq 0.05$) contrasts are reported. These results are presented in footnotes, in order to alleviate the reading of the text. Only significant effects are documented in tables (means and standard deviations).

5.9 Results of Experiment A

5.9.1 Reaction Times (RT)

Under all conditions (see **table 5.3**), schizophrenic patients showed significantly slower reaction times than controls (Group: $F(1,32)=13.25$, $p=.001$). From low to high provisional memory load conditions reaction times enlarged in both groups by comparable amounts (Provisional Memory: $F(1,32)=8.23$, $p=.007$). Furthermore, RT were significantly faster with low compared to high ambiguity of experimental stimuli (Ambiguity: $F(1,32)=37.69$, $p=.0001$). An augmented frequency of response errors in the patient group with high provisional memory load and low ambiguity had been documented in **table 5.2**.

Table 5.3 Provisional Memory and Ambiguity Effects on Reaction Times

		Memory	Ambiguity
Patients	Low	724.5 (181.0)	729.3 (177.5)
	High	874.5 (265.2)	869.8 (217.1)
Controls	Low	545.7 (163.7)	515.1 (149.6)
	High	599.0 (163.7)	629.6 (211.2)

unit is ms; SD in parentheses

5.9.2 Event-Related Potentials

5.9.2.1 Grand Averages of the Patient and the Control Groups

As **figures 5.3** and **5.4** reveals the involvement of the subjects in task execution resulted in a pattern of successive positive and negative potential shifts. Following the WS, positive deflections, identifiable as an LPC due to their centro-parietal maxima (C3, C4, P3, P4), occur in both groups. Following this WS-LPC at fronto-central electrodes near the sagittal axis (F3, F4, C3, C4, P3, P4) but also at the left-frontal lead F7, a negativity arises which increases in amplitude towards the IS. This negativity with a central (C3, C4) predominance is defined as a CNV. The putative subcomponents of the CNV, iCNV and tCNV, cannot clearly be differentiated by visual inspection of the grand averages. Following the IS a positivity similar in shape and topography to the WS-LPC arises, labeled IS-LPC. After the IS-LPC a second negative potential shift can be observed: the PINV. The PINV has a clear right-frontal predominance in healthy control subjects, and a clear left fronto-central maximum in schizophrenic patients.

As can be seen from **figures 5.3** and **5.4** there is no overall diminution of potential amplitudes in schizophrenic patients when compared to controls. Instead, both centro-parietal LPC are larger in the grand averages of the patients, as compared to those of the healthy control subjects. Furthermore, the left fronto-central PINV is clearly larger in patients compared to controls. In addition, **figures 5.3** and **5.5** show that the regularly delivered auditory probes (clicks) evoke potential deflections mainly with a positive component (see *chapter 7*).

5.9.2.2 Initial and Terminal Contingent Negative Variation (iCNV, tCNV)

Table 5.4 summarizes the results of repeated measures ANOVA for the iCNV and the tCNV. Concerning the *iCNV* there are no statistically significant overall amplitude differences between groups. For both groups iCNV is larger over the left than the right hemisphere, as the significant main effect **Hemisphere** (see **table 5.6**) indicates. The significant **Electrode** effect of the iCNV results from significant contrasts between (F7, F8)[27], (T3, T4)[28], and (P3, P4)[29], *each compared to* (C3, C4). **Table 5.5** reveals that iCNV amplitudes are largest at (C3, C4) and (F3, F4), with *smaller* values at (P3, P4),(F7, F8), and (T3, T4). Enhanced Provisional Memory load results in enhanced iCNV amplitudes, as indicated by the significant Provisional Memory effect, but must be differentiated topographically. The significant interaction of **Provisional Memory by Electrode** results from a smaller enhancement with higher memory load at (F7, F8)[30], at (T3, T4)[31], and at (P3, P4)[32], each compared to (C3, C4) (see **table 5.5**). Only the frontal (F3, F4) iCNV enhancement with higher memory load is comparable to the central (non-significant contrast). This interaction remains significant after normalization of the data.

To summarize: schizophrenic patients and healthy controls do not differ in iCNV amplitudes, the iCNV being predominant over the left hemisphere and at fronto-central sites ((F3, F4), (C3, C4)) in both groups. High compared to low provisional memory load enhances the iCNV in both groups. This effect is most pronounced at frontal ((F3, F4)) and central sites ((C3, C4)). The results of normalized data suggest the exclusion of an underlying amplitude effect.

---

[27] $F(1,36)=8.76$, $p=.005$
[28] $F(1,36)=42.38$, $p=.0001$
[29] $F(1,36)=6.25$, $p=.02$
[30] $F(1,36)=13.76$, $p=.0007$
[31] $F(1,36)=12.28$, $p=.001$
[32] $F(1,36)=7.72$, $p=.009$

Table 5.4 Experimental Effects on iCNV and tCNV

		iCNV		tCNV	
Effect	df1, df2	F	p	F	p
M	1,36	4.17	.05	4.96	.03
M * G	1,36	1.34	.25	0.47	.50
A	1,36	1.12	.30	0.06	.81
A* G	1,36	1.41	.24	4.91	.06
H	1,36	7.88	.008	8.79	.005
H * G	1,36	0.87	.36	0.01	.91
E	4,144	10.85	.0001	29.29	.0001
E * G	4,144	1.41	.25	3.94	$^{G}.02^{1.2}$
M	1,36	1.89	.18	0.01	.92
M * A	1,36	0.06	.80	0.83	.37
M * H	1,36	0.10	.75	0.10	.75
M * H * G	1,36	0.86	.35	1.63	.21
A * H	1,36	1.56	.22	1.89	.18
A * H * G	1,36	0.01	.93	0.01	.93
M * E	4,144	5.83	$^{G}.001^{2.1}$	3.20	$^{A}.03^{2.2}$
M * E * G	4,144	0.70	.55	0.46	.72
A * E	4,144	0.35	.77	0.15	.90
A * E * G	4,144	0.47	.69	1.24	.30
H * E	4,144	1.22	.30	1.92	.12
H * E * G	4,144	1.00	.38	1.02	.39
M * A * H	1,36	0.07	.80	0.12	.73
M * A * H * G	1,36	0.99	.33	1.38	.25
M * A * E	4,144	1.32	.27	1.15	.33
M * A * E * G	4,144	0.71	.50	1.94	.14
M * H * E	4,144	1.70	.17	1.34	.26
M * H * E * G	4,144	1.17	.32	1.23	.30
A * H * E	4,144	1.59	.20	0.99	.41
A * H * E * G	4,144	0.10	.96	0.43	.77
M * A * H * E	4,144	1.06	.36	1.73	.16
M * A * H * E * G	4,144	2.06	.12	0.74	.55
Group	1,36	0.00	.10	0.82	.37

Table legend: M=Provisional Memory factor (low vs high), A=Ambiguity factor (clear vs ambiguous), H=Hemisphere factor (left vs right), E=Electrode factor (F3,F4),(F7,F8),(C3,C4), (T3,T4), (P3,P4), G=Group factor (patient vs control group); univariate repeated measures ANOVA, Huynh-Feldt corrected p values; $^{G(G)}$=Generator effect, $^{A(A)}$= Amplitude effect as indicated by results of normalization; $^{1.2}$=F(4,144)=3.49,p=.03; $^{2.1}$=F(4,144)=2.92,p=.03; $^{2.2}$=F(4,144)=2.26, p=.08

ANOVA results for the tCNV, presented in **table 5.4**, parallel those obtained for the iCNV. Again, there are no overall group differences in tCNV amplitudes, which are larger over the left than the right hemisphere, as indicated by the significant **Hemisphere** main effect (see **table 5.6**).

Table 5.5 Provisional Memory by Electrode Interaction on iCNV and tCNV

		(F3,F4)	(F7,F8)	(C3,C4)	(T3,T4)	(P3,P4)
iCNV	Low	-0.70 (1.9)	-0.30 (2.4)	-0.69 (1.4)	0.20 (0.9)	-0.54 (1.3)
	High	-1.33 (2.0)	-0.24 (1.7)	-1.63 (2.2)	0.04 (1.0)	-0.95 (1.5)
tCNV	Low	-1.50 (2.4)	-0.63 (2.7)	-2.53 (2.3)	-0.80 (1.5)	-2.19 (1.7)
	High	-2.23 (2.5)	-0.69 (2.0)	-3.26 (2.8)	-0.44 (1.3)	-2.65 (2.2)

unit is µV; SD in parentheses

Table 5.6 Hemisphere Effect on iCNV and tCNV

	Left Hemisphere	Right Hemisphere
iCNV	-0.84 (1.5)	-0.39 (1.1)
tCNV	-1.98 (1.7)	-1.41 (1.6)

unit is µV; SD in parentheses

As statistically confirmed by the significant **Electrode** main effect, tCNV has a central (C3, C4) amplitude maximum with *smaller* values at (F3, F4)[33], (F7, F8)[34], (T3, T4)[35], and (P3, P4)[36], *each compared to* (C3, C4). Differences between parietal and central positions are smallest (see **table 5.7**).

The interaction of **Group by Electrode** is due to significant contrasts between all fronto-temporal amplitudes ((F3,F4)[37], (F7,F8)[38], (T3,T4)[39]) on the one hand, and the central amplitudes (C3, C4) on the other hand ((P3, P4) not differing significantly from (C3, C4)). As can be seen in **table 5.7**, healthy control subjects have larger tCNV amplitudes over centro-parietal leads than schizophrenic patients. A difference in the underlying generator structure between the groups is suggested by the significant results of normalized data.

Table 5.7 Group by Electrode Interaction on tCNV

	(F3,F4)	(F7,F8)	(C3,C4)	(T3,T4)	(P3,P4)
Patients	-1.76 (2.2)	-0.94 (2.5)	-2.10 (2.0)	-0.21 (1.1)	-1.88 (1.7)
Controls	-1.97 (2.4)	-0.39 (1.8)	-3.59 (2.5)	-0.41 (1.4)	-2.97 (2.0)

unit is µV; SD in parentheses

---

[33] $F(1,36)=19.02$, $p=.0001$
[34] $F(1,36)=33.30$, $p=.0001$
[35] $F(1,36)=86.09$, $p=.0001$
[36] $F(1,36)=4.97$, $p=.03$
[37] $F(1,36)=8.16$, $p=.007$
[38] $F(1,36)=7.28$, $p=.01$
[39] $F(1,36)=5.56$, $p=.02$

Only at lateral frontal sites (F7, F8) is the tCNV of schizophrenic patients slightly larger than the tCNV of controls. Separate ANOVA for the two groups indicate significant Electrode effects for either group, with larger F-values for the control[40] as compared to the patient[41] group. This confirms the larger amplitude differences between the electrode (pairs) in controls than in patients. Since this result remains significant after normalization of the data, different underlying source configurations in patients and controls are suggested. Like iCNV amplitudes, tCNV amplitudes are augmented with high compared to low provisional memory load, as indicated by the significant **Provisional Memory** effect. Again, a topographical differentiation appears warranted due to the significant interaction of **Provisional Memory by Electrode**. This interaction results from significant contrasts of provisional memory effects at (F7, F8)[42], (P3, P4)[43], *each compared to* (C3, C4). Differences between (T3, T4)[44] and (C3, C4) just fall short of significance. Hence, the provisional memory effect is most pronounced for the fronto-central leads F3, F4, C3, C4 and significantly smaller for the remaining leads (see **table 5.5**). Non-significant results of normalized data indicate that this effect is due to an amplitude augmentation with high provisional memory load.

Results on tCNV can be summarized as follows: There are no overall group differences in tCNV amplitudes, but larger amplitudes in controls than in patients can be found for the centro-parietal leads, suggesting a "shallower" posterior-anterior tCNV gradient in patients compared to controls, which is compatible with a difference between groups in the generator structure of the tCNV. With a left-hemisphere, central predominance of the tCNV in both groups high compared to low provisional memory load results in maximal enlargements over fronto-central regions. Since the provisional memory by electrode interaction becomes non-significant after data normalization, this result is compatible with an amplitude effect.

Results on iCNV and tCNV reveal both topographical similarities and differences for the two CNV components. A fronto-central predominance is present for the iCNV, a central one for the tCNV. Both components are larger over the left than over the right hemisphere, and both components exhibit maximal augmentations due to high provisional memory load at (F3,F4) and (C3,C4). For the iCNV this experimental effect is compatible with a change in the underlying generator structure. Only with respect to the tCNV do the two experimental groups differ in their scalp distribution. This may be a group difference in the generator structure of that component, since this effect remains significant after normalization of the data.

---

[40] $F(4,72)=26.52$, $p=.0001$
[41] $F(4,72)=7.29$, $p=.003$
[42] $F(1,36)=7.03$, $p=.01$
[43] $F(1,36)=4.75$, $p=.04$
[44] $F(1,36)=3.54$, $p=.07$

5.9.2.3 *Post-Imperative Negative Variation (PINV)*

Table 5.9 summarizes the univariate ANOVA effects for the two successive PINV parameters, PINV-1 and PINV-2 (2.0-3.0, resp. 3.0-4.0 s after IS onset, excluding probe EP). Since the results are almost identical for the two parameters, only experimental effects on the first measure, PINV-1, will be described in detail here.

The significant **Group** effect indicates overall larger PINV amplitudes in schizophrenic patients than in healthy controls (see also **figures 5.3** and **5.4**).

Table 5.8 Hemisphere by Group Interaction on PINV

	Left Hemisphere	Right Hemisphere
Patients	-1.71 (1.8)	-0.65 (1.7)
Controls	0.29 (1.6)	-0.50 (1.8)

unit is µV; SD in parentheses

This "overall" pattern must be specified by the significant **Hemisphere by Group** interaction: Both groups exhibit reverse hemispheric asymmetries of the PINV (see **table 5.8**). In schizophrenic patients PINV is larger over the left than over the right hemisphere, in controls PINV is larger over the right than over the left hemisphere. Within-group ANOVA confirm significant hemispheric asymmetries in patients[45] and a trend in controls[46] (see **figures 5.3** and **5.4**).

Table 5.9 Experimental Effects on PINV-1 and PINV-2

		PINV-1		PINV-2	
Effect	**df1,df2**	**F**	**p**	**F**	**p**
M	1,36	11.51	.002	7.44	.01
M * G	1,36	5.19	.03	4.36	.04
A	1,36	21.16	.0001	36.95	.0001
A * G	1,36	2.88	.10	2.55	.12
H	1,36	0.40	.53	0.22	.64
G * G	1,36	18.23	^{G}.0001$^{1.1}$	19.59	^{G}.0001$^{1.2}$
E	4,144	29.79	.0001	36.66	.0001
E * G	4,144	7.61	^{G}.0004$^{2.1}$	8.62	^{G}.0002$^{2.2}$
M * A	1,36	1.79	.19	2.51	.12
M * A * G	1,36	1.99	.17	2.40	.13
M * H	1,36	0.92	.34	0.60	.44
M * H * G	1,36	1.14	.29	3.11	.09
A * H	1,36	1.32	.26	2.38	.13
A * H * G	1,36	8.87	^{A}.005$^{3.1}$	10.25	^{A}.003$^{3.2}$

[45] $F(1,18)=36.48$, $p=.0001$
[46] $F(1,18)=3.95$, $p=.06$

Table 5.9 continued

		PINV-1		PINV-2	
Effect	**df1,df2**	**F**	**p**	**F**	**p**
M * E * G	4,144	1.22	.30	0.86	.44
A * E	4,144	13.31	[G].0001[4.1]	18.95	[G].0001[4.2]
A * E * G	4,144	3.64	[A].03[5.1]	3.79	[A].02[5.2]
H * E	4,144	1.52	.22	2.50	.07
H * E * G	4,144	5.60	[A].002[6.1]	5.57	[A].003[6.2]
M * A * H	1,36	2.84	.10	3.39	.07
M * A * H * G	1,36	0.28	.60	0.31	.58
M * A * E	4,144	0.09	.91	0.06	.95
M * A * E * G	4,144	1.51	.23	1.56	.22
M * H * E	4,144	0.94	.44	0.72	.58
M * H * E * G	4,144	1.85	.12	2.62	[A].04[7.2]
A * H * E	4,144	3.43	[A].02[8.1]	3.82	[A].01[8.2]
A * H * E * G	4,144	0.86	.46	0.87	.46
M * A * H * E	4,144	0.93	.44	1.31	.27
M * A * H * E* G	4,144	0.94	.44	1.07	.37
Group	1,36	4.40	.04	4.65	.04

*Table legend: **M**=Provisional Memory factor, **A**=Ambiguity factor, **H**=Hemisphere factor,**E**=Electrode factor, **G**=Group factor; univariate ANOVA; Huynh-Feldt corrected p-values; average mastoids reference; baseline is tCNV;* [G(G)]=Generator effect,[A(A)]=Amplitude effect as indicated by results of normalization; [1.1]=F(1,36)=19.50, p=.0001; [1.2]=F(1,36)=22.15, p=.0001; [2.1]=F(4,144)=6.51, p=.0007; [2.2]=F(4,144)=7.29, p=.0004; [3.1]=F(1,36)=3.87, p=.06; [3.2]=F(1,36) 3.91, p=.06;[4.1]=F(4,144)=3.43, p=.02; [4.2]=F(4,144)=4.32,p=.01;[5.1]=F(4,144)=2.30, p=.09;[5.2]=F(4,144)=.75,p=.06;[6.1]=F(4,144) =2.06,p=.11;[6.2]=F(1,144)=2.39,p=.07;[7.2]=F(4,144)= 2.21, p=.07;[8.1]=F(4,144)=2.06, p=.09;[8.2]=F(4,144)=2.24, p=.07

The significant results of normalized data indicate that the group difference in hemisphere asymmetry is in accordance with a difference in the generator structures of the PINV.

A significant effect of the **Electrode** factor results from a frontal maximum of the PINV in both groups. Both frontal (F3, F4)[47] and fronto-lateral (F7, F8)[48] PINV amplitudes are significantly larger than the central (C3, C4), which do not differ from temporal (T3, T4) but are significantly larger than parietal (P3, P4)[49] PINV amplitudes. Again, this "overall" pattern must be differentiated for the groups, due to the significant **Electrode by Group** interaction. This interaction results from significant contrasts of (F3, F4)[50], (F7, F8)[51] and (T3, T4)[52], each contrasted with (C3, C4), which does not differ significantly from (P3, P4).

[47] F(1,36)=28.34, p=.0001
[48] F(1,36)=15.31, p=.0008
[49] F(1,36)=32.91, p=.0001
[50] F(1,36)=4.81, p=.03
[51] F(1,36)=15.00, p=.0004
[52] F(1,36)=16.93, p=.0002

The amplitude differences between frontal (F3, F4, F7, F8) and central (C3, C4) electrodes are larger in healthy controls than in schizophrenic patients, who have substantial PINV amplitudes also at central leads.

Figure 5.3a Schizophrenic Patients and Healthy Controls —Low Ambiguity—

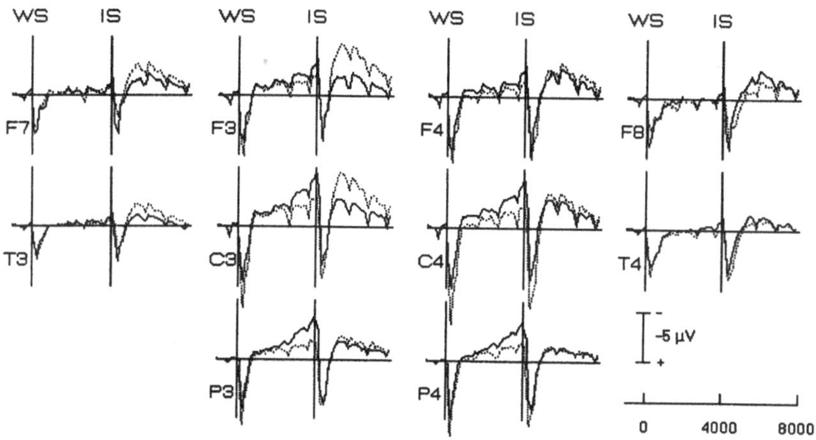

Groups: ______ = Healthy Controls; .... = Schizophrenic Patients

Interestingly, the topographical maximum of the PINV is at (F3, F4) in patients and at (F7, F8) in healthy controls (see **table 5.10**). The within-group ANOVA confirm the significance of the electrode factor in both groups, with a larger F-value in controls[53] compared to patients[54], indicating larger amplitude differences between electrodes in controls. Since the Electrode by Group interaction remains significant after data normalization, different source configurations are suggested to be present in the two groups during PINV. Different topographical patterns in the two groups are also suggested by the significant **Hemisphere by Electrode by Group** interaction (see **table 5.11**). Over the range of the different electrodes there are different lateralizational

[53] $F(4,72)=25.75$, $p=.0001$
[54] $F(4,72)=11.73$, $p=.0001$

patterns in the two groups. Lateralizational differences between (F3, F4)[55] and (P3, P4)[56] on the one hand and (C3, C4) on the other hand differ between the two groups, which do not differ significantly at fronto-temporal ((F7, F8), (T3, T4)) leads.

Table 5.10 Electrode by Group Interaction on PINV

	(F3,F4)	(F7,F8)	(C3,C4)	(T3,T4)	(P3,P4)
Patients	-2.27 (2.7)	-1.59 (2.3)	-1.57 (2.3)	-0.76 (1.3)	-0.28 (1.3)
Controls	-0.77 (1.7)	-1.78 (1.7)	0.90 (2.3)	-0.67 (1.3)	1.82 (2.3)

unit is µV; SD in parentheses

Figure 5.3b Schizophrenic Patients and Healthy Controls

—High Ambiguity—

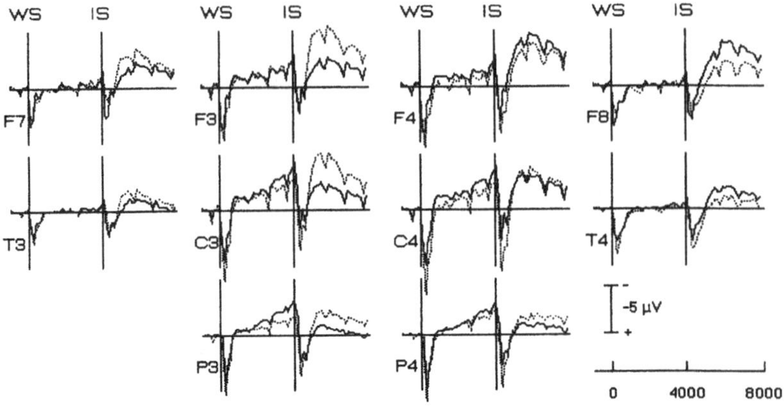

Groups: ______ = Healthy Controls; .... = Schizophrenic Patients

[55] F(1,36)=4.15, p=.05
[56] F(1,36)=6.26, p=.02

Table 5.11 Hemisphere by Electrode by Group Interaction on PINV-1

	(F3,F4)	(F7,F8)	(C3, C4)	(T3,T4)	(P3,P4)
Patients					
Left	-2.89 (2.7)	-2.16 (2.5)	-2.20 (2.2)	-1.23 (1.5)	-0.09 (1.2)
Right	-1.64 (2.7)	-1.02 (2.3)	-0.94 (2.4)	-0.29 (1.3)	0.64 (1.5)
Controls					
Left	0.01 (2.2)	-1.20 (1.3)	1.32 (2.8)	-0.50 (1.4)	1.83 (2.2)
Right	-1.55 (2.3)	-2.36 (1.7)	0.48 (2.6)	-0.85 (1.7)	1.81 (2.6)

unit is µV; SD in parentheses

As the within-group ANOVA reveal, a significant Hemisphere by Electrode interaction is present only in controls[57], but not in patients[58]. According to the results of normalized data, this effect arises from amplitude differences between the groups.

Except for parietal (P3, P4) leads, patients exhibit a left-frontal, controls a right-frontal predominance of the PINV. However, at (F3, F4) the asymmetry in the control group (right > left) is relatively larger than that in patients (left > right). At parietal (P3, P4) leads, an asymmetry of the PINV is lacking in controls only.

To summarize: Topographical differences between schizophrenic patients and controls can be summarized as follows: A broad left-hemisphere PINV maximum in patients contrasts with a right-frontal PINV maximum in controls. Both groups exhibit a frontal PINV predominance, however, patients at (F3, F4) leads and controls at (F7, F8) leads. The electrode effect is larger in controls than in patients, suggesting a "broader" distribution in the latter group. The Group by Hemisphere and the Group by Electrode interactions remain significant after normalization of the data, suggesting group differences in the generator structure of the PINV.

Underlying the significant **Ambiguity** effect is a PINV enhancement due to ambiguous as compared to clear IS (see **figure 5.3**). As suggested by the significant **Ambiguity by Electrode** interaction the effect must be differentiated topographically. This interaction stems from significant contrasts of (F3, F4)[59] and (P3, P4)[60], *each contrasted against* (C3, C4). In both groups ambiguous IS enlarge the PINV most pronounced at frontal (F3, F4) and least pronounced at parietal (P3, P4) leads. The ambiguity effect at (F7, F8) is only descriptively (but not significantly) stronger than that at (C3, C4), and comparable at the latter positions with (T3, T4). Significant results of normalized data suggest that generator structures differ between the conditions.

---

[57] $F(4,72)=4.30$, $p=.02$
[58] $F(4,72)=1.56$, $p=.21$
[59] $F(1,36)=21.95$, $p=.0001$
[60] $F(1,36)=14.75$, $p=.0005$

The significant **Ambiguity by Electrode by Group** interaction indicates, however, that the last mentioned effect must again be differentiated for the two groups. Concerning this interaction only the contrast (F3, F4)[61] versus (C3, C4) is significant, while the contrast (P3, P4)[62] versus (C3, C4) just falls short of significance (non-significant contrasts also for (F7, F8) and (T3, T4) versus (C3, C4)).

Table 5.12 Ambiguity by Electrode by Group Interaction on PINV

	(F3,F4)	(F7,F8)	(C3, C4)	(T3,T4)	(P3,P4)
Patients					
Clear	-1.96 (2.6)	-1.38 (2.1)	-1.39 (2.4)	-0.71 (1.1)	-0.35 (1.3)
Ambiguous	-2.57 (2.7)	-1.79 (2.5)	-1.75 (2.2)	-0.81 (1.5)	0.21 (1.3)
Controls					
Clear	0.05 (2.1)	-1.35 (1.2)	1.24 (2.4)	-0.50 (1.3)	1.81 (2.3)
Ambiguous	-1.59 (1.6)	-2.20 (1.3)	0.56 (2.3)	-0.85 (1.4)	1.84 (2.5)

unit is µV; SD in parentheses

Table 5.12 reveals that the ambiguity effect is larger at frontal (F3, F4) than at central (C3, C4) sites and that this difference between electrodes is more pronounced in the control[63] than in the patient[64] group. Except for (P3, P4), slightly larger ambiguity effects can be seen in controls than in patients at almost all of the remaining sites; only at (P3, P4) a PINV augmentation can be seen in patients but is absent in controls. The non-significant results of normalized data suggest that an amplitude effect cannot be excluded from this interaction effect.

Another topographical differentiation of the ambiguity effect is documented by the significant **Ambiguity by Hemisphere by Group** interaction. **Table 5.13** illustrates that the PINV enhancement in controls due to ambiguous IS is larger over the right than over the left hemisphere, while the reverse is true of the patient group. The within-group ANOVA, however, show a significant Ambiguity by Hemisphere interaction only for the control[65] group, but not for patients[66]. An amplitude effect cannot be excluded for this interaction, since the normalized data reveal non-significant results.

A final topographical differentiation of the Ambiguity effect is the significant **Ambiguity by Hemisphere by Electrode** interaction (see **table 5.14**).

[61] $F(1,36)=7.62, p=.009$
[62] $F(1,36)=3.91, p=.06$
[63] $F(4,72)=13.29, p=.0001$
[64] $F(4,72)=1.99, p=.16$
[65] $F(1,18)=6.03, p=.02$
[66] $F(1,18)=2.85, p=.10$

Left-right differences in the ambiguity effect are different at frontal (F3 versus F4) and central (C3 versus C4) positions[67], since amplitude enhancements due to ambiguous IS are comparable for the two central positions but are larger at F4 compared to F3.

Table 5.13 Ambiguity by Hemisphere by Group Interaction on PINV

	Clear		Ambiguous	
	Left	Right	Left	Right
Patients	-1.47 (1.7)	-0.57 (1.8)	-1.96 (1.9)	-0.73 (1.7)
Controls	0.46 (1.7)	0.03 (1.8)	0.13 (1.7)	-1.02 (2.0)

unit is μV; SD in parentheses

Table 5.14 Ambiguity by Hemisphere by Electrode Interaction

	(F3,F4)	(F7,F8)	(C3, C4)	(T3,T4)	(P3,P4)
Clear					
Left	-1.01 (2.9)	-1.43 (1.9)	-0.20 (3.0)	-0.74 (1.4)	0.85 (1.9)
Right	-0.90 (2.6)	-1.31 (1.9)	0.04 (2.7)	-0.47 (1.4)	1.30 (2.1)
Ambiguous					
Left	-1.97 (3.0)	-1.93 (2.3)	-0.69 (3.1)	-0.99 (1.7)	0.90 (2.2)
Right	-2.30 (2.5)	-2.06 (2.4)	-0.51 (2.6)	-0.67 (1.8)	1.20 (2.3)

unit is μV; SD in parentheses

The overall significant **Provisional Memory** effect on PINV indicates an enhancement of the PINV with high provisional memory load. This effect must be differentiated by groups, since the interaction of **Provisional Memory by Group** is significant, too.

As the within-group ANOVA reveal, only schizophrenic patients[68] but not control subjects[69] exhibit an enhancement in PINV amplitudes from low to high provisional memory load conditions (see **table 5.15** and **figure 5.4**). Only for PINV-2 does this effect enter into a higher-order interaction, as the significant **Provisional Memory by Hemisphere by Electrode by Group** interaction, for which an amplitude effect cannot be rejected by the normalization of the data, indicates. This interaction rises from a significant contrast between the fronto-lateral (F7, F8) and central (C3, C4) leads in the group-specific lateralization of the Provisional Memory effect[70].

---

[67] $F(1,36)=16.50$, $p=.0003$
[68] $F(1,18)=15.60$, $p=.0009$
[69] $F(1,18)=0.64$, $p=.43$
[70] $F(1,36)=11.01$, $p=.002$

While the augmentation of the PINV in schizophrenic patients is relatively weak at the F8 lead, compared to the F7, C4, and C3 leads, healthy controls exhibit a weak PINV augmentation at C3 and a somewhat stronger PINV diminution at F7, with almost no changes at F8 and C4, and with high as compared to low provisional memory load.

Figure 5.4a Schizophrenic Patients and Healthy Controls
—Low Provisional Memory Load—

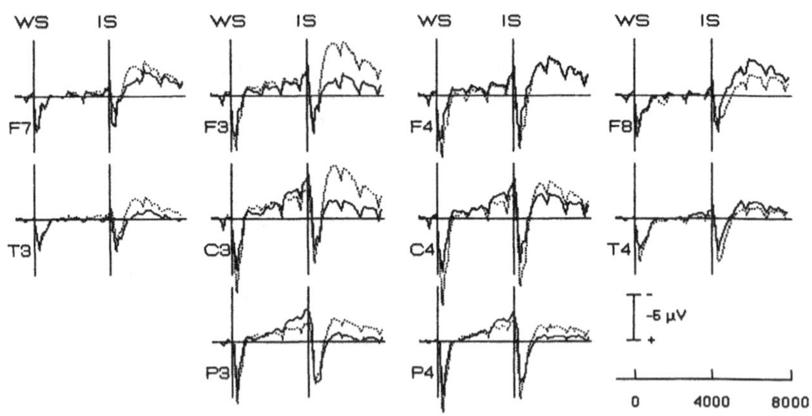

Groups: ______ = Healthy Controls; .... = Schizophrenic Patients

Table 5.15 Provisional Memory by Group Interaction on PINV

	Low Memory Load	High Memory Load
Patients	-0.74 (1.6)	-1.62 (1.9)
Controls	-0.01 (1.5)	-0.10 (1.6)

unit is µV, SD in parentheses

Figure 5.4a Schizophrenic Patients and Healthy Controls —High Provisional Memory Load—

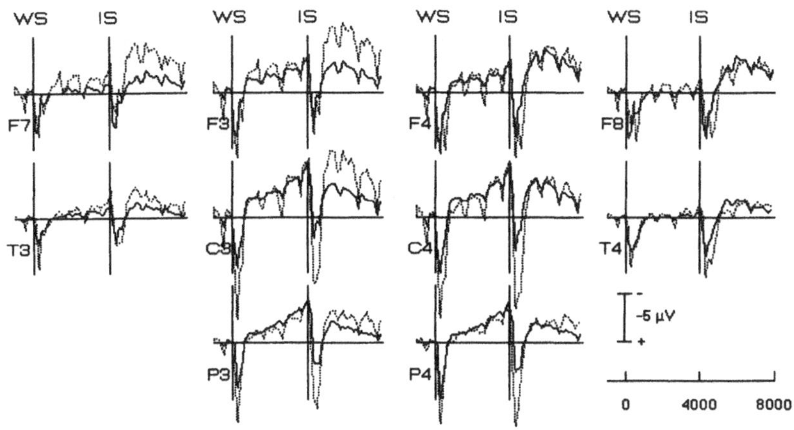

Groups: ______ = Healthy Controls; .... = Schizophrenic Patients

Table 5.16 Provisional Memory by Hemisphere by Electrode by Group Interaction on PINV-2

	(F3,F4)	(F7,F8)	(C3, C4)	(T3,T4)	(P3,P4)
Patients					
Low Memory Load					
Left	-2.25 (2.3)	-1.70 (2.2)	-1.58 (2.3)	-0.81 (1.3)	0.36 (1.4)
Right	-1.57 (2.3)	-1.33 (1.7)	0.64 (2.4)	-0.40 (1.0)	0.88 (1.6)
High Memory Load					
Left	-3.36 (3.2)	-2.62 (2.7)	-2.49(2.6)	-1.54 (1.7)	0.43 (1.0)
Right	-2.21 (3.0)	-1.51 (2.6)	-1.49(2.5)	-0.78 (1.6)	0.23 (1.3)
Controls					
Low Memory Load					
Left	0.04 (2.2)	-1.60 (1.4)	1.64 (2.7)	-0.64 (1.6)	2.15 (2.1)
Right	-1.76 (2.3)	-2.71 (1.8)	0.57 (2.6)	-0.90 (1.8)	2.04 (2.3)
High Memory Load					
Left	-0.05 (2.4)	-1.07 (1.6)	-1.31 (2.7)	-0.41 (1.3)	1.68 (2.3)
Right	-1.66 (2.4)	-2.71 (1.8)	-0.45 (2.7)	-1.12 (1.7)	1.46 (2.9)

unit is µV; SD in parentheses

The two conditions effects may be summarized as follows: In both groups ambiguous compared to clear IS matching conditions result in augmented PINV amplitudes, particularly at frontal leads and presumably because of a change in the generator structure. The frontal accentuation of the ambiguity effect is stronger in controls than in patients, but this interaction is more compatible with an amplitude effect. The groups differ in the lateralization of the ambiguity effect, which clearly predominates over the right hemisphere in controls but is less asymmetric in patients. Again, this effect is non-significant for the normalized data. From low to high provisional memory load a statistically significant enhancement in PINV amplitudes is only found in the patient group.

5.9.2.4 Late Positive Complexes (LPC)

Since the LPC following the WS and IS are related to the evaluation of those stimuli that are varied in order to manipulate the PINV, it seems meaningful to scrutinize the experimental effects and the topographies of the LPC.

The WS-LPC is scarcely affected by the experimental task variations, but its amplitudes require some topographical differentiation (see **table 5.16**). The significant **Hemisphere** factor reveals that WS-LPC peak amplitudes are larger over the right than over the left hemisphere (see **table 5.17**). Furthermore, as indicated by the significant **Electrode** main effect, WS-LPC peak amplitudes are larger over central (C3, C4) than over parietal (P3, P4)[71] or frontal (F3, F4)[72] positions; in addition fronto-lateral (F7, F8)[73] and temporal (T3, T4)[74] amplitudes are significantly smaller than the central amplitudes (see **table 5.18**).

The only condition-related effect on the WS-LPC peak amplitude is the **Provisional Memory by Hemisphere** interaction (see **table 5.17**). While there is a diminution of peak amplitudes from the low to the high memory level over the left hemisphere, a slight enhancement can be observed over the right hemisphere with the change of task conditions.

A significant **Electrode** effect can be found for the WS-LPC peak latency, too (see **table 5.19**). The WS-LPC peaks earliest over central (C3, C4)and parietal (P3, P4) regions, which do not differ significantly[75]. While frontal (F3, F4)[76] latencies do not differ from central, at fronto-lateral (F7, F8)[77] and temporal (T3, T4)[78] positions peak latencies occur significantly later.

Table 5.20 reveals the that enhancement of provisional memory load has different effects on peak latencies in patients and controls, as statistically sup-

[71] $F(1,36)= 10.25, p=.0003$
[72] $F(1,36)=18.50, p=.0001$
[73] $F(1,36)= 60.52, p=.0001$
[74] $F(1,36)= 107.85, p=.0001$
[75] $F(1,36)=0.81, p=.38$
[76] $F(1,36)=1.79, p=.19$
[77] $F(1,36)=7.98, p=.008$
[78] $F(1,36)=7.11, p=.01$

Table 5.16 Experimental Effects on WS-LPC

		Amplitude		Latency	
Effect	**df1, df2**	**F**	**p**	**F**	**p**
M	1,36	0.34	.57	1.54	.22
M * G	1,36	0.79	.38	1.64	.21
A	1,36	3.91	.06	0.02	.88
A * G	1,36	0.05	.82	0.13	.73
H	1,36	22.88	.0001	2.63	.11
H * G	1,36	1.96	.17	0.36	.55
E	4,144	28.51	.0001	4.65	.01
E * G	4,144	1.67	.19	2.61	.08
M * A	1,36	0.20	.66	0.36	.55
M * A * G	1,36	2.15	.15	0.60	.44
M * H	1,36	7.99	[G].008[1]	0.81	.37
M * H * G	1,36	0.15	.70	2.89	.10
A * H	1,36	0.62	.44	0.01	.94
A * H * G	1,36	0.06	.81	0.17	.69
M * E	4,144	2.78	.07	0.13	.95
M * E * G	4,144	1.86	.16	2.38	.06
A * E	4,144	0.70	.51	0.40	.74
A * E * G	4,144	0.55	.60	0.81	.49
H * E	4,144	0.20	.86	5.48	.0009
H * E * G	4,144	0.30	.78	2.51	.05
M * A * H	1,36	1.41	.24	0.43	.52
M * A * H * G	1,36	0.79	.38	0.09	.76
M * A * E	4,144	0.72	.51	1.10	.36
M * A * E * G	4,144	1.85	.16	0.34	.83
M * H * E	4,144	1.47	.24	1.59	.19
M * H * E * G	4,144	1.83	.17	0.65	.61
A * H * E	4,144	1.25	.29	0.14	.96
A * H * E * G	4,144	2.05	.15	2.16	.08
M * A * H * E	4,144	0.42	.59	1.63	.18
M * A * H * E * G	4,144	0.75	.43	2.09	.10
Group	1,36	2.18	.15	0.21	.65

M=*Provisional Memory factor,* ***A***=*Ambiguity factor,* ***H***=*Hemisphere factor,* ***E***=*Electrode factor,* ***G***=*Group factor; Results of univariate ANOVA, Hunynh-Feldt corrected p values;* [G]*=Generator effect as indicated by results of normalization;* [1]$=F(1,36)=6.37, p=.02$

Table 5.17 Provisional Memory by Hemisphere Interaction on WS-LPC Peak Amplitude

	Low Memory Load	High Memory Load
Left	5.37 (2.4)	5.23 (2.1)
Right	6.42 (2.2)	6.52 (2.1)

unit is μV, SD in parentheses

Table 5.18 Electrode Effect on WS-LPC Peak Amplitude

(F3,F4)	(F7,F8)	(C3,C4)	(T3,T4)	(P3,P4)
6.47 (2.9)	4.39 (2.2)	7.94 (3.0)	6.68 (2.6)	4.39 (1.4)

unit is µV, SD in parentheses

ported by the significant **Provisional Memory by Electrode by Group** interaction. However, there is no significant contrast underlying this effect. Descriptively, there is a temporo-parietal enhancement and a centro-frontal diminution in peak latencies in healthy controls with high compared to low provisional memory load, while the delay of peaking is pronounced over frontal and central, but absent over temporo-parietal positions in patients. At frontal ((F3, F4), (F7, F8)) leads patients exhibit clearly longer latencies than controls, while the reverse holds for the centro-parietal leads.

Table 5.19 Hemisphere by Electrode by Group Interaction on WS-LPC Peak Latency

	(F3,F4)	(F7,F8)	(C3,C4)	(T3,T4)	(P3,P4)
Patients					
Left	297.8(57.5)	319.3(78.6)	290.3(33.6)	322.9(69.1)	282.0(33.9)
Right	336.7(87.5)	365 (100.1)	304.1(55.6)	315.7(62.2)	289.0(45.3)
Controls					
Left	298.6(67.4)	293.6(76.7)	296.1(38.2)	302.8(50.8)	317.2(51.4)
Right	300.8(55.9)	330.8(90.0)	307.2(53.3)	321.7(60.0)	292.8(40.7)

unit is ms, SD in parentheses

Table 5.20 Provisional Memory by Electrode by Group Interaction on WS-LPC Peak Latency

	(F3,F4)	(F7,F8)	(C3,C4)	(T3,T4)	(P3,P4)
Patients					
low	310.1(60.1)	328.4(72.8)	291.5(34.4)	317.8(58.0)	285.7(34.0)
high	324.3(64.3)	355.9(86.6)	303.3(46.4)	320.8(42.6)	286.2(35.5)
Controls					
low	302.6(63.1)	316.8(87.4)	304.2(48.9)	308.8(45.0)	299.0(39.7)
high	296.7(61.3)	307.5(75.2)	299.1(39.3)	315.7(55.2)	311.3(48.5)

unit is µV, SD in parentheses

Table 5.21 Experimental Effects on IS-LPC

		Amplitude		Latency	
Effect	**df1, df2**	**F**	**p**	**F**	**p**
M	1,36	2.92	.10	0.57	.46
M * G	1,36	3.09	.08	0.01	.99
A	1,36	58.16	.0001	0.33	.57
A * G	1,36	13.11	.0009	0.04	.84
H	1,36	20.10	.0001	2.14	.13
H * G	1,36	4.43	^{G}.04[1]	0.13	.72
E	4,144	35.98	.0001	3.16	.03
E * G	4,144	0.97	.39	12.61	.0001
M * A	1,36	0.06	.81	0.85	.36
M * A * G	1,36	0.34	.56	4.74	.04
M * H	1,36	0.22	.64	0.32	.58
M * H * G	1,36	0.61	.44	0.32	.57
A * H	1,36	0.46	.50	0.28	.60
A * H * G	1,36	0.26	.61	0.21	.65
M * E	4,144	2.70	.06	1.25	.30
M * E * G	4,144	1.45	.24	0.75	.53
A * E	4,144	12.15	^{G}.0001[2]	8.59	.0001
A * E * G	4,144	1.13	.32	2.12	.09
H * E	4,144	1.92	.13	4.79	.001
H * E * G	4,144	0.69	.55	1.74	.15
M * A * H	1,36	0.86	.36	6.50	.02
M * A * H * G	1,36	3.87	.06	2.31	.14
M * A * E	4,144	1.06	.36	2.04	.09
M * A * E * G	4,144	1.90	.14	0.95	.43
M * H * E	4,144	1.83	.14	1.56	.19
M * H * E * G	4,144	0.41	.76	1.13	.34
A * H * E	4,144	0.62	.64	2.22	.08
A * H * E * G	4,144	1.06	.38	0.56	.68
M * A * H * E	4,144	1.30	.28	1.50	.21
M * A * H * E * G	4,144	1.13	.34	0.24	.90
Group	1,36	0.06	.81	2.63	.11

M=Provisional Memory factor, A=Ambiguity factor, H=Hemisphere factor, E=Electrode factor, G=Group factor; Results of univariate ANOVA, Huynh-Feldt corrected p-values; $^{G(G)}$=Generator effect as indicated by results of normalization; [1]=F(1,36)=5.13, p=.03; [2]=F(1,36)=5.67, p=.001

The significant **Hemisphere by Electrode** interaction comprises significant contrasts between fronto-lateral[79] and parietal[80] positions on the one hand and central positions on the other hand (see **table 5.19**).

This interaction, however, must be differentiated for the two groups, as the significant interaction of **Hemisphere by Electrode by Group** reveals (see **table**

[79] F(1,36)=5.90, p=.02

[80] F(1,36)=5.28, p=.03

5.19). Concerning this interaction, only the contrast between frontal (F3, F4) and central (C3, C4) positions approaches conventional significant level[81]. While in controls peak latencies do not substantially differ between these positions, in patients right-frontal latencies are larger than left-frontal, but left and right central latencies do not differ to that extent.

Table 5.21 summarizes the results for the IS-LPC. In both groups the IS-LPC is larger over the right than over the left hemisphere, as the significant **Hemisphere** effect indicates. This difference is, however, larger in patients than in controls, as statistically confirmed by the significant interaction of **Hemisphere by Group** (see **table 5.22**). The interaction effect turns out significant after normalization of the data, suggesting different underlying source configurations in the two groups.

The significant **Electrode** main effect reveals that IS-LPC peak amplitudes are largest at the central leads. The significant contrasts between the parietal[82], frontal[83], fronto-lateral[84], and temporal[85] leads on the one hand and the central leads on the other hand underline these amplitude differences. This effect can be followed in **table 5.24** by comparing the columns pertaining to the different electrode levels.

Ambiguous IS result in IS-LPC *diminutions* in both groups, as the significant **Ambiguity** effect shows. This amplitude reduction, however, is larger in controls than in schizophrenic patients, as confirmed by the significant **Ambiguity by Group** interaction.

The within-group ANOVA reveal a significant Ambiguity effect in patients[86] which is even larger in controls[87] (see **table 5.23**).

Table 5.22 Hemisphere by Group Interaction on IS-LPC Peak Amplitude

	Left Hemisphere	Right Hemisphere
Patients	6.45 (2.2)	7.95 (2.7)
Controls	6.75 (2.2)	7.29 (2.4)

unit is µV, SD in parentheses

This "overall" amplitude reduction has a topography which does not distinguish groups, as attested by the significant interaction of **Ambiguity by Electrode.** The ambiguity effects are largest at the central (C3, C4) and parietal

[81] F(1,36)=3.41, p=.06
[82] F(1,36)=6.84, p=.01
[83] F(1,36)=9.83, p=.006
[84] F(1,36)=60.45, p=.0001
[85] F(1,36)=127.23, p=.0001
[86] F(1,18)=8.50, p=.009
[87] F(1,18)=59.85, p=.0001

(P3, P4) leads (which do not differ significantly), but significantly smaller at frontal (F3, F4)[88], fronto-lateral (F7, F8)[89], and temporal (T3, T4)[90] leads, when compared with the central ambiguity effects.

Since the normalized data reveal a significant interaction effect, too, a change in the underlying generator structure can be assumed.

In addition to these significant ambiguity effects, the interaction of **Provisional Memory by Group** *approaches* the conventional significance level. Although this interaction is statistically only a trend, it is of theoretical interest, since it precedes a differential provisional memory effect in PINV amplitudes. As was the case for the PINV, only schizophrenic patients respond to increased loads on provisional memory functions with an amplitude enhancement in the IS-LPC, an effect which is virtually absent in the control group (see **table 5.23**).

Table 5.23 Ambiguity by Group *and* Provisional Memory by Group Interactions on IS-LPC Peak Amplitude

	Ambiguity		**Provisional Memory**	
	Clear	Ambiguous	Low	High
Patients	7.48 (2.4)	6.92 (2.3)	6.56 (2.8)	7.84 (2.8)
Controls	7.80 (2.2)	6.24 (2.2)	7.03 (2.4)	7.01 (2.1)

unit is µV, SD in parentheses

Table 5.24 Ambiguity by Electrode Interaction on IS-LPC Peak Amplitude

	(F3,F4)	**(F7,F8)**	**(C3,C4)**	**(T3,T4)**	**(P3,P4)**
Clear	8.40 (3.2)	5.60 (2.7)	10.07 (3.4)	5.09 (1.6)	9.02 (3.5)
Ambiguous	7.25 (3.3)	5.03 (2.9)	8.58 (3.3)	4.59 (1.7)	7.47 (3.1)

unit is µV, SD in parentheses

The latency aspect of the IS-LPC shows topographical differences, as indicated by the significant **Electrode** factor. The contrasts between frontal[91], fronto-lateral[92], temporal[93], and parietal[94] leads on the one hand, and central (C3, C4) leads on the other hand are significant (see **table 5.25**). This effect must, however, be distinguished in a two-fold way.

---

88 $F(1,36)=4.03$, $p=.05$
89 $F(1,36)=15.71$, $p=.0003$
90 $F(1,36)=30.53$, $p=.0001$
91 $F(1,36)=5.28$, $p=.03$
92 $F(1,36)=4.41$, $p=.04$
93 $F(1,36)=7.15$, $p=.01$
94 $F(1,36)=8.37$, $p=.007$

Table 5.25 Electrode by Group Interaction on IS-LPC Peak Latency

	(F3,F4)	(F7,F8)	(C3,C4)	(T3,T4)	(P3,P4)
Patients	414.2(68.4)	425.9(89.6)	413.3(58.1)	414.7(53.8)	387.4(37.7)
Controls	372.7(54.9)	350.4(55.2)	407.2(63.0)	360.4(50.5)	438.8(63.6)

unit is ms, SD in parentheses

Firstly, the significant interaction of **Electrode by Group** suggests differences in the latency patterns between the two groups. This interaction comprises significant contrasts between the frontal[95], fronto-lateral[96], temporal[97], and parietal[98] leads and the central leads. While frontal and central IS-LPC peak roughly simultaneously in patients, peak amplitudes are delayed at central compared to frontal leads in controls. Compared to central latencies fronto-lateral latencies are delayed in patients, but occur earlier in controls. The reverse holds for parietal latencies which are reduced in patients but increased in controls, when compared to the central leads. Temporal latencies are similar to central latencies in patients, but reduced in controls (see **table 5.25**).

Secondly, the significant **Hemisphere by Electrode** interaction indicates, that while peak latencies are smaller over the right than over the left hemisphere at central, temporal, and parietal leads, this pattern cannot be found at frontal[99] and fronto-lateral[100] positions (see **table 5.26**).

Table 5.26 Hemisphere by Electrode Interaction on IS-LPC Peak Latency

	(F3,F4)	(F7,F8)	(C3,C4)	(T3,T4)	(P3,P4)
Left	391.6(64.3)	382.8(87.6)	424.6(72.5)	390.3(75.7)	426.3(69.2)
Right	393.3(75.2)	393.4(85.6)	395.9(60.9)	384.9(60.9)	399.9(59.7)

unit is ms, SD in parentheses

Ambiguous, as compared to clear IS alter the peak latencies at different leads to a different extent, as indicated by the significant **Ambiguity by Electrode** interaction. While ambiguous IS are associated with latency diminutions at central leads, prolongations of peak latencies can be seen at frontal[101], fronto-lateral[102] sites. However, at the parietal leads the latency

[95] $F(1,36)=4.68$, $p=.04$
[96] $F(1,36)=10.84$, $p=.002$
[97] $F(1,36)=8.09$, $p=.005$
[98] $F(1,36)=8.37$, $p=.007$
[99] $F(1,36)=9.30$, $p=.004$
[100] $F(1,36)=11.10$, $p=.002$
[101] $F(1,36)=4.20$, $p=.05$
[102] $F(1,36)=4.34$, $p=.04$

diminution is even stronger than at the central leads[103]. (Almost) no change can be seen at the temporal leads, but this does not differ significantly from the central change (see **table 5.27**).

Table 5.27 Ambiguity by Electrode Interaction on IS-LPC Peak Latency

	(F3,F4)	(F7,F8)	(C3,C4)	(T3,T4)	(P3,P4)
Clear	387.9(67.4)	384.6(85.6)	414.1(63.5)	387.4(59.0)	430.9(67.7)
Ambiguous	397.0(76.6)	393.6(87.5)	406.5(75.4)	387.8(67.1)	395.3(62.1)

unit is ms, SD in parentheses

There is a group-specific pattern of the ambiguity by provisional memory interaction, as revealed by the significant **Provisional Memory by Ambiguity by Group** interaction on IS-LPC peak latencies. The within-group ANOVA show a significant Provisional Memory by Ambiguity interaction only in the patient[104], but not in the control group[105]. While in patients there is a weak enhancement of peak latencies within the low ambiguity condition from the low memory load to the high memory load condition, the same variation is associated with a latency reduction under the high ambiguity condition. On a much smaller scale the reverse pattern can be seen in the control group.

Table 5.28 Provisional Memory by Ambiguity by Group Interaction on IS-LPC Peak Latency

	Low Memory Load		High Memory Load	
	Clear	Ambiguous	Clear	Ambiguous
Patients	409.8 (52.3)	416.3 (61.8)	417.6 (68.3)	399.1 (64.5)
Controls	392.3 (41.4)	384.2 (63.2)	382.5 (38.0)	384.6 (63.8)

unit is ms, SD in parentheses

5.10 Discussion of Results

5.10.1 Reaction Times

Delayed motor responses, a common finding in schizophrenic patients (King, 1991), was replicated in the present study. Enlarged response times under high provisional memory conditions may indicate that additional stages of response selection were required under this condition. Instead of associations

[103] F(1,36)=6.86, p=.01
[104] F(1,18)=5.89, p=.03
[105] F(1,18)=0.67, p=.42

between IS and response hand that were stable over the trials under low provisional memory conditions, under high provisional memory load conditions contingencies between IS and response side had to be created *depending on* the trial-specific WS type. Assuming that ambiguous IS would not go undetected by the subjects as physically deviating, longer response times under the high ambiguity condition were also expected and may indicate additional time needed to cope with conflicting response tendencies.

5.10.2 Initial and Terminal CNV

Groups did not differ in their iCNV amplitudes. Storage and recall of conditional probabilities (Rockstroh et al., 1989) and orientation (Rohrbaugh & Gaillard, 1983), as putatively reflected by the iCNV (see *chapter 3*), therefore appear not to be impaired in the present sample of schizophrenic patients. The significant enhancement of the iCNV particularly at frontal (F3, F4) and central (C3, C4) positions with the high provisional memory load is in line with results suggesting sensitivity of this component to the information value of the WS (McCarthy & Donchin, 1978). Since the Provisional Memory by Electrode interaction turned out significant after normalization of the data, a *qualitative* topographical change of the generator structure of the iCNV is suggested. The non-significant interaction between the provisional memory and the group factor suggests that schizophrenic patients are as sensitive as controls to the experimental variation of this aspect of task demands.

Terminal CNV was enhanced in a similar way as the iCNV under the condition of high provisional memory load in both groups, with this effect extending over frontal, central, and parietal leads. This suggests that higher provisional memory load was associated with an intensified preparedness for IS evaluation and responding. CNV has been shown to be positively related to task complexity (McCallum, 1986). Schizophrenic patients exhibited neither reduced CNV amplitudes nor less responsiveness to this task variation. This result contrasts with reports of generally reduced CNV amplitudes and reduced reactivity in schizophrenic patients (cf. Cohen, 1990). One explanation may be the psychopathological state of the patients who were sufficiently remitted in order to participate in a two-session, two hours each, EEG task. Indeed, CNV amplitude increases after the onset of neuroleptic therapy or with remission have been reported (Knott et al., 1976; Rizzo et al 1983; Verhey et al., 1984).

It is interesting to note that iCNV and tCNV yielded rather similar results. In particular, the same left-hemisphere predominance and a similar topography of the provisional memory effect (maximal at frontal and central leads) were found for the two components. Both components have been suggested to reflect different and independent, as well as independently inducable processes (Rohrbaugh & Gaillard, 1983). Since the interval for the parametrization of the iCNV was chosen rather late with respect to the large WS-LPC, it is possible that an overlap of iCNV and, to a lesser extent, tCNV was present during the iCNV interval. Alternatively, it seems plausible that a "true" provisional memory effect on iCNV was carried over to the tCNV, "mimicing" a provisional memory by electrode interaction for that component. In *chapter 3* it has been argued that many of the experiments designed to disentangle the assumed "stimulus anticipation" and "response preparation" proportions of the tCNV yielded equivocal results since the experimental variations targeted at modulating the tCNV also affected the preceding components (especially the iCNV). An experimental variation of the tCNV may therefore be secondary to an analogous variation of the iCNV. This interpretation might explain the fronto-central predominance of the memory effect on the centro-parietal tCNV. It is more plausible that intensified provisional memory requirements had an impact on stimulus storage and contingency formation (assumed to be reflected by the iCNV) than on the preparation of response and stimulus evaluation (assumed to be reflected by the tCNV), since neither response, nor IS features were varied by the provisional memory factor.

It is therefore tentatively concluded that the provisional memory effect on the tCNV is secondary to the provisional memory effect on the iCNV.

5.10.3 WS-LPC and IS-LPC

The groups did not differ significantly in the overall amplitude or the scalp distribution of the WS-LPC. This result conflicts with the results reported in the literature concerning the P300 (see *chapter 3*). Even the reported increase of the *visual* P300 amplitudes with clinical improvement following neuroleptic treatment (Duncan et al., 1987) should not amount to *larger* than normal peak amplitudes in schizophrenic patients, with the Group main effect reaching an F value of still 2.18 ($p=.15$). However, since this group difference is not statistically substantiated, further speculations will not be undertaken.

The WS-LPC showed a broad topographical distribution. Its central predominance is compatible with the maximal superposition at the geometrical center of an array of distributed sources (Lutzenberger et al., 1987). Despite this, a right-hemisphere preponderance could be observed. This might suggest that memory storage and updating of information, presumably underlying the P300/LPC (cf. Donchin & Coles, 1988), was more pronounced right- than left-hemispherically in the present experiment, compatible with the rather spatial character of the experimental stimuli.

The IS-LPC, like the WS-LPC, was larger over the right than over the left hemisphere, also possibly reflecting the rather spatial nature of the experimental stimuli. This lateralizational pattern holds for both groups, although with larger hemisphere differences in schizophrenic patients. Further similarities between WS- and IS-LPC are the central predominance and the larger amplitudes in patients than in controls, which have been discussed for the WS-LPC.

The IS-LPC peak amplitudes proved to be more sensitive than the WS-LPC to the experimental task variations. Ambiguous, when compared to clear IS resulted in IS-LPC amplitude decreases primarily at central and parietal leads, and this effect was much larger in the control than in the patient group. This may be a rather trivial effect of the different relative frequencies of the clear and ambiguous IS, since P300 is sensitive to stimulus probabilities (cf. Donchin & Coles, 1988). Yet, other interpretations appear more plausible. One of the determinants of the P300 according to the Triarchic Model (Johnson, Jr., 1986) is equivocation, which affects information transmission. Equivocation is a function of the aposteriori uncertainty about having correctly perceived a given event (see *chapter 3*). This uncertainty should have been intensified with the presentation of ambiguous IS, which was expected to result in increased response uncertainty. The larger amplitude reduction in controls than in patients might indicate a greater difference in perception-related (un)certainty about stimulus identification between the clear and the ambiguous IS in controls. A similar interpretation is supported by the "Context Closure" model of the P300 by Verleger (1988). Since the ambiguous IS were less integratable into a meaningful context than the clear IS, an amplitude decrease is compatible with a diminished context closure. Patients exhibited a reduced sensitivity to this condition effect when compared with controls. This differential sensitivity of the groups might indicate a *generally* decreased sensitivity of the patient group to experimental task variations. This interpretation is, yet, not plausible since the preceding (and some of the following) ERP components revealed no hint to a decreased sensitivity in patients. Similarly, a generally impaired stimulus integration into the task context (closure) cannot be assumed in patients, because there was no overall IS-LPC amplitude diminution in this group.

5.10.4 PINV

Ambiguous IS enhanced PINV amplitudes in both groups by comparable amounts, suggesting the experience of ambiguous stimulus-stimulus or stimulus-response contingencies to be a *general* antecedent condition for the development of a PINV, "stressing" no schizophrenia-specific processing characteristics. The significant Ambiguity by Electrode interaction, which remained significant after normalization of the data, indicates a topographical maximum of the Ambiguity effect at frontal (F3, F4) leads which is compatible with a shift of the underlying generator structure towards these regions with ambiguous compared to clear IS. As indicated by the significant Ambiguity by Electrode by Group interaction, the frontal predominance of the Ambiguity effect was more pronounced in healthy controls, with larger effects in schizophrenic patients at parietal (P3, P4) leads. This "broader" topography of the Ambiguity effect in schizophrenic patients is also indicated by the Ambiguity by Electrode interaction of the within-group ANOVA which was larger in controls ($p=.0001$) than in patients ($p=.16$). The Ambiguity by Electrode by Group interaction vanished, however, after normalization of the data, suggesting *no* qualitative group differences in the anterior-posterior distribution of the ambiguity effect between groups. Altogether these results indicate a similarity of the responses to ambiguous IS in the two groups, both with respect to magnitude and the non-lateralized topography of this effect.

A frontal predominance (Fz) has also been reported for the PINV augmentation from "constant lack of control" to "random lack of control" conditions by Kathmann et al. (1990). The topographical similarity between these two experimental effects suggest common factors to be present with both of them. This possibility will be discussed in *chapter 9*.

While the frontal predominance of the Ambiguity effect underlines the assumed generality of this effect, its lateralization, which differed between groups, might be seen to demarcate group differences in the reaction to ambiguous IS. The significant Ambiguity by Hemisphere by Group interaction revealed a larger right-sided PINV augmentation in controls and a larger left-sided augmentation in patients. Yet, this interaction effect turned out non-significant after normalization of the data, not excluding a simple amplitude effect. Since healthy controls exhibited under both experimental conditions a right-sided predominance of the PINV and schizophrenic patients a left-sided, the effect of ambiguous IS would be largest at the side of maximum PINV amplitudes. This "common rule" of the lateralization of the Ambiguity effect would again indicate the similarity of the response of both groups to ambiguous IS. The within-group ANOVA, however, reveal a significant Ambiguity by Hemisphere interaction only in controls ($p=.02$), but not in patients ($p=.10$), suggesting that the "common rule" mentioned above would not exist. Since the Ambiguity by Hemisphere interaction was present in patients at least as a statistical trend, it appears none the less warranted to conclude that in both groups ambiguous IS augmented the PINV over the side of its topographical maximum, confirming that both groups responded in comparable ways to ambiguous IS.

PINV amplitudes were also affected by the Provisional Memory factor, with larger PINV under conditions of enhanced memory load. The significant Provisional Memory by Group interaction and the within-group Provisional Memory effects (significant only for the patient group) indicate that this effect could be observed *only* in schizophrenic patients. Hence, it is likely that the provisional memory load specifically activates processes underlying or contributing to the PINV of schizophrenic patients, but not of controls. In this respect the Provisional Memory factor differs from the Ambiguity factor. According to the "hypofrontality" hypothesis of schizophrenia (e.g. Andreasen et al., 1992) dysfunctions of the prefrontal cortex may be assumed to be present in schizophrenic patients and warrant the deduction that electrophysiological responses to experimental variations of PFC involvement in task execution should differ between the two groups. This is especially true of delay tasks whose theoretical importance for research on schizophrenia has recently been emphasized by Parks & Holzman (1992, 1993).

However, envisaging the abundant functional interconnections between the PFC and other cortical and subcortical structures, most noteably the basal ganglia and the temporo-limbic system including the hippocampus (cf. Fuster, 1989), disentangling PFC contributions to task performance from those of other brain structures is most difficult. Furthermore, experimentation (e.g. Rubin et al., 1991; Siegel et al., 1993; Weinberger et al., 1993) and argumentation (e.g. Buchsbaum, 1990; Frith, 1992; Frith & Done, 1988; Weinberger 1987, 1991, 1994) concerning the brain abnormalities underlying schizophrenia is increasingly taking into account these interconnections. The selective enhancement of the PINV in schizophrenic patients with high provisional memory load should be interpreted within this context and underlines the role of ERP research in linking neurophysiological and cognitive abnormalities in schizophrenia.

Since the PINV is considered to reflect "*response* uncertainty" (Cohen, 1990), the significant augmentation of PINV amplitudes in schizophrenic patients by high compared to low provisional memory load is noteworthy in that this result suggests that the mere execution of a motor response (which was the same under all conditions) is not the only factor contributing to the development of a PINV in schizophrenic patients.

With the present experimental task which included a condition of an ambiguous matching of the experimental stimuli, a PINV could be observed in healthy subjects. While in this way one of the aims of experiment A was achieved, allowing a *comparative* investigation of the functional significance of the PINV (see *chapter 7*) and its topography (see *chapter 8*) in schizophrenic patients *and* healthy controls, an experimental design is added to those described in *chapter 3.4* which induces a PINV in subjects that do not exhibit a PINV under standard conditions of the two-stimulus paradigm. Relationships between this matching task and the "loss of control" experiments as well as the possible meaning of the ambiguity factor will be discussed in detail in *chapter 9*.

In accordance with a large number of published studies (see *paragraph 3.5*) larger PINV amplitudes were observed in schizophrenic patients compared to healthy controls. PINV has been suggested to reflect "response uncertainty". The present results confirm an increased sensitivity of schizophrenic patients to develop response uncertainty. This sensitivity was not altered by the clearness or ambiguity of the matching process, since the Ambiguity factor did not enter into a statistical interaction with the group factor, but was selectively modulated in schizophrenic patients by the provisional memory requirements of the task, suggesting that the involvement of the prefontal cortex or the "network" between prefrontal cortex and temporo-limbic structures such as the hippocampus, which are both considered to be dysfunctional in schizophrenia, to be crucial for the development of response uncertainty.

As revealed by the significant Hemisphere by Group interaction, averaged over the experimental conditions schizophrenic patients exhibited a left-sided predominance of the PINV (p=.0001) and healthy controls a right-sided (p=.06). These different asymmetries are compatible with the assumption of different generator structures in the two groups, since the Hemisphere by Group interaction remained significant after normalization of the data.

Left-hemisphere overactivity as well as left-hemisphere dysfunctions (Gur, 1978) have repeatedly been reported for schizophrenic patients. Furthermore, an *over*activity restricted to the left PFC has been observed during the Wisconsin Card Sorting Task in schizophrenic patients, but not in healthy controls (Berman & Weinberger, 1990). None the less the interpretation of these topographical results has to be taken with quite some reservation. Firstly, the literature on hemispheric asymmetries in general (Bradshaw & Nettleton, 1981; Moscovitch, 1979) and with respect to schizophrenia (e.g. Liddle et al., 1992) does not provide sufficiently generalizable interpretational premises. Secondly, the reported results from other studies refer to the neurophysiological activities *during* task execution, while the PINV reported here reflects neurophysiological activity *following* the execution of a task.

The possible meaning of these lateralizational results will be discussed in *chapter 9*, together with the topographical results from multi-channel recordings (see *chapter 8*).

In sum, experiment A reached its goal in determining an experimental condition under which a PINV can be observed in schizophrenic patients and *healthy controls. This constitutes the basis for exploring aspects of the functional significance of the PINV by exploiting the "probe"-technique (see chapter 7) and for a comprehensive investigation of the PINV topography (see chapter 8) in schizophrenic patients and healthy subjects.*

Chapter 6 The Impact of the Motor Response on the PINV

Summary of Chapter 6 So far, PINV in schizophrenic patients has been observed only under task conditions involving a motor response. Following the neuropsychological theory of schizophrenia by Frith (1987, 1992; Frith & Done, 1988) corollary discharges, sent from the motor to the perceptual monitoring system, should be impaired in schizophrenic patients during the motor response, triggering the PINV as a putative manifestation of response uncertainty (Cohen, 1990). Following this hypothesis, the present experiment integrated a nogo imperative stimulus (IS) into the delayed matching-to-sample task structure which comprised an IS that could be matched clearly or ambiguously with the Warning Stimulus (WS). If the PINV is primarily related to impaired monitoring during of the motor response, the nogo IS should diminish or abolish PINV. A replication of the PINV enlargement due to ambiguous IS, as well as group-specific hemispheric PINV asymmetries were expected.

Schizophrenic patients showed significantly larger PINV amplitudes than controls. A significant diminution of the PINV especially at centro-parietal sites was found under the nogo condition. While a PINV augmentation with ambiguous IS was replicated, the topographical differences between patients and controls were dissimilar to those of experiment A.

The results confirm the main hypothesis of this experiment and suggest the presence of two topographically distinguishable subcomponents of the PINV. In its posterior (centro-parietal) proportions, PINV depends upon the presence or absence of a motor response, possibly reflecting "action uncertainty", while the anterior proportions may reflect "performance uncertainty".

6.1 The Theoretical Background of Experiment B

The reviews of *chapter 3* showed that PINV has been observed so far only under experimental conditions requiring a motor response. In *chapter 5* an experimental task was presented which revealed that the PINV of schizophrenic patients was specifically sensitive to the "stressing" of provisional memory functions, which are considered to be based on the functional integrity of the PFC and the communication between PFC and temporo-limbic structures.

A common feature of both, the studies reviewed in *chapter 3* and the experiment of *chapter 5*, is the inclusion of a motor response (a button press) into

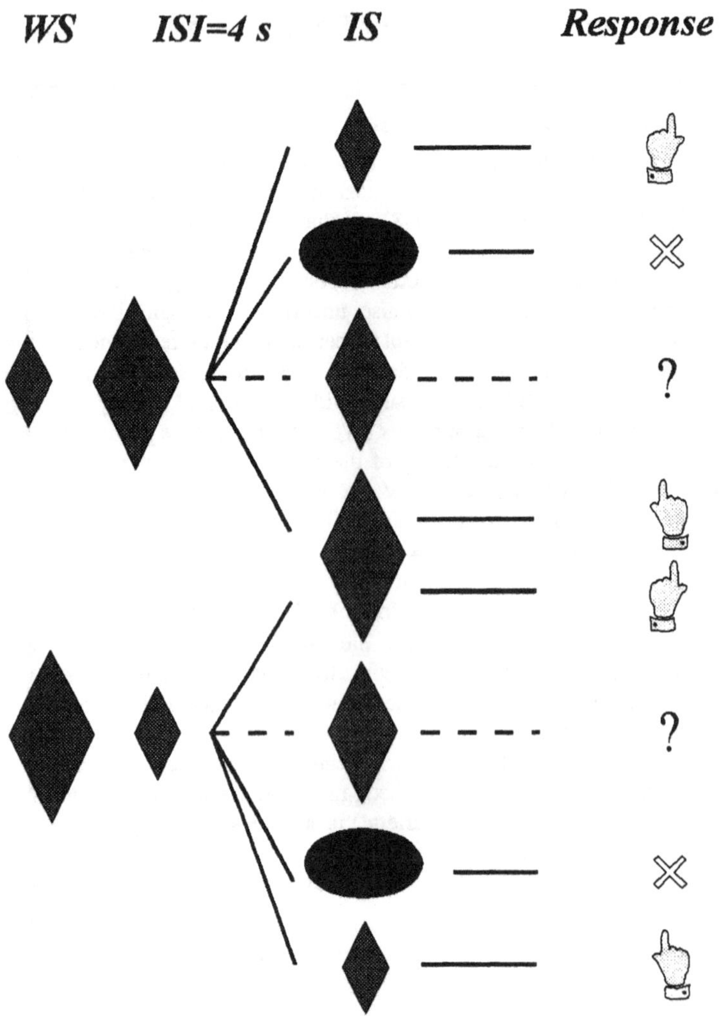

Figure 6.1

the task structure. To my knowledge, there is currently *no* PINV study which investigated *whether PINV occurs also when no motor response has to be given by the patient.*

There are reasons to assume that PINV as indicator of response uncertainty is linked to a motor response. These reasons will be considered in the following.

One of the 3 disorders of schizophrenia suggested in the neuro-psychological theory of Frith (1992; Frith & Done, 1988) concerns *self-monitoring*. The concept of self-monitoring is based on the distinction between the execution of an action and the perception of the (self-generated) intention to initiate the action. This distinction itself rests upon the assumption that, in parallel to efferent muscle commands, the perceptual system receives information about intention to give these efferent commands. The terms "corollary discharge" (Sperry, 1950) and "re-afference copy" (Von Holst & Mittelstaedt, 1950) have been coined to describe the message sent from the motor to the perceptual system in case of self-initiated movements. This concept elucidates how an organism is capable of distinguishing self-initiated from externally initiated perceptual changes. The term "self-monitoring" refers to the perception of corollary discharges.

A typical situation exemplifying the necessity for self-monitoring (perception of corollary discharges) is the execution of smooth pursuit eye movements (SPEM). If we follow an object (target) moving in front of a stationary background with our eyes, the retinal image of the background moves in the direction opposite to the eye movement. It is apparent that much perceptual confusion would arise if we were not able to recognize our own smooth pursuit eye movement in order to explain the movement of the retinal background image.

In animals, the firing of prefrontal units after a motor response has been associated with corollary discharges (Watanabe, 1986). The frontal eye fields are suspected to be one of the sources of oculomotor corollary discharges by Frith (1992) and others (cf. Fuster, 1989, p. 92, p. 130). The example of smooth pursuit eye movements (SPEM) was chosen on purpose, since SPEM disorders have frequently been observed in schizophrenic patients (Levy et al., 1993) and have been suggested to result from frontal eye field dysfunctions (Levin, 1984). The repeated observation of effectively reduced or even lacking SPEM disorders in schizophrenic patients when the target moves in front of a dark or dimmed background (Klein & Andresen, 1991; Pivik et al., 1988; Thaker et al., 1989; Tomer et al., 1981) may lend support to the hypothesis that SPEM disorders are related to a disorderd perception of oculomotor corollary discharges (compensated by an invisible or dark background).

More direct evidence in favor of disordered self-monitoring was presented by Malenka et al. (1982). In these studies subjects had to track a target jumping between the middle and the lateral sides of a video screen by adjusting a continuously moving (not "jumping") cursor with a joystick. While in the control condition directions of joystick and cursor movement were identical, in the experimental condition the cursor moved in the direction opposite to that of the joystick. Since only the centre and lateral areas of the screen were visible, the movement of the cursor was invisible until it reached one of the two lateral positions. Hence, no exteroceptive feedback about the direction of the cursor movement was provided. Every tenth trial subjects were instructed that the mapping of joystick and cursor direction would change (same versus opposite) in the following trial. During this task schizophrenic patients exhibited an impaired ability to recognize and correct (reverse) false, but an increased tendency to reverse correct movements, when no exteroceptive cues were given. The authors interpreted this impairment as suggestive of "an impaired ability to monitor ongoing behavior by means of internal, self-generated cues" (Malenka et al., 1982). Similar results were obtained by Frith and Done (1989), Mlakar et al. (1994), and Malenka et al. (1986)[106].

As mentioned above, PINV in schizophrenic patients has so far been observed *only* after the execution of a motor response. Since motor responses during CNV tasks are generally retarded and frequently preceded by reduced CNV amplitudes, impairments of "willed actions" in the terminology of Frith (as presumably reflected by reduced tCNV amplitudes) may be assumed. A deficit in central monitoring, on the other hand, means that "schizophrenic patients are lacking a clear, distinct and solid representation of internally initiated mental activity" (Mlakar et al., 1994). Hence, it is plausible that "patients fail to appreciate having completed their task" (Cohen, 1990), and a PINV, as a putative indicator of response uncertainty, should occur.

For these reasons it is particularly interesting to investigate, whether PINV does occur in schizophrenic patients under task conditions which require response preparation but not response execution. If PINV is related to a deficient monitoring of motor commands, abolishing these commands by the task structure should reduce PINV amplitudes.

6.2 Hypotheses

The following presentation of research hypotheses will be supplemented by a number of questions which do not have the quality of hypotheses but are exploratory in nature. When hypotheses and questions are simultaneously treated statistically (leapfrog strategy), the distinction between the two kinds of tests should be made explicit. Therefore, hypotheses and questions will be stated separately in a confirmatory and an exploratory part. Results from experiment A will be taken into consideration *unless* they do not contradict results reported in the literature (e.g. no reduced CNV amplitudes).

---

[106] A recent study of Kopp & Rist (1994) using a similar design, did not replicate these results. The authors attempt to explain this divergence with the memory- instead of stimulus-driven structure of the tasks of the 3 studies cited above. However, "a lack of clear representations of the appropriate actions" as supposed by Kopp & Rist (1994) to be present in the schizophrenic patients should have affected also the probability of initiating correct responses. This was not the case in the study of Malenka et al. (1982).

Confirmatory Part
CNV
Hypothesis 1a: Schizophrenic patients show diminished tCNV amplitudes. *Reason:* Published results on diminished CNV in schizophrenia; literature presented in *paragraph 3.3*. *Statistical form:* Group main effect.
Hypothesis 1b: The parieto-frontal tCNV gradient is steeper in controls than in patients. *Reason:* Published results on relative CNV augmentations at frontal leads in schizophrenic patients; literature presented in *paragraph 3.3*. *Statistical form:* Group by Electrode interaction.
PINV
Hypothesis 2a: Ambiguous imperative stimuli augment the PINV amplitudes in schizophrenic patients and induce a PINV in healthy controls. *Reason:* Hypothesized association between ambiguous contingencies and PINV; expected replication of the observation of a PINV and of the Ambiguity effect in experiment A; literature and arguments presented in *paragraphs 3.5-3.7* and *5.1* *Statistical form:* Ambiguity main effect[107].
Hypothesis 2b: Ambiguous imperative stimuli exert their strongest PINV augmenting impact at frontal leads. *Reason:* expected replication of the "frontal" topography of the Ambiguity effect reported for experiment A. *Statistical form:* Ambiguity[107] by Anteriority (see *paragraph 6.6*) interaction with a significant apriori contrast (F3,F4) versus (C3,C4).
Hypothesis 2c: The frontal PINV augmentation due to ambiguous IS will be larger in controls than in patients. *Reason:* expected replication of the Ambiguity by Electrode by Group interaction reported for experiment A. *Statistical form:* Ambiguity[107] by Anteriority by Group interaction with a significant planned contrast (F3,F4) versus (C3,C4)
Hypothesis 3: Nogo imperative stimuli abolish or significantly diminish the PINV. *Reason:* A deficit of schizophrenic patients in central monitoring (cf. Frith, 1992) of the response should lead to a failure of appreciating task completion and, hence, a PINV; in controls, those processes triggered by the inclusion of ambiguous IS should be suspended by nogo IS; literature reported in *paragraph 2.1.2, 3.5, 3.6, and 6.1*. *Statistical form:* Nogo main effect[107].
Hypothesis 4: The two subject groups exhibit different PINV asymmetries. *Reason:* expected replication of the different laterlization of the PINV in experiment A. *Statistical form:* Laterality (see *paragraph 6.6*) by Group interaction and significant hemisphere effects in the within-group ANOVA[108].

---

107 Since the conditions clear, ambiguous, and nogo will be different levels of one and the same Condition factor, the statistical counterpart of these research hypotheses is a significant main effect Condition and significant apriori contrasts clear vs. ambiguous, respectively clear vs. nogo.

108 Since different electrode montages are used in experiments A and B an exact replication of the results from experiment A is not possible. For this reason the hypothesis is not entirely stringent.

Hypothesis 5: PINV will be predominant at frontal leads. *Reason:* expected replication of the frontal predominance of the PINV in experiment A. *Statistical form:* Anteriority (see *paragraph 6.6*) main effect with significant apriori contrasts (F3,F4) versus (C3,C4)[108].

LPC

Hypothesis 6: Schizophrenic patients exhibit diminished LPC amplitudes following the warning and the imperative stimuli[109]. *Reason:* published results on diminished P300 amplitudes in schizophrenic patients; literature reported in *paragraph 3.9*. *Statistical form:* Group main effect.

Hypothesis 7: Ambiguous imperative stimuli will diminish IS-the LPC amplitudes. *Reason:* expected replication of the IS-LPC amplitude reduction due to ambiguous IS observed in experiment A; literature presented in *paragraph 3.8*. *Statistical form:* Ambiguity main effect[107].

Hypothesis 8a: Nogo imperative stimuli will enlarge the IS-LPC amplitudes. *Reason:* diminished equivocation with nogo compared to clear/ambiguous IS should enlarge the IS-LPC amplitudes according to the Triarchic Model by Johnson, Jr. (1986), and NoGo-IS have been shown to augment P300 amplitudes (Simson et al., 1977, Roberts et al., 1994); literature reported in *paragraph 3.8* and *6.1*. *Statistical form:* Nogo main effect[107].

Hypothesis 8b: Under the nogo condition the IS-LPC enlargement will be maximal centrally, shifting the topographical LPC maximum anteriorly, compared to the clear condition. *Reason:* published topography of the Nogo-P300 (e.g. Roberts et al., 1994); literature reported in 6.1, *Statistical form:* Nogo[107] by Anteriority (see *paragraph 6.6*) interaction with significant apriori contrasts between the central versus parietal and central versus frontal coronal electrode rows (see *paragraph 6.6*).

RT

Hypothesis 9: Reaction Times will be longer in patients than in controls. *Reason:* expected replication of the RT Group main effect of experiment A and results reported in the literature (cf. King, 1991). *Statistical form:* Group main effect.

Hypothesis 10: Reaction Times will be longer under the ambiguous than under the clear condition. *Reason:* expected replication of the RT Ambiguity main effect of experiment A. *Statistical form:* Condition[107] main effect.

Exploratory Part

If the nogo imperative stimulus reduces PINV amplitudes: Will this reduction have a distinct topography ? Also, will one of the two groups "benefit" more than the other from the PINV reducing impact of the nogo imperative stimulus?

---

[109] Although diminished LPC amplitudes in schizophrenic patients were not found in experiment A, they have frequently been reported in the literature (see, for instance, the CNV shape typical of schizophrenic patients according to Timsit-Berthier et al. (1973) in *chapter 3.5*).

6.3 The Laboratory

Experiment B was conducted in the psychophysiological laboratory of the University of Konstanz in the State Hospital Reichenau, which is described in *paragraph 4.2*.

6.4 Recruitment of Subjects and Assessment of Clinical Variables in the Patient Group

Recruitment of schizophrenic patients and healthy controls was done in the same manner as described in *paragraph 4.3*. The clinical variables assessed were the Brief Psychiatric Rating Scale (BPRS) and the medication of the day of the testing session as described in *paragraph 4.4*.

Fourteen schizophrenic patients and 14 healthy control subjects participated in the experiment. The patients met a DSM-III-R diagnosis of a schizophrenic disorder. Due to excessive artifact contamination data of 3 patients and 1 control subject were rejected from data analysis. **Table 6.1** summarizes the socio-demographic features of both subject groups.

Thirteen of the 14 DSM-III-R schizophrenic patients who participated in experiment B were recruited from the research ward of the State Hospital Reichenau. Another schizophrenic patient was recruited from a ward for long-term chronically ill patients from the same hospital. For years this patient was known on the research ward and fulfilled DSM-III-R criteria for a schizophrenic disorder. During the time of testing all schizophrenic patients were receiving neuroleptic medication (mean=184.5 mg/day CPZ equivalent; SD=93.3) with mean BPRS psychopathological ratings of 44.3 (SD=10.8; range=28-62).

All healthy control subjects who participated in experiment B denied having had any psychiatric or neurologic disease or taking any medication.

Table 6.1 Socio-Demographic Features of the Patient and the Control Group Participating in Experiment B

	Patients	Controls
N	11	13
Age (SD; Range)	30.8 (9.14; 20-54)	29.3 (9.21; 20-48)
Male : Female	10 : 1	11 : 2
Smoker : Non-Smoker	7 : 4	8 : 5
Education (Years)		
<8	1	0
8	2	3
10	4	8
13	4	2
Right-Handedness	10	12

6.5 Testing Procedure

The testing procedure of experiment B corresponds to the general scheme described in *paragraph 4.5*. After a subject had acceded to participate, the time of his/her experimental session was arranged. Prior to the experimental session the subject was shown the laboratory. Also, the written consent had to be given before beginning with the experiment. After the instruction, the headset was adjusted by which the acoustical probes were delivered.

Each subject received DM 25,- for his/her participiation, and an additional financial bonus of DM 10,- was promised in case of fast and correct responding. None of the subjects terminated the experiment by him/herself.

6.6 The Task Presentation

Figure 6.1 illustrated schematically the experimental design. Since the basic experimental design is described in *paragraph 4.1*, it will only briefly be described here.

As WS a pair of horizontally arranged green diamonds was presented against a black background for 100 ms. These diamonds had clearly distinguishable sizes. The lateral arrangement of the two diamonds varied randomly from trial to trial. Four seconds after WS onset (SOA=4.0 s) a single green diamond was presented as IS in the middle of the monitor for 100 ms during 78% of the trials. This single green diamond was in 23.5% (48 trials) of all trials identical with the small WS diamond and in 23.5% of all trials with the large WS diamond. During further 29.5% (60 trials) of all trials the IS diamond had a size lying intermediate between those of the WS diamonds, but the occurence of these middle-size diamonds was not introduced to the subjects who were instructed to match the IS diamond with the identical WS diamond by pressing the left or right hand response button. During the remaining 23.5% of the trials (48 trials) a red ellipse was presented. Subjects were instructed that the red ellipse would terminate the trial and that no response had to be given to the red ellipse. Due to its color and shape the ellipse IS was clearly distinguishable from the diamond IS.

To sum: Under the "clear" condition the IS diamonds were identical with one of the WS diamonds, while under the "ambiguous" condition IS of a size lying inbetween those of the WS diamonds were presented. Finally, the "nogo" condition comprised trials with red ellipses.

Via a headphone auditory probe stimuli (clicks of 20 ms duration and 80 dB intensity) were delivered during the pretrial-baseline, the iCNV, tCNV, or PINV segments in one out of 6 possible pairwise and equally distributed combinations. Subjects were instructed to ignore these task-irrelevant auditory stimuli[110].

[110] Results on probe-evoked potentials will be presented in *chapter 7*.

After the electrode application, described in *paragraph 4.5.1*, each experimental session began with the execution of calibration eye movements for the off-line EOG correction. The calibration task is described in *paragraph 4.5.2*.

Prior to the 204 trials of the main experimental block, 20 practice trials without probe stimuli, intermediate-size or ellipse IS were conducted in order have the subjects become acquainted with the task.

Figure 6.2 Electrode Montage in Experiment B

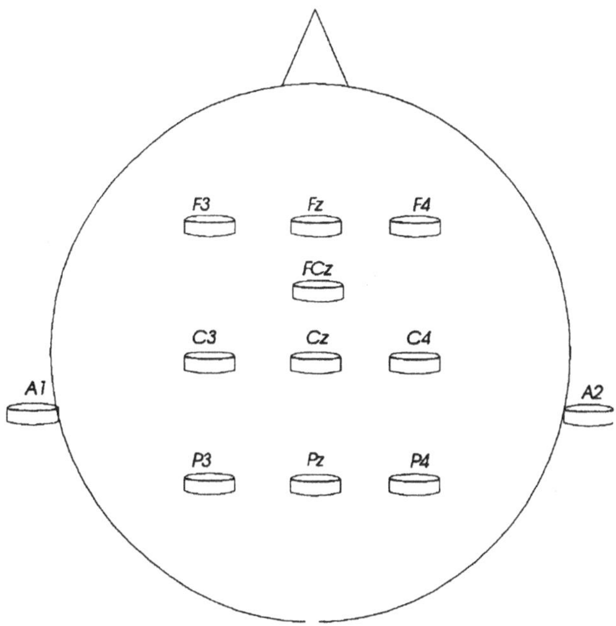

6.7 Apparatus and Physiological Recordings

The equipment of the laboratory where experiment B took place was described in *paragraph 4.2*. Experiment B was conducted with the GENTASK version for 386 PCs. The amplifier settings are listed in table 4.1. (see *paragraph 4.6.3*). EEG electrodes were affixed to the positions illustrated in **figure 6.2.** At first, Cz was determined as geometrical centre between nasion and inion and the left and right preauricular points as described in Lutzenberger et al., 1985. Then, electrode positions F3, Fz, F4, C3, C4, P3, Pz, P4, A1, and A2 (mastoids) were determined according to the international 10-20-system as described in *paragraph 4.5*. FCz served as the common recording reference, and a ground electrode was placed at the forehead.

In addition to these EEG electrodes, 4 EOG electrodes (*not* illustrated in **figure 6.2**) were affixed to the outer canthi and 2-3 cm above and below the left eye in order to record eye movements. Data of experiment B were corrected for horizonal and vertical saccades and for blinks with correction coefficients determined from the calibration trials.

6.8 Data Reduction and Data Analysis

The EEG data recorded by the DCMES system were at first corrected for DC shifts with the program DCAUS (MES, Munich; see *paragraph 4.2*). For this DC correction a fourth-order polynom is fitted to the mean DC levels of a 1.0 s pretrial baseline interval in order to compute each trial as a difference from the statistically fitted DC level.

The subsequent steps in data analysis were performed with the analysis programs described in detail in *paragraph 4.2*. The following processing steps were accomplished (in that order): Transformation of data format from the DCMES to the Konstanz raw data format; categorization of each trial according to its combination of stimulus and response event marker and computation of RT; initial baseline subtraction; re-referencing of raw data on an average reference; computation of EOG correction coefficients from the calibration trials; application of the EOG correction coefficients, i.e. EOG correction; condition-wise computation of single subject averages; averaging over predefined conditions, including only trials with correct responses as defined below, re-referencing of the average reference data on the average mastoid reference; computation of mean amplitudes over predefined intervals for statistical analyses; computation of peak amplitudes and latencies over predefined intervals for statistical analyses. Details of the EOG correction are described in *paragraph 4.7.3* (programs REGREYE and CORREYE).

Table 6.2 presents mean relative frequencies of correct, false, missing or commisional responses and of artifact-free EEG trials, separated for experimental groups and conditions. Correct, false, or missing responses can be differentiated for the condition “clear“, while for the condition “ambiguous“ only correct or missing responses can be distinguished (due to the ambiguity of the IS “correct“ or “false“ is not defined). Response errors under the nogo condition are comission errors, i.e. pressing the response button.

Mean amplitude values for each subject, condition, and channel were determined for 1 tCNV, and 3 PINV intervals. Mean amplitudes for the interval from 3.0 to 3.5 s following WS onset served as tCNV parameter (i.e. the last 500 ms prior to the possible tCNV probe). PINV-1 was defined as the mean amplitude of the 1.0-1.5 s interval following IS onset (i.e. the last 500 ms prior to the possible PINV probe), and PINV-2 and PINV-3 were defined as the mean amplitudes of the interval 2.0-2.5 s and 2.5-3.0 s, following IS onset. In addition, peak amplitude and latency values for the LPC following the IS were determined as positive peaks within the range of 320-560 ms following the IS. This range includes the range of condition and group specific peak latencies at Pz with a buffer of plus/minus 200 ms. Finally, grand averages were computed for each group, distinguishing the conditions clear, ambiguous, and nogo and including only trials with correct responses.

Table 6.2 Mean Relative Frequencies of Correct Responses and of Artifact-Free EEG Trials

(a) Relative Frequencies of Correct Responses

Condition	Clear	Ambiguous	NoGo
Patients	.93	.97	.99
Controls	.93	.99	.99

Under the ambiguous condition responses with the left or the right hand are correct, under the nogo condition withholding of the response is correct.

(b) Relative Frequencies of Artifact-Free EEG Trials

Condition	Clear	Ambiguous	NoGo
Patients	.73	.77	.80
Controls	.78	.81	.90

The ERP parameters were statistically analyzed with the GLM-Procedure of SAS. *Univariate* analyses will be reported for reasons described in *paragraph 4.7.4*. Within a mixed-model repeated measures ANOVA the between-subject factor "GROUP" (patients versus controls) and the within-subject factors "CONDITION" (referring to the IS types clear, ambiguous and nogo), "LATERALITY" (distinguishing the left (F3-C3-P3), middle (Fz-Cz-Pz), and right (F4-C4-P4) sagittal electrode rows), and "ANTERIORITY" (distinguishing the frontal (F3-Fz-F4), central (C3-Cz-C4), and parietal (P3-Pz-P4) coronal electrode rows) were analyzed. Reaction times were statistically analyzed by repeated meaures ANOVA, including the "GROUP" and "CONDITION" factors. Following the ANOVA, planned contrasts specified the origin of a given significant effect including an experimental factor with more than 2 levels. In experiment B, the levels "ambiguous" and "nogo" of the CONDITION factor are contrasted with the level "clear"; the levels "right" (F4, C4, P4) and "left" (F3, C3, P3) of the LATERALITY factor are contrasted with the level "middle" (Fz,

Cz, Pz); the levels "frontal" (F3, Fz, F4) and "parietal" (P3, Pz, P4) of the ANTERIORITY factor are contrasted with the level "central" (C3, Cz, C4). Huynh-Feldt-corrected p-values are reported for all main and interaction effects including a within-subject factor with more than 2 levels in order to control for possible violations of the sphericity assumption of split-plot designs (see *paragraph 4.7.4*). Planned contrasts are accomplished in order to specify the origin of a significant effect including an experimental factor with more than 2 levels. Only significant ($p \leq 0.05$) contrasts will be reported. These results will be presented in footnotes, in order to alleviate the reading of the text. Only significant effects will be documented in tables (means and standard deviations).

6.9 *Results of Experiment B*

6.9.1 Reaction Times (RT)

Groups differed significantly in their mean RT, as indicated by the significant **Group** effect ($F(1,22)=9.50$, $p=.005$), with longer RT in the patient group. The significant **Condition** effect indicates that ambiguous compared to clear IS were associated with an increase in RT ($F(1,22)=31.87$, $p=.0001$).

Table 6.3 Reaction Times

	Clear	Ambiguous
Patients	680.7 (172)	854.8 (184)
Controls	510.5 (86)	653.7 (168)

RT in ms; SD in parentheses

This RT increase was similar in the two groups, as the nonsignificant interaction of **Condition by Group** ($F(1,22)=0.54$, $p=.47$) reveals. **Table 6.3** shows the mean RT under the varying experimental conditions.

6.9.2 Event Related Potentials

6.9.2.1 Grand Averages of the Patient and the Control Group

Figure 6.3 (a-c) shows the grand averages of schizophrenic patients and healthy controls for the 3 experimental conditions. The grand averages reveal sequences of positive and negative deflections similar to those already described in *paragraph 5.9.2.1*, and will not be described in greater detail for this reason. Briefly, a centro-parietally predominant WS-LPC can be observed under all conditions. It is followed by a broadly distributed iCNV. Towards the IS the amplitudes of the centro-parietal tCNV increase. Following the IS, the IS-LPC can be seen. This component is strongly augmented under the nogo condition. After the IS-LPC, the broadly distributed PINV can be seen. The PINV is clearly reduced under the nogo, compared to the clear condition.

ERP amplitudes are similar in schizophrenic patients and controls. However, patients exhibit a larger PINV than controls under the clear and ambiguous conditions.

In both groups click-evoked potentials show up during the pretrial-baseline, the iCNV, the tCNV, and the PINV, with largest amplitudes at Cz. These click EP have a rather weak negative component (N100) and a clearly visible positive component (P200). Together they can be identified as vertex potentials (see *chapter 7*).

6.9.2.2 Terminal Contingent Negative Variation (tCNV)[111]

Table 6.4 summarizes the results on tCNV. **Groups** did not differ in the tCNV amplitudes. The significant **Laterality** effect rises from significant contrasts between the left[112] and the right[113] sagittal rows, each contrasted with the mid-sagittal row.

While tCNV is largest over the sagittal midline, left hemisphere and right hemisphere amplitudes are significantly smaller. This effect can be followed in **table 6.5** by comparing the middle with the outer columns.

Only the significant contrast between the frontal and the central coronal row[114] underlies the significant **Anteriority** effect. As expected for visual stimulation, tCNV is largest parietally and smallest frontally with intermediate amplitudes at central electrodes. This effect can be followed in **table 6.4** by comparing the rows for the frontal, central, and parietal amplitude means under the 3 experimental conditions.

The significant interaction of **Laterality by Anteriority** is due to a significant difference between the left versus midline sagittal row concerning the difference between the parietal and central coronal row[115]. While the amplitudes at C3 and Cz are almost identical, amplitudes at Pz are larger than amplitudes at P3 (see **table 6.6**).

Unexpectedly —because the type of IS could not be predicted from the WS— the scalp distribution of the tCNV differed between the conditions.

---

[111] Since the type of the WS was not experimentally varied in this experiment, the iCVN will not be analyzed.

[112] $F(1,22)= 5.31, p=.03$

[113] $F(1,22)=12.51, p=.002$

[114] $F(1,22)=52.00, p=.0001$

[115] $F(1,22)=22.70, p=.0001$

Table 6.4 Experimental Effects on tCNV

Effect	df1, df2	F	p
C	2,44	1.22	.30
C * G	2,44	0.95	.39
L	2,44	6.63	.003
L * G	2,44	0.32	.73
A	2,44	31.92	.0001
A * G	2,44	1.16	.31
C * L	4,88	6.85	^{G}.0002[1]
C * L * G	4,88	0.76	.54
C *A	4,88	5.32	^{G}.0007[2]
C * A * G	4,88	0.49	.74
L * A	4,88	5.46	.002
L * A * G	4,88	0.57	.63
C * L * A	8,176	6.14	^{G}.0001[3]
C * L * A * G	8,176	0.44	.82
Group	1,22	1.18	.29

C*=Condition (clear, ambiguous, nogo);* ***L****=laterality (left, middle right sagittal row);* ***A****=Anteriority (frontal, central, parietal coronal row);* ***G****=group (patient vs control subjects); univariate F-values, Huynh-Feldt adjusted;* $^{G(G)}$=Generator effect as indicated by results of normalization; [1]=F(4,88)=6.85, p=.0002; [2]=F(4,88)=5.32, p=.0007; [3]=F(4,88)=6.14, p=.0001;

Table 6.5 Condition by Laterality by Anteriority Interaction on tCNV

	Left	Middle	Right
Clear			
Frontal	-1.00 (2.7)	-1.11 (2.7)	0.02 (2.5)
Central	-2.86 (2.7)	-3.00 (3.0)	-1.96 (1.8)
Parietal	-2.72 (2.1)	-3.79 (2.4)	-2.56 (1.8)
Ambiguous			
Frontal	-1.34 (3.3)	-1.36 (3.0)	0.04 (4.3)
Central	-2.96 (2.5)	-3.06 (3.0)	-2.57 (2.7)
Parietal	-3.41 (2.3)	-4.05 (2.7)	-3.02 (2.3)
NoGo			
Frontal	-1.21 (3.1)	-0.73 (3.6)	-0.45 (3.5)
Central	-3.63 (3.0)	-3.36 (2.5)	-2.91 (2.9)
Parietal	-3.68 (2.2)	-4.29 (2.8)	-3.43 (2.5)

unit is µV; SD in parentheses; for the sake of conciseness those subordinate interaction and main effects which can easily be followed in this table will not be presented in separate tables

Table 6.6 Laterality by Anteriority Interaction on tCNV

	Left	Middle	Right
Frontal	-1.19 (2.9)	-1.07 (2.8)	-0.13 (3.2)
Central	-3.15 (2.4)	-3.14 (2.9)	-2.48 (2.2)
Parietal	-3.27 (2.0)	-4.04 (2.4)	-3.01 (1.9)

unit is μV, SD in parentheses

Condition dependent scalp distributions are revealed by the significant **Condition by Laterality** interaction. This effect is due to a significant difference at the left- versus mid-sagittal row concerning the ambiguity effect (clear vs ambiguous)[116], and differences between the left-[117]and right-[118] versus mid-sagittal rows concerning the nogo effect (clear vs nogo; see **table 6.7**).

Table 6.7 Condition by Laterality Interaction on tCNV

	Left	Middle	Right
Clear	-2.20 (2.3)	-2.63 (2.5)	-1.50 (1.7)
Ambiguous	-2.57 (2.4)	-2.82 (2.7)	-1.85 (2.7)
NoGo	-2.84 (2.5)	-2.79 (3.0)	-2.27 (2.6)

unit is μV, SD in parentheses

Amplitude augmentations from the clear to the ambiguous condition are larger over the left hemisphere than along the mid-sagittal axis, which does not differ from right-sided leads in this effect. Furthermore, amplitude augmentations from clear to nogo are least pronounced at the mid-sagittal leads when compared to the left- and right-sided leads. The significant interaction of **Condition by Anteriority** rises from a significant contrast between the frontal and the central coronal row in the nogo *effect* (i.e. nogo versus clear difference)[119]. Differences in tCNV amplitudes between the frontal and the central coronal row are larger under the nogo than under the clear condition, while the central versus parietal difference remains similar under both conditions. In other words: the nogo effect on tCNV is least pronounced frontally (see **table 6.8**).

No significant contrast is associated with the **Condition by Laterality by Anteriority** interaction.

---

[116] $F(1,22)=4.64$, $p=.04$
[117] $F(1,22)=6.88$, $p=.02$
[118] $F(1,22)=9.19$, $p=.006$
[119] $F(1,22)=11.43$, $p=.003$

Table 6.8 Condition by Anteriority Effect on tCNV

	Frontal	Central	Parietal
Clear	-0.70 (2.5)	-2.61 (2.4)	-3.02 (1.9)
Ambiguous	-0.89 (3.2)	-2.86 (2.6)	-3.49 (2.3)
NoGo	-0.80 (3.2)	-3.30 (3.0)	-3.80 (2.4)

unit is µV, SD in parentheses

To summarize: tCNV is most pronounced over the sagittal midline and larger at centro-parietal than at frontal leads. Groups do not differ in the tCNV amplitudes or its topography. Unexpectedly, the experimental conditions are associated with amplitude augmentations especially at left-sided (ambiguity, nogo) and right-sided (nogo) leads. The nogo effect is stronger at centro-parietal as compared to frontal sites.

6.9.3.2 Post-Imperative Negative Variation (PINV)[120]

Table 6.9 Experimental effects on PINV-1 to PINV-3

		PINV-1		PINV-2		PINV-3	
Effect	**df1, df2**	**F**	**p**	**F**	**p**	**F**	**p**
C	2,44	8.69	.003	19.16	.0001	18.47	.0001
C * G	2,44	4.16	.04	1.90	.17	2.41	.11
L	2,44	2.14	.13	8.36	.0008	12.46	.0001
L * G	2,44	6.85	${}^{G}.003^{1.1}$	6.00	${}^{G}.006^{1.2}$	7.74	${}^{G}.001^{1.3}$
A	2,44	9.22	.0005	20.88	.0001	18.27	.0001
A * G	2,44	0.66	.52	1.38	.26	1.62	.21
C * L	4,88	4.75	${}^{G}.003^{2.1}$	1.97	.12	3.76	${}^{G}.01^{2.3}$
C * L * G	4,88	3.36	${}^{A}.02^{3.1}$	0.81	.51	0.94	.44
C * A	4,88	5.61	${}^{G}.0008^{4.1}$	3.32	${}^{G}.03^{4.2}$	2.39	.09
C * A * G	4,88	0.69	.59	0.45	.70	0.21	.86
L * A	4,88	3.79	.008	3.77	.008	2.50	.06
L * A * G	4,88	2.95	${}^{A}.03^{5.1}$	2.33	.06	1.13	.35
C * L * A	8,176	0.97	.45	2.19	${}^{G}.05^{6.2}$	2.28	${}^{G}.04^{6.3}$
C*L*A*G	8,176	1.39	.21	1.32	.25	1.22	.30
Group	1,22	29.02	.0001	17.82	.0004	20.20	.0002

[120] Secondary analyses of the course of the PINV (by using the 3 PINV segments as levels of a within-subject ANOVA factor "SEGMENT") and of the influence of the tCNV on the PINV (by separately analysing the PINV data with a pretrial versus tCNV baseline) will be published by Klein et al. (submitted).

Table legend: *C=Condition factor; L=Laterality factor; A=Anteriority factor; G=Group factor; PINV-1=1.0-1.5s; PINV-2=2.0-2.5s;PINV-3=2.5-3.0s (each after IS onset; F=uni-variate F-values, p=Huynh-Feldt corrected p values; average mastoids referene, baseline is tCNV;* [G(G)]=Generator effect,[A(A)]=Amplitude effect as indicated by results of normalization; [1.1]=F(2,44)=6.02,p=.007;[1.2]=F(2,44)=7.72,p=.001;[1.3]=F(2,44) =10.12, p=.0003;[2.1]= F(4,88)=5.23, p=.001;[2.3]=F(4,88)=3.79,p=.007;[3.1]=F(4,88)=1.47, p=.22;[4.1]=F(4,88)=3.90,p=.008;[4.2]=F(4,88)=3.47,p=.02;[5.1]=F(4,88)=2.15, p=.08;[6.2]=F(8, 176)=3.25, p=.004; [6.3]=F(8,176)=3.61, p=.001

Results for the 3 PINV segments are documented in **table 6.9.** For all 3 PINV segments the main effect **Anteriority** rises from significantly larger frontal than central amplitudes[121].

While the PINV is smaller at parietal than at central sites, this difference is significant only for the second PINV segment[122] (see **table 6.10**).

The **Laterality** main effect, which is significant for the PINV segments 2 and 3, stems from a significant contrast between the right and the midline sagittal row[123]. PINV is larger over the right hemisphere than along the sagittal midline and over the left hemisphere (see **table 6.10**). For all three PINV segments, however, the interaction of **Laterality by Group** reaches statistical significance.

Table 6.10 Laterality by Anteriority Interaction on PINV

	Left	Middle	Right
PINV-1			
Frontal	-0.65 (2.6)	-1.08 (2.8)	-1.72 (1.8)
Central	-0.28 (2.8)	-0.16 (3.0)	-0.59 (2.1)
Parietal	-0.09 (1.8)	0.54 (2.5)	-0.18 (2.3)
PINV-2			
Frontal	-0.55 (2.6)	-1.05 (2.8)	-2.09 (2.0)
Central	0.30 (2.8)	0.52 (2.7)	-0.46 (2.0)
Parietal	0.79 (1.7)	1.05 (2.1)	0.39 (1.8)
PINV-3			
Frontal	-0.24 (2.7)	-0.62 (3.1)	-1.82 (2.3)
Central	0.75 (2.7)	0.86 (2.8)	-0.25 (2.1)
Parietal	1.09 (1.8)	1.10 (2.3)	0.46 (1.8)

unit is µV, SD in parentheses

[121] PINV-1: F(1,22)=8.12, p=.009; PINV-2: F(1,22)=23.43, p=.0001; PINV-3: F(1,22)=27.26, p=.0001
[122] F(1,22)=5.74, p=.03
[123] PINV-2: F(1,22)=13.16, p=.002; PINV-3: F(1,22)=20.10, p=.0002

Figure 6.3a PINV in Schizophrenic Patients and Healthy Controls —Clear Condition—

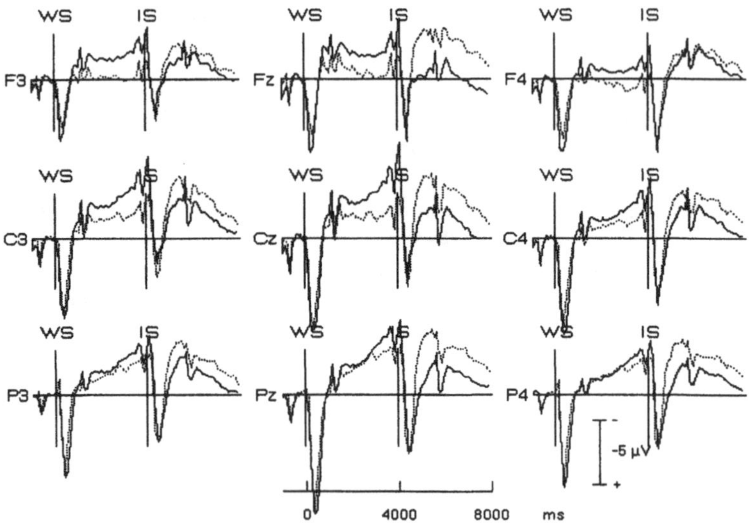

Groups: ______ = Healthy Controls; .... = Schizophrenic Patients

Table 6.11 Laterality by Group Interaction on PINV

	Left	Middle	Right
PINV-1			
Patients	-2.05 (1.4)	-2.52 (1.4)	-2.03 (1.3)
Controls	1.12 (1.7)	1.52 (1.9)	0.19 (1.7)
PINV-2			
Patients	-1.23 (1.9)	-1.74 (1.7)	-1.72 (1.4)
Controls	1.37 (1.8)	1.73 (1.8)	0.12 (1.4)
PINV-3			
Patients	-0.97 (2.0)	-1.65 (2.2)	-1.67 (1.7)
Controls	1.81 (1.6)	2.21 (1.4)	0.43 (1.3)

unit is µV; SD in parentheses

This interaction is due to significant contrasts between the left and the mid-sagittal row[124] and the right and the mid-sagittal row[125] concerning the group differences. Different generator structures are suggested by the significant results of the normalized data.

Figure 6.3b PINV in Schizophrenic Patients and Healthy Controls —Ambiguous Condition—

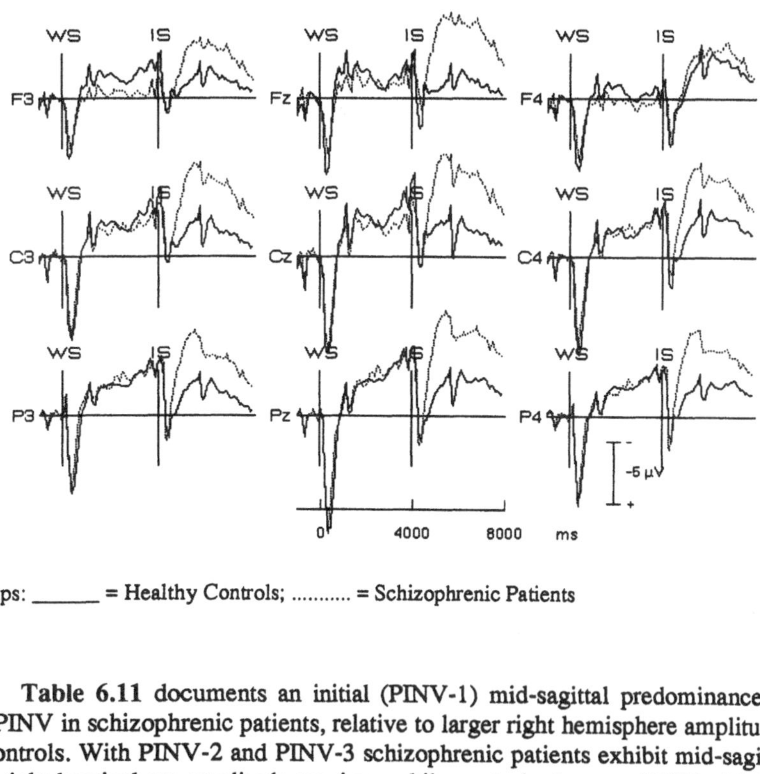

Groups: ______ = Healthy Controls; .... = Schizophrenic Patients

Table 6.11 documents an initial (PINV-1) mid-sagittal predominance of the PINV in schizophrenic patients, relative to larger right hemisphere amplitudes in controls. With PINV-2 and PINV-3 schizophrenic patients exhibit mid-sagittal and right hemisphere amplitude maxima, while controls show a PINV which is maximally pronounced over the right hemisphere.

---

124 PINV-1: $F(1,22)=4.94$, $p=.04$; PINV-2: $F(1,22)=4.62$, $p=.04$; PINV-3: $F(1,22)=6.94$, $p=.02$

125 PINV-1: $F(1,22)=12.97$, $p=.002$; PINV-2: $F(1,22)=14.01$, $p=.001$; PINV-3: $F(1,22)=19.16$, $p=.0002$

For the patient group, the within-group ANOVA reveals no significant Laterality effect for the first two PINV segments and a trend for PINV-3 [126].

In controls the Laterality effect is significant for all 3 PINV segments[127]

Figure 6.3c PINV in Schizophrenic Patients and Healthy Controls —Nogo Condition—

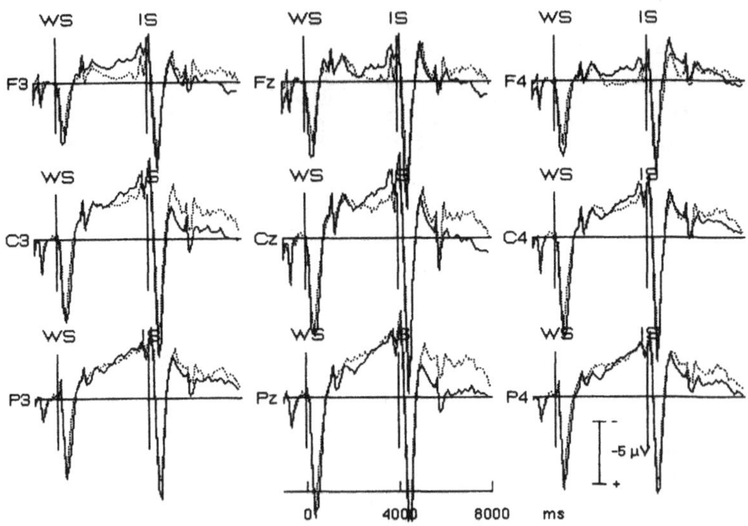

Groups: ______ = Healthy Controls; .... = Schizophrenic Patients

These effects rise from significant contrasts between the right and the mid-sagittal row[128], which does not differ significantly from the left-sagittal row.

The significant interaction of **Laterality by Anteriority** results from differences between the left and the mid-sagittal row concerning differences between the parietal and the central coronal row. This contrast is significant only

[126] PINV-1: F(2,20)=1.13, p=.34; PINV-2: F(2,20)=1.40, p=.27; PINV-3: F(2,20)=3.30, p=.06

[127] PINV-1: F(2,24)=8.08, p=.003; PINV-2: F(2,20)=14.21, p=.0002; PINV-3: F(2,24)=16.82, p=.0001

[128] PINV-1: F(1,12)=15.25, p=.002; PINV-2: F(1,12)=34.20, p=.0001; PINV-3: F(1,12)=51.68, p=.0001

for the first PINV segment[129]. Amplitude differences between left and mid-sagittal leads are smaller along the central than along the parietal coronal row (see **table 6.10**).

Although the interaction of **Laterality by Anteriority by Group** is significant, no planned contrast reaches statistical significance.

To summarize: While the PINV is maximal over frontal regions in both groups, a right-hemisphere predominance in controls contrasts with a less asymmetric PINV topography in patients, which favors the sagittal midline initially (PINV-1) and the mid-sagittal and right-sided leads during the subsequent segments (PINV-2, PINV-3). These differences are compatible with different generator structures in the two groups, since they turn out significant after normalization of the data.

For all 3 PINV segments, the **Condition** main effect is significant. These main effects arise from the significant contrasts between the experimental (ambiguous, nogo) and the control conditions (clear). While ambiguous compared to clear IS result in a significant overall increase of PINV amplitudes[130], the nogo IS reduce PINV ampitudes[131] (see **table 6.12**). Only for the first PINV segment this effect has to be differentiated for the two groups.

Table 6.12 Condition by Group Interaction on PINV-1 and Condition Effect on PINV-2 and PINV-3

	Clear	Ambiguous	NoGo
PINV-1			
Patients	-2.50 (1.2)	-4.03 (2.9)	-0.06 (2.5)
Controls	1.03 (1.7)	0.57 (2.4)	1.32 (1.8)
PINV-2			
Both Groups	-0.38 (2.0)	-1.46 (2.7)	1.44 (2.6)
PINV-3			
Both Groups	0.13 (2.1)	-1.21 (2.8)	1.52 (2.6)

unit is µV; SD in parentheses

This **Condition by Group** interaction on PINV-1 is due to a significant contrast between the clear and the nogo condition only[132]. While PINV amplitudes are larger under the ambiguous than under the clear condition in both groups, the PINV reduction under the nogo compared to the clear condition is larger in patients than in controls (see **table 6.12**). The within-group ANOVA for

[129] $F(1,22)=6.50, p=.02$

[130] PINV-1: $F(1,22)=4.77, p=.04$; PINV-2: $F(1,22)=6.45, p=.02$; PINV-3: $F(1,22)=10.81, p=.003$

[131] PINV-1: $F(1,22)=8.98, p=.007$; PINV-2: $F(1,22)=23.46, p=.0001$; PINV-3: $F(1,22)=14.09, p=.001$

[132] $F(1,22)=5.58, p=.03$

PINV-1, however, reveal a significant Condition (nogo) effect only in the patient group[133], but not in controls[134].

Since the Condition by Group interaction is non-significant for the subsequent PINV segments, the further ambiguity augmentation and nogo diminution of the PINV are similar in the two groups (see **table 6.12**).

For the 1st and the 3rd PINV segments the interaction of **Condition by Laterality** is significant, in both cases rising from a significant difference between the left and mid-sagittal row in the ambiguity *effect* (clear vs ambiguous)[135]. The PINV enlargement from the clear to the ambiguous condition is more pronounced along the right- and mid- than along the left-sagittal rows (see **table 6.13**). For the 1st PINV segment right versus mid-sagittal rows differ in the ambiguity effect, too[136]. This is likewise due to the pronounced ambiguity effect at all mid-sagittal leads. The Condition by Laterality interaction is compatible with a change in the generator structure, as indicated by the results of normalized data.

Concerning the 1st PINV segment, the last mentioned interaction must be differentiated according to the groups, as indicated by the significant interaction of **Condition by Laterality by Group**. The differences between the left and mid-sagittal rows in the ambiguity effect are not the same in patients and controls[137]. While ambiguous IS augment the PINV along all sagittal rows in patients, only mid- and right-sagittal amplitudes are enhanced in controls, with a slight PINV diminution over the left hemisphere giving rise to the significant planned contrast mentioned.

An amplitude effect cannot be rejected as the results of normalized data are non-significant. The within-group ANOVA indicate a significant Condition by Laterality interaction in controls[138], but only a trend in patients[139], confirming the "broader" topography of the ambiguity effect for the first PINV segment in patients statistically. Altogether ambiguous IS augment the PINV along all sagittal rows in patients, but only over the right- and mid-sagittal leads in controls.

A further topographical differentiation of the Condition effect is indicated by the significant interaction of **Condition by Anteriority** for the 1st and the 2nd PINV segment, due to significant differences between the frontal and the central coronal row concerning the nogo effect[140]. The PINV amplitude diminution by the nogo IS is more pronounced at central (and parietal) than at frontal leads (see

---

[133] $F(2,20)=7.47$, $p=.01$
[134] $F(2,24)=0.87$, $p=.42$
[135] PINV-1: $F(1,22)=17.64$, $p=.0004$; PINV-3: $F(1,22)=14.18$, $p=.001$
[136] $F(1,22)=5.06$, $p=.03$
[137] $F(1,22)=12.77$, $p=.002$
[138] $F(4,48)=5.90$, $p=.0006$
[139] $F(4,72)=2.82$, $p=.06$
[140] PINV-1: $F(1,22)=19.47$, $p=.0002$; PINV-2: $F(1,22)=12.08$, $p=.002$

table 6.14) and is compatible with a change in the generator structure of the PINV.

Table 6.13 Condition by Laterality Interaction on PINV-1 and PINV-3 and Condition by Laterality by Group Interaction on PINV-1

	Left	Middle	Right
PINV-1			
Clear	-0.73 (2.2)	-0.21 (3.0)	-0.83 (2.0)
Ambiguous	-1.22 (3.4)	-1.60 (4.3)	-1.80 (3.2)
NoGo	0.93 (2.4)	0.98 (2.6)	0.14 (2.0)
PINV-3			
Clear	0.32 (2.2)	0.61 (2.7)	-0.53 (1.7)
Ambiguous	-0.44 (2.9)	-1.18 (3.5)	-2.01 (2.6)
NoGo	1.73 (2.6)	1.89 (3.1)	0.93 (2.4)
PINV-1			
Patients			
Clear	-2.43 (1.4)	-2.81 (1.3)	-2.27 (1.5)
Ambiguous	-3.67 (2.6)	-4.74 (3.4)	-3.69 (3.1)
NoGo	-0.04 (2.7)	-0.01 (3.1)	-0.14 (2.0)
Controls			
Clear	0.71 (1.8)	1.98 (2.2)	0.39 (1.5)
Ambiguous	0.85 (2.5)	1.06 (2.9)	-0.20 (2.5)
NoGo	1.76 (1.8)	1.81 (2.0)	0.39 (2.0)

unit is µV; SD in parentheses

Table 6.14 Condition by Anteriority Interaction on PINV-1 and PINV-2

	Frontal	Central	Parietal
PINV-1			
Clear	-1.11 (2.4)	-0.59 (2.9)	-0.07 (2.2)
Ambiguous	-1.96 (3.3)	-1.75 (4.0)	-0.90 (2.6)
NoGo	-0.38 (2.2)	1.19 (2.8)	1.25 (2.0)
PINV-2			
Clear	-1.32 (2.2)	-0.27 (2.5)	0.45 (1.7)
Ambiguous	-2.48 (3.0)	-1.45 (3.2)	-0.45 (2.5)
NoGo	0.11 (2.9)	1.97 (3.2)	2.23 (2.2)

unit is µV; SD in parentheses

The interaction of **Condition by Laterality by Anteriority**, which is significant for the 2nd and the 3rd PINV segments, originates from differences at left versus mid-sagittal[141] and right versus mid-sagittal[142] rows concerning the parietal versus central coronal difference in the Ambiguity effect (see **table 6.15**).

Table 6.15 Condition by Laterality by Anteriority Interaction on PINV-2 and PINV-3

	Left	Middle	Right
PINV-2			
Clear			
Frontal	-0.84 (2.4)	-0.93 (2.8)	-2.19 (1.8)
Central	-0.16 (2.7)	-0.05 (3.2)	-0.70 (1.9)
Parietal	-0.35 (1.7)	0.93 (2.1)	-0.07 (1.7)
Ambiguous			
Frontal	-1.56 (3.2)	-2.31 (3.9)	-3.56 (2.9)
Central	-1.10 (3.4)	-1.31 (3.9)	-1.94 (2.7)
Parietal	-0.00 (2.5)	-0.75 (2.7)	-0.75 (2.8)
NoGo			
Frontal	0.74 (3.0)	0.10 (3.6)	-0.51 (2.5)
Central	2.18 (3.5)	2.46 (3.8)	1.26 (2.8)
Parietal	2.03 (2.1)	2.83 (2.5)	1.84 (2.5)
PINV-3			
Clear			
Frontal	-0.36 (2.5)	-0.21 (2.9)	-1.67 (1.9)
Central	0.52 (2.7)	0.82 (3.2)	-0.22 (2.0)
Parietal	0.79 (2.0)	1.23 (2.3)	0.31 (1.9)
Ambiguous			
Frontal	-1.20 (3.1)	-2.02 (4.2)	-3.50 (3.0)
Central	-0.58 (3.4)	-0.92 (3.9)	-1.79 (2.8)
Parietal	0.44 (2.6)	-0.59 (3.4)	-0.73 (2.6)
NoGo			
Frontal	0.85 (3.2)	0.38 (3.9)	-0.27 (2.9)
Central	2.30 (3.2)	2.63 (3.8)	1.27 (2.7)
Parietal	2.02 (2.0)	2.66 (2.6)	1.80 (2.4)

units is µV; SD in parentheses

Compared to the rather homogeneous enlargement of PINV amplitudes along the central coronal row there is only a substantial PINV augmentation for the parietal coronal row at Pz present for PINV-2 and PINV-3 and additionally at P4 for PINV-3.

Hence, the ambiguity effect on PINV is mainly restricted to Pz concerning the parietal coronal row, but is present at all central and frontal leads. Again, this effect is significant for the normalized ERP data, indicating that ambiguous

[141] PINV-2: $F(1,22)=5.36$, $p=.03$; PINV-3: $F(1,22)=4.40$, $p=.05$

[142] PINV-2: $F(1,22)=4.61$, $p=.04$; PINV-3: $F(1,22)=4.24$, $p=.05$

compared to clear IS are associated with a change in the generator structure of the PINV.

To summarize: PINV amplitudes are larger under the ambiguous than under the clear condition, particularly over the right hemisphere and at mid-sagittal leads. The Ambiguity effect at the sagittal rows distinguishs between the groups, being present at left-hemisphere leads only in patients. The nogo effect is associated primarily with an amplitude diminution along the central and parietal coronal rows and weaker effects at the frontal leads. The topography of this effect does not differentiate groups. Ambiguous and nogo IS appear to activate different generators when compared to the clear IS, as suggested by the significant results of normalized data.

6.9.3.3 Late Positive Complex (LPC)

The results for the IS-LPC are documented in **table 6.16**. The IS-LPC peak amplitudes differ between the sagittal rows, as attested by the significant **Laterality** effect.

Table 6.16 Experimental Effects on IS-LPC

		Amplitude		Latency	
Effect	**df1,df2**	**F**	**p**	**F**	**p**
C	2,44	43.76	.0001	38.83	.0001
C* G	2,44	0.23	.70	1.50	.24
L	2,44	24.65	.0001	0.88	.40
L * G	2,44	0.17	.83	0.37	.63
A	2,44	8.36	.005	5.04	.02
A * G	2,44	0.49	.53	4.00	.04
C * L	4,88	31.47	.0001[1]	0.77	.52
C * L * G	4,88	0.53	.68	0.39	.76
C * A	4,88	13.84	.0001[2]	5.16	.001
C * A * G	4,88	0.29	.82	1.62	.18
L * A	4,88	3.48	.03	1.83	.13
L * A * G	4,88	0.19	.86	1.75	.15
C * L * A	8,176	0.74	.57	0.50	.82
C * L * A * G	8,176	0.35	.85	0.75	.62
Group	1,22	0.68	.42	0.27	.61

***C**=Condition factor, **L**=Laterality factor, **A**=Anteriority factor**G**=Group factor; Results of univariate ANOVA, Hunynh-Feldt corrected; average mastoids reference; baseline is tCNV;* [1]=F(4,88)=6.78, p=.0001; [2]=F(4,88)=2.44, p=.05

This effect originates from significant contrasts between the left and the mid-sagittal row[143] and the right and the mid-sagittal row[144]. The peak amplitudes are largest along the mid-sagittal row and smaller over the left and right hemispheres (see **table 6.17**).

The significant **Anteriority** main effect rises only from a significant contrast between the frontal and the central sagittal rows[145]. The IS-LPC is largest at parietal and central leads and smallest at frontal positions.

Both main effects are subordinate to the significant **Laterality by Anteriority** interaction, with significant contrasts between the left and the mid-sagittal[146] and the right and the mid-sagittal[147] rows concerning the parieto-central difference. The amplitude diminution from the mid-sagittal electrodes Cz and Pz to the lateral leads is steeper along the parietal than along the central coronal row (see **table 6.17**).

Table 6.17 Laterality by Anteriority Interaction on IS-LPC Peak Amplitude

	Left	Middle	Right
Frontal	6.20 (3.5)	8.06 (5.4)	6.93 (4.7)
Central	8.29 (3.9)	10.06 (4.9)	8.97 (4.2)
Parietal	8.85 (3.7)	11.64 (4.7)	8.72 (4.2)

unit is µV; SD in parentheses

The significant **Condition** main effect is due to significant contrasts between the ambiguous[148] and the nogo[149] conditions on the one hand and the clear condition on the other hand. Compared to the clear condition, IS-LPC peak amplitudes are smaller following ambiguous, but larger following nogo IS (see **tables 6.18** and **6.19**). The significant **Condition by Laterality** interaction is due to a topographical differentiation of the nogo effect. The amplitude augmentation due to the nogo IS is larger along the mid-sagittal row than over the left[150] and the right[151] hemisphere. The interaction of **Condition by Anteriority** is significant, because the nogo effect on the IS-LPC amplitudes is more pronounced along the central coronal row than along the frontal[152] and the parietal[153]coronal rows. In

[143] $F(1,22)=44.67$, $p=.0001$
[144] $F(1,22)=45.34$, $p=.0001$
[145] $F(1,22)=14.52$, $p=.001$
[146] $F(1,22)=5.56$, $p=.03$
[147] $F(1,22)=12.00$, $p=.002$
[148] $F(1,22)=14.43$, $p=.001$
[149] $F(1,22)=43.95$, $p=.0001$
[150] $F(1,22)=32.07$, $p=.0001$
[151] $F(1,22)=53.73$, $p=.0001$
[152] $F(1,22)=34.36$, $p=.0001$
[153] $F(1,22)=4.73$, $p=.05$

addition, the amplitude diminution due to ambiguous IS (compared to clear IS) is larger at central than at parietal[154] leads, but of the same magnitude at central and frontal positions. Both interaction effects are compatibel with a change in the generator structure as indicated by the significant results of normalized data. They are documented in **table 6.18** and **6.19**.

Table 6.18 Condition by Laterality Interaction on IS-LPC Peak Amplitudes

	Left	Middle	Right
Clear	6.34 (3.8)	8.00 (4.8)	7.21 (4.2)
Ambiguous	5.01 (3.9)	6.12 (4.7)	5.64 (3.9)
NoGo	12.00 (4.5)	15.63 (5.9)	11.77 (4.8)

unit is µV; SD in parentheses

Table 6.19 Condition by Anteriority Interaction on IS-LPC Peak Amplitudes

	Frontal	Central	Parietal
Clear	6.06 (4.6)	7.33 (4.8)	8.15 (4.4)
Ambiguous	4.51 (4.0)	5.53 (4.9)	6.74 (4.4)
NoGo	10.62 (5.7)	14.45 (5.4)	14.31 (5.3)

unit is µV; SD in parentheses

IS-LPC *latencies* differ between the coronal rows, as the significant **Anteriority** main effect indicates. The effect rises from significant contrasts between the ambiguous[155] and the nogo[156] conditions, when compared to the clear condition. These differences, however, are not the same in the two groups, since the interaction of **Anteriority by Group** is significant due to different latency differences in the two groups between the frontal and the central coronal row[157].

Table 6.20 Anteriority by Group Interaction on IS-LPC Peak Latencies

	Frontal	Central	Parietal
Patients	380.6 (71.3)	380.3 (59.1)	383.7 (57.5)
Controls	366.7 (40.5)	397.0 (47.0)	412.6 (46.2)

unit is ms; SD in parentheses

[154] $F(1,22)=5.74, p=.03$
[155] $F(1,22)=39.52, p=.0001$
[156] $F(1,22)=17.90, p=.0003$
[157] $F(1,22)=7.58, p=.01$

The within-group ANOVA, yet, reveal a significant Anteriority effect only in the control group[158] but not in patients[159].

Table 6.21 Condition Effect on IS-LPC Peak Latency

Clear	Ambiguous	NoGo
383.4 (61.3)	347.6 (49.6)	430.7 (54.2)

unit is ms; SD in parentheses

While in controls frontal IS-LPC peak latencies are shorter than central, they do not differ in patients. This differentiation holds also for the IS-LPC peak latencies under the nogo condition.

The **Condition** main effect on the IS-LPC peak latency is significant, too. While ambiguous compared to clear IS diminish the latencies significantly[160], nogo IS augment the peak latency[161] (see **table 6.21**). There is no significant contrast for the **Condition by Anteriority** interaction.

6.10 Discussion

6.10.1 Reaction Times (RT)

In experiment B the often reported (e.g., King, 1991) slowing of motor responses in schizophrenic patients was replicated. This effect has already been discussed in *paragraph 5.10.1*.

6.10.2 Terminal Contingent Negative Variation (tCNV)

The tCNV was most pronounced over the sagittal midline and larger at left- than at right-sided leads (see **table 6.6**). The larger left- than right-sided tCNV confirms findings reported before, e.g. by Klein et al. (1994), Roberts et al. (1994), and in experiment A. Schizophrenic patients did not differ from healthy controls in their overall or condition-specific tCNV amplitudes, although smaller CNV amplitudes can be seen in **figures 6.3a-c**, especially at fronto-lateral leads. CNV amplitudes have been reported frequently to be reduced in schizophrenic patients (e.g. Callaway, 1970; Tecce & Cole, 1976; Timsit-Berthier et al., 1973). The present results suggest that globally diminished tCNV amplitudes or atypical spatial generation of the tCNV can hardly be considered a general sign,

158 $F(2,24)=11.30, p=.001$
159 $F(2,20)=0.05, p=.87$
160 $F(1,21)=39.52, p=.0001$
161 $F(1,21)=17.90, p=.0003$

comparable to a "trait marker", in schizophrenic patients. Rather, state-dependent or task-dependent influences may explain the present results. *Firstly*, the patients participating in this study were well beyond acute stages of their illness. Several results suggest that following acute phases neuroleptic treatment and remission may be associated with trends towards the normalization of CNV (e.g. Knott et al., 1976; Rizzo et al 1983). Only longitudinal studies, however, could clarify, whether the lack of group differences in CNV were due to a temporary "recovery" of the CNV. *Secondly*, promising a financial bonus in the case of fast and correct responses in the instruction may have motivated schizophrenic patients particularly, since their economical situation is frequently worse than that of healthy subjects. CNV is sensitive to motivational factors (cf. Rockstroh et al., 1989) and improved task performance after providing sufficient financial reinforcement has been observed in schizophrenic patients (Summerfelt et al., 1991).

The tCNV distribution, including the often reported *frontal shift* (relative frontal enlargement) of the tCNV (Cohen et al., 1991; Rockstroh et al., 1993; Wagner et al., 1993; Van den Bosch, 1983, 1984) did not differ between groups. Under the rather complex experimental delayed matching-to-sample task of the present experiments (compared to the standard two-stimulus paradigm used in most reports on reduced CNV in schizophrenia, e.g. Van den Bosch, 1983); the topographical shift of the tCNV towards frontal areas in schizophrenic patients was not seen. This frontal predominance has been discussed by Rockstroh et al. (1994) as indicating frontal compensatory effort for impaired parietal processing of the visual WS. One may speculate that the rather demanding but perhaps not too difficult delayed matching-to-sample task of the present experiment motivated schizophrenic patients optimally, thereby increasing the centro-parietally predominant tCNV and the posterior-anterior amplitude gradient.

6.10.3 Late Positive Complex (LPC)

The LPC following the visual IS had the expected parietal, mid-sagittal predominance. The effects of varying task conditions on IS-LPC included both a significant diminution of peak amplitudes following ambiguous IS, and augmented peak amplitudes following nogo IS (both compared to clear).

Reduced peak amplitudes following ambiguous IS have been reported for the IS-LPC data of experiment A. Several explanations, depending on the theoretical model of P300, can be given for this result. *Firstly*, task "closure" which has been suggested to underlie the P300 (Verleger, 1988) can be expected to be less evoked by ambiguous when compared to clear IS. *Secondly*, "information transmission" may be expected to be less for ambiguous than for clear IS and should result in diminished peak amplitudes according to the model of Johnson, Jr. (1986). The same argument may hold for the large LPC following the nogo IS: The red ellipses associated with the instruction to withhold the motor response should be apt to "close" the current perceptual epoch efficiently. Furthermore, with these clearly distinguishable IS, information transmission should be maximal according to the model of Johnson (1986).

Nogo IS increased the IS-LPC especially at central positions, thereby shifting the parietal LPC maximum of the clear condition to the more balanced centro-parietal maximum of the nogo condition. The amplitude augmentation and topographical shift is in accordance with other studies including nogo stimuli (Karlin et al., 1970; Pfefferbaum & Ford, 1988; Pfefferbaum et al., 1985; Simson et al., 1977).

While larger LPC amplitudes following nogo IS have been suggested to reflect enhanced CNV resolution (Simson et al., 1977), recent results (Roberts et al., 1994) support the notion that the augmentation of the nogo P300 reflects sensorimotor inhibition. This interpretation is in agreement with the interpretation of positive ERP such as the P300 as widespread cortical inhibition (Rockstroh & Elbert, 1990; Rockstroh et al., 1992) or dampening of excited motor systems by means of corticothalamic inhibition following the nogo signal (Roberts et al., 1994). Following Brunia (1993; see *chapter 7*) such inhibitory processes must be expected to be selective, and the selection should plausibly favor those cortical regions which are most engaged in response preparation. As has been argued in *chapter 3* and discussed by Brunia (1993), MI and SI regions exhibiting increased excitability during the centro-parietally dominant tCNV are among the cortical structures preparing for response execution and should be inhibited most by the processing of the nogo signal. Furthermore, preparatory unit activity, has been observed in prefrontal, precentral motor (MI), premotor, supplementary motor and posterior parietal areas in the monkey (cf. Watanabe, 1986). These activities are evidently reflected by the centro-parietal tCNV. However, while the tCNV was largest parietally, the IS-LPC increase due to nogo stimuli was largest centrally. This result might suggest that following the nogo command mainly the areas that prepare the motor response are inhibited, and these are located more anteriorly.

Since no differences between schizophrenic patients and healthy controls were found for the IS-LPC, functions reflected by this component may be assumed to be unimpaired in the patient group. As has been discussed in *chapter 3* evidence for temporal stability of a reduced *visual* P300 in schizophrenic patients is weak, and recovery of this component with clinical improvement can be assumed (Duncan et al., 1987). The schizophrenic patients participating in this experiment were well beyond acute stages of their illness and this may possibly explain their unimpaired P300 regulation.

6.10.4 Post-Imperative Negative Variation (PINV)

In experiment B, a larger PINV was observed in schizophrenic patients than in healthy controls. This result confirms numerous published reports (see *paragraph 3.5*) and the results of experiment A. Since PINV has been suggested to reflect response uncertainty (cf. Cohen, 1990), schizophrenic patients appear more susceptible than healthy subjects of developing uncertainty about the correctness of their response. A discussion of this effect shall not be repeated here (see *paragraph 5.10.4*).

In experiment B a PINV could again be observed in healthy subjects. This results replicates the observation reported for experiment A and suggests that the basic task design, including ambiguity in WS-IS and IS-response contingencies, is capable of inducing a PINV in this group. The possible meaning of this result will be thoroughly discussed in *chapter 9*.

Similarly to the non-lateralized topography of the PINV in experiment A, a frontal predominance of this component was found in experiment B. This topography of the PINV has been reported by some authors (e.g. Elbert et al., 1982; Kathmann et al., 1990; Lutzenberger et al., 1981), but has not always been found (e.g. Van den Bosch, 1983; Roth et al., 1979). Some studies reveal different fronto-central gradients of the PINV in schizophrenic patients (larger at frontal sites) and healthy controls (larger at central sites) (Eikmeier et al., 1992; Wagner et al., submitted). This was not found for the samples of experiment B.

As was the case in experiment A, the lateralization of the PINV, averaged over the experimental conditions, *differed* between groups (interaction Laterality by Group), and this effect was compatible with a difference in the generator structure of the PINV in patients and controls. In contrast to experiment A, *no* left-sided predominance of the PINV was present in patients (with a non-significant Laterality effect in the within-group ANOVA), while the right-sided predominance of the PINV in controls, which was present only as a trend in experiment A, now turned out highly significant. One *might* consider the choice of the planned contrasts (left versus right in experiment A; left versus mid-sagittal and right versus mid-sagittal in experiment B) as a trivial explanation for the divergency. But it is clear that the selection of contrasts has no impact on the within-group Laterality simple effects; furthermore a post hoc *re*definition of the contrasts (mid-sagittal versus left and right versus left) did not reveal significant results for the left versus right *contrast* of the Laterality by Group interaction[162]. Also, the left versus right *contrasts* of the patient within-group Laterality effect revealed no significant results for the first and second PINV segment[163]. For PINV-3 this *contrast* was significant[164], however, indicating a larger right- than left-sided PINV in the patient group, contrary to expectations (see also **table 6.11**).

The significant Condition main effect (contrast clear versus ambiguous) together with the non-significant clear versus ambiguous contrast of the Condition

162 PINV-1: $F(1,22)=2.67$, $p=.12$; PINV-2: $F(1,22)=1.84$, $p=.19$; PINV-3: $F(1,22)=1.62$, $p=.12$

163 PINV-1: $F(1,10)=0.0$, $p=.96$; PINV-2: $F(1,10)=1.55$, $p=.24$

164 PINV-3: $F(1,10)=4.80$, $p=.05$; it should be noted that the within-group Laterality effect itself fell short of significance even for this segment ($F(2,20)=3.30$, $p=.06$)

by Group interaction indicates that ambiguous compared to clear IS augmented the PINV in both groups to a similar extent. This result replicates the corresponding effect of experiment A and suggests that ambiguity during the matching of experimental stimuli is a general factor of the PINV.

For the second and third PINV segments this effect was rather homogeneous at frontal and central leads, but was present parietally only at Pz (interaction of Condition by Laterality by Anteriority). Hence, the anterior-posterior distribution of the effect of ambiguous compared to clear IS does not exactly correspond to the topography of the Ambiguity effect in experiment A, which was maximum at frontal (F3, F4) leads. In order to check whether the failure to replicate the result of experiment A was related to differences between experiments A and B in the electrode montage, an ANOVA for the F3, F4, C3, C4, P3, P4 configuration of the data of experiment B was accomplished, testing the Condition by Anteriority interaction and the clear versus ambiguous planned contrast, which both included the comparison between the frontal (F3, F4) and the central (C3, C4) electrode pairs. Neither the Condition by Anteriority interactions ($Fs \leq 1.48$) nor the planned contrasts ($Fs \leq 2.33$) were statistically significant. Accordingly, a different electrode montage cannot explain the different topographies of the Ambiguity effect in experiments A and B. It must remain unclear at present, whether procedural differences between the experiments such as the interspersed nogo IS or the different schedule for the presentation of the probe stimuli can explain this divergency in the topography of the Ambiguity effect. Despite these discrepancies, this non-lateralized topography of the Ambiguity effect (Condition by Laterality by Anteriority interaction) suggests a change in the generator structure to take place following ambiguous compared to clear IS, a result which has also been reported in experiment A.

The lateralization of the Ambiguity effect was not consistent over the three PINV segments. For the *first* PINV segment the Ambiguity effect was stronger at mid-sagittal (Fz, Cz, Pz) compared to right-sided (F4, C4, P4) leads, where PINV was larger than over the left hemisphere (F3, C3, P3). While the lateralization of the Ambiguity effect turned out non-significant for the *second* PINV segment, a mid-sagittal to right-sided maximum was present towards the end of the period (*PINV-3*). Since the interaction effects for PINV-1 and PINV-3 remained significant after normalization of the data, a change in the generator structure of the PINV with ambiguous compared to clear IS is again suggested. Due to the different electrode montage, including the mid-sagittal leads in experiment B, these results are not entirely comparable with those of experiment A.

However, contrary to experiment A, the lateralization of the Ambiguity effect did not consistently differ between groups. In experiment B the Condition by Laterality by Group interaction was significant only for the first PINV segment. This interaction, together with the within-group ANOVA, indicates a clear mid-sagittal and right-sided maximum of the Ambiguity effect in controls ($p=.0006$), with a statistical trend for the *same* topographical change in patients ($p=.06$). While a (trend towards a) right-sided predominance of the Ambiguity effect was not found for the patient group of experiment A, the patient groups of experiment A and B correspond in the less "restricted" topography of the Ambiguity effect. In other words, in both experiments patients exhibited a less "focal" and "broader" (lateralized) topography of the Ambiguity effect when compared to healthy controls. This corresponds to the "overall" topography of the PINV in patients and controls, which is topographically more limited (to the right side) in controls than in patients.

The "overall" topography of the PINV and the topography of the Ambiguity effect of experiment B are dissimilar to those of experiment A, but correspond to the notion about functional asymmetries, stating that "patients with schizophrenia may be less completely lateralized than normal individuals" (Weinberger, 1994). The multi-channel recordings of experiment C may clarify which of the two lateralizational patterns (of experiments A and B) can be replicated. In any event, so far the lateralization of the PINV in schizophrenic patients appears *consistent only in its divergency* from a clear right-frontal maximum in healthy controls but offers the possibility of heterogeneity *between* patient *samples*.

The key question of experiment B focused on the effects on PINV by eliminating the motor response from the task structure by "nogo" IS. Under the nogo compared to the clear condition a diminution of the first and second PINV segment was present at central and parietal leads and to a lesser extent also at frontal leads[165]. This is a striking result since it suggests a topographical *complementarity* of the Ambiguity and the Nogo effects, with a rather anterior PINV topography and augmenting effect of the ambiguous, and a rather posterior PINV topography and diminishing effect of nogo IS, each compared to clear IS. The centro-parietal PINV reduction under the nogo condition does not differ in magnitude between the groups, and is compatible with a change in the generator structure (significant after normalization of the data). A functional differentiation of subcomponents of the PINV within the total topographical field has been suggested by Birbaumer et al. (1986), when referring to the PINV in healthy subjects. Birbaumer et al. (1986) associated a frontal PINV with uncertainty about response-outcome contingencies and a centrally dominant PINV with uncertainty about stimulus characteristics. Tentatively applying this differentiation to the present results would suggest that centro-parietal PINV reduction is due to a diminished uncertainty about having correctly identified the nogo IS. Since the nogo IS was by color and shape certainly better discernible from the clear and ambiguous IS than the clear and ambiguous IS among one another, equivocation defined as the amount of loss of information during the stimulus presentation due to an aposteriori uncertainty *about having correctly perceived a given event* (cf. Johnson, Jr., 1986), should be minimal following the nogo IS. The posterior predominance of the PINV diminution due to nogo IS is therefore compatible

[165] Although for PINV-3 the Condition by Anteriority interaction was present only as a trend, the included contrast for the difference between the frontal versus the central coronal row in the nogo *effect* remained significant ($F(1,22)=8.09$, $p=.009$).

with the notion that the posterior proportions of the PINV reflect uncertainty about stimulus characteristics (Birbaumer et al., 1986). The augmentation of the IS-LPC amplitudes following nogo compared to clear IS support this interpretation. According to the Triarchic Model of the P300 (Johnson, Jr., 1986; see *paragraph 6.10.4*) reduced equivocation means augmented information transmission, with augmented P300 amplitudes as a consequence[166]. Following this reasoning the centro-parietal reduction of the PINV by the nogo IS may indicate the sensitivity of the posterior proportions of the PINV to uncertainty about stimulus characteristics, as suggested by Birbaumer et al. (1986).

Another interpretation of the nogo effect refers to the signal function of the nogo IS, to withhold the response. This signal function may be expected to trigger the inhibition of excited motor structures, and this is confirmed by the behavioral (reaction) data. Furthermore, the amplitude augmentation of the *anteriorly shifted* IS-LPC has been considered to reflect sensorimotor inhibition or dampening of excited motor systems via corticothalamic inhibition (Roberts et al., 1994; see also *paragraph 6.10.3*). Excitation of motor systems is generally considered to be reflected by the tCNV. However, while the tCNV was largest parietally, the IS-LPC increase due to nogo stimuli was largest centrally. This result might suggest that following the nogo command, the rather anteriorly located cortical areas preparing the motor response are mainly inhibited. Hence, the centro-parietal PINV could be considered a sign of ongoing motor preparation. This interpretation would suggest that the posterior proportions of the PINV reflect a non-inhibited tCNV, lending support to the interpretation of the PINV as "delayed CNV resolution" (Mantanus et al., 1981; Roth et al., 1977) or "′prolongated duration′ CNV" (Timsit-Berthier et al. (1987), see *paragraph 3.5*). It should, however, be recalled that negativity during the PINV interval *exceeded* negativity during the tCNV intervall (see **table 6.11**). Hence, PINV concepts which stress the persistence of processes that arise during the ISI cannot fully explain the data.

The posterior topography of the nogo effect suggests a differentiation of different components of "response uncertainty" and, hence, PINV, coupled to different topographical proportions of this component. From experiment A it became clear that the quality of the matching process (clear versus ambiguous) affected the PINV primarily at frontal sites (chapter 8 will report on two replications (experiments C and D) of this effect which was, unfortunately, not seen in experiment B). Furthermore, a frontal predominance has been reported for the PINV augmentation from the "constant lack of control" to the "random lack of control" condition by Kathmann et al. (1990). In both instances, consistent or

[166] It could be argued that the IS-LPC *augmentation* following nogo compared to clear IS is a consequence of the lower subjective probability of the nogo IS. Although this hypothesis cannot be ruled out at present, it is important to note that the ambiguous IS, which occured also with lower probability than the clear IS, was associated with *diminished* IS-LPC amplitudes. Hence, subjective probability yields no *simple* explanation for the present IS-LPC results.

reliable contingencies (between WS and IS, and IS and response in the present experiment, between response and response consequences in the study of Kathmann and collegues) were untied experimentally (by ambiguous IS or random lack of control), presumably triggering uncertainty about the correctness of the response. Assuming that subjects generate a representation of the task structure which comprises *reliable contingencies* between the experimental stimuli and the response (a necessary condition for stimulus evaluation, response preparation, and a *defining* condition of aspired *correct* responding), a divergency between this representation and the perception of *dis*contingencies in concrete trials may be an "antecedent condition"[167] for the development of an anterior PINV. This anterior PINV may reflect ruminations about the appropriateness of the response *vis-à-vis* untied contingencies, and hence "performance uncertainty". In contrast, the posterior proportions of the PINV, which were less affected by the aforementioned experimental manipulations, are especially affected by the requirement to respond versus to withhold the response. It is therefore conceivable that these posterior proportions of the PINV reflect "action uncertainty", i.e. ruminations about the appropriateness of responding *per se*, suspended by the nogo instruction.

Unfortunately, still another interpretation of the nogo effect cannot be ruled out with certainty at present. Although the red nogo ellipse provided the unambiguous instruction not to respond, the implications for the task structure were *twofold*. *Firstly*, an abolishing of the motor response, as discussed above; but *secondly*, an abolishing of the matching process itself. "No motor response" and "no-matching" are two different consequences of the same nogo IS, and both could have been responsible for the PINV diminution under the nogo condition. However, if the "no-matching" consequence were the crucial aspect of the nogo IS, this effect should have had the same topography like the other experimental effect related to the matching process: the Ambiguity effect. But this is not the case: *Firstly*, in this experiment ambiguous IS changed the PINV lateralization, while the nogo IS had no lateralized effect. *Secondly*, in experiment A ambiguous IS augmented PINV amplitudes maximally at frontal leads (this will be seen in experiments C and D, too; see *chapter 8*). While this effect could not be replicated within the present experiment, at least the nogo effect was *least* pronounced over frontal regions in experiment B. Had the nogo IS provided primarily a "no-matching" effect, PINV diminution should not have been most prominent at centro-parietal sites. In conclusion, the predominant consequence of the nogo IS was presumably the response inhibition and not the abolishing of the matching process.

[167] Donchin & Coles (1988) give an explanation of this term: "For example, the claim that the amplitude of the P300 is inversely proportional to the subjective probability of the eliciting events, (...), is a statement about the conditions antecedent to the elicitation of the P300. (...) Such statements are necessary precursors of a theory of the P300." (p.358)

It is certainly a matter of further experimentation to attempt to clarify which of these hypotheses is correct. In any event, the present results suggest a double dissociation between anterior and posterior proportions of the PINV.

How do these results relate to the deficit in self-monitoring, proposed by Frith (1987, 1992, 1994; Frith & Done, 1988) and suggested to be related to the PINV by Cohen (1990)?

A lack of corollary discharges has been suggested to underlie some of the symptoms of schizophrenic patients (Frith, 1992). In the early version by Frith (1987) the corollary discharge or re-afference copy hypothesis pertains to motor processes. While the extension to hallucinatory experiences by Frith (1992) is still somewhat speculative, direct supportive evidence has been given for motor responses with the central error correction paradigm (e.g. Malenka et al. 1982; Mlakar et al., 1994; see *paragraph 6.1*). Deficits in the internal monitoring of actions are related to deficient information transmission between prefrontal and hippocampal sites (Frith, 1992), and the assumed "stressing" of the "information networking" between these sites has been shown to be associated with the PINV in schizophrenic patients (experiment A).

A deficit of schizophrenic patients in the generation of corollary discharges during termination of the task by the motor response should result in a failure of appreciating "having completed their task" (Cohen, 1990), with the development of the PINV as a consequence.

Removing the motor response from the task structure, as in the present nogo condition, should circumvent impaired corollary discharge generation, avoiding the triggering of response uncertainty and PINV. In this sense, the smaller PINV under the nogo condition is compatible with the hypothesis that impaired corollary discharges during the motor response in CNV standard tasks are triggering response uncertainty and PINV in schizophrenic patients.

To sum: Experiment B replicated the observation of a PINV in healthy subjects under experimental conditions that comprised ambiguity in matching the IS with the WS. Again larger PINV amplitudes were observed in schizophrenic patients, suggesting a heigthened sensitivity of this group to the development of response uncertainty. While the left-sided predominance of the PINV in the patient group of experiment A could not be replicated in experiment B, the right-sided predominance of the PINV in healthy controls was more pronounced in experiment B compared to experiment A. The possible meaning of these results and of the divergency between experiment A and B will be discussed in chapter 9, after presentation of the results of multi-channel recordings in chapter 8.

Experiment B added insight into the functional meaning of the PINV in schizophrenia in that the contribution of motor responding to PINV could be specified.

Although the elucidation of the exact meaning of the PINV diminution following nogo IS requires further experimentation, we can conclude that PINV in schizophrenia is not entirely induced by processes associated with the motor response or deficits such as missing corollary discharges, since a small negative shift still remained in this group under the nogo condition. These results are intriguing for the discussion as to what extent PINV indicates response uncertainty. It is tempting to differentiate "action uncertainty" associated with the posterior, nogo-sensitive proportion of the PINV, from "performance uncertainty", associated with the frontal, nogo-insensitive but ambiguity-sensitive proportion of this component.

Chapter 7 Does the PINV Reflect Facilitation or Dysfacilitation ?

Summary In experiments A and B acoustical probe stimuli were presented during the pretrial baseline, CNV, and PINV in order to test the facilitatory nature of the negative potential shifts and the question whether schizophrenic patients exhibit atypical modulation of the probe EP. The results on probe EP will be reported in *chapter 7* subsequent to a brief introduction into the brain structures asssumed to be responsible for the generation of negative and positive potential shifts and a survey of studies using the probe technique. In both experiments, EP elicited by probe stimuli delivered during the PINV (in particular the N100 component) were smaller than those during the CNV. Also, in both experiments schizophrenic patients and healthy controls did not differ in the probe EP components. These results suggest normal modulation of the probe EP in schizophrenic patients and a dysfacilitatory nature of the PINV, compared to the CNV. Methodological issues are discussed.

7.1 Theoretical Background of the Investigation of Acoustical Probe-EP in Experiments A and B

One approach to the functional meaning of ERP and their deviances in clinical populations considers neurophysiological mechanisms. This chapter will deal with one aspect of the functional significance of CNV and PINV in schizophrenic patients and healthy control subjects.

The neurophysiological mechanisms of ERP generation have mainly been investigated in non-human organisms. Differences in experimental conditions between nun-human and human ERP research, higher encephalization of the human cortex (Rebert et al., 1993), and the scarcity of simultaneous cortical unit and scalp recordings may hamper the generalization of results from non-human to human organisms. With these considerations in mind, the next paragraph will describe the brain structures assumed to be *functionally*[168] involved in ERP generation. In *paragraph 7.2* a model will be presented which states that positive and negative potential shifts indicate cortical facilitation and dysfacilitation. This hypothesis will be explored by presenting probe stimuli during the development of negative potential shifts, these studies are reported in *paragraph 7.8* and *7.9*.

[168] Brain structures functionally involved in ERP generation should not be mistaken as biophysical generators of the potential shift.

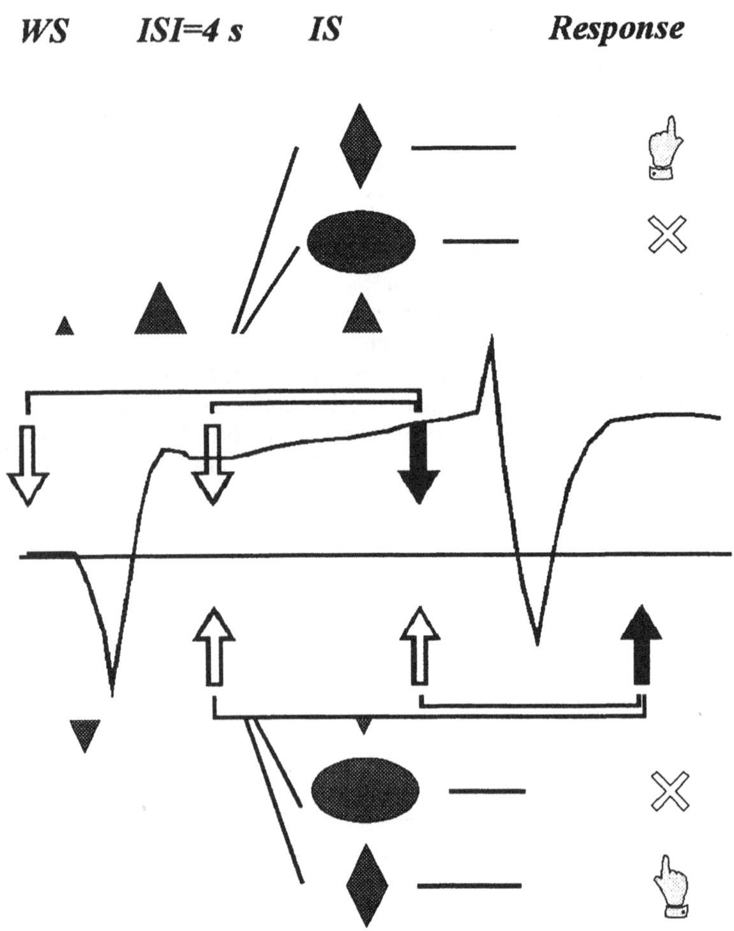

Figure 7.1

7.1.1 Brain Structures Involved in ERP Generation

Excitatory post-synaptic potentials generated in apical dendrites of radially (vertically) arranged pyramidal neurons in deep cortical layers, innervated by thalamic afferents and axon collaterals may be considered responsible for the generation of *negative* potentials (Elbert, 1993; Mitzdorf, 1991). Pharmacological studies have revealed that surface-*negative* slow potential shifts are associated with cortical presynaptic release of Acetylcholine (ACh; Marczynski, 1978). Thereby, ACh appears to work more like a neuro-modulator than like a neurotransmitter, in that cortical neurons, "primed" with ACh in subthreshold concentrations, exhibit a reduced threshold for incoming volleys and exciting aminoacids (Marczynski, 1978).

EEG synchronization can be associated with surface-positive slow potential shifts. During EEG synchronization release of cortical GABA and inhibitory post-synaptic potentials (IPSP) of pyramidal cells can be observed (cf. Marczynski, 1978). Therefore, surface-*positive* potentials have been suggested to result from a hyperpolarizing inhibition of pyramidal cells and the electrotonic spreading of inhibitory post-synaptic potentials (Marczynski, 1978). The latter part of this conclusion may be questioned, since there are only few inhibitory cortical neurons; on the other hand, these neurons may be very effective since inhibitory synapses terminate on the cell bodies or proximate dendrites and not on distal dendrites (see also *chapter 8*). Summation of excitatory post-synaptic potentials (EPSP) and IPSP appear to result in some of the SCP observed in humans (Rockstroh et al., 1989).

The regulation of these *cortical* processes is certainly not confined to cortical neurons, since glial cell activity appears to be involved in the regulation of slow potential shifts as well (Bauer et al., 1993). Furthermore, subcortical brain structures are considered to be involved, as has been exemplified in *chapters 3* and *6* for the CNV generation and has recently been emphasized by Elbert (1993) and Rebert et al. (1993). According to Rockstroh et al. (1989) subcortical brain structures assumed to be involved in SCP generation include:

(1) the <u>M</u>esencephalic <u>R</u>eticular <u>F</u>ormation (MRF; part of the <u>A</u>scending <u>R</u>eticular <u>A</u>ctivation <u>S</u>ystem (ARAS),
(2) the <u>R</u>eticular nucleus of the thalamus (R),
(3) the <u>B</u>asal <u>G</u>anglia (BG),
(4) and the midline <u>I</u>ntralaminar <u>N</u>uclei of the <u>T</u>halamus (INT).

(MRF) Enhanced activity of the MRF is associated with negative, reduced activity of the MRF with positive cortical potential shifts. Electrical stimulation of the MRF induces behavioral arousal and EEG desynchronization (Moruzzi & Magoun, 1949, cited after Skinner, 1978). Electrical ARAS stimulation completely abolishes the cortical release of GABA (Marczynski, 1978). These results suggest a relation between cortical activity and brainstem activity, a

relation which is further supported (a) by a study of McCallum et al. (1973), who observed similar potentials from brainstem and vertex recordings in a neurosurgical patient during a CNV task with a 1 s interstimulus interval (ISI), and (b) by the observed linkage between negative potential shifts in midbrain structures and enhanced multiple-unit activity (Rebert 1972, 1976).

(R) The influence of the MRF on cortical potential shifts is mediated by its influence on the reticular nucleus of the thalamus (*R*, Skinner & Yingling, 1976, 1977). Through this structure thalamic afferents and efferents have to pass. Electrical stimulation of the ARAS is associated with large positive slow potentials in R, reflecting tonic hyperpolarizing inhibition of R (Marczynski, 1978). The inhibitory influence of MRF on R is further supported by the observation of the abolishing of spontaneous firing of R units following MRF stimulation (Skinner, 1978). During sensory information processing an inhibitory influence of R on thalamic nuclei occurs (Skinner & Yingling, 1977). For instance, during EEG synchronization enduring bursts of high-frequency firing can be measured in R cells, while in the surrounding thalamic nuclei activity is lacking (cf. Skinner, 1978). These effects are topologically specific, since stimulation of a given part of R inhibits only the information transmission to the adjacent thalamic nucleus, leaving the other thalamic nuclei unaffected (Yingling & Skinner, 1976). Inhibition of R may be mediated pharmacologically by catecholaminergic projections (Marczynski, 1978). Slow potential shifts during CNV and RP tasks have been observed, among others, in dorsomedial and centromedian thalamic nuclei. Non-specific thalamic fibers synapse with dendritic trees of pyramidal neurons, the depolarization of which modulates the excitability of the neuron and the neuronal networks (see below and *paragraph 8.2.2*).

The projections from R to the thalamic nuclei appear to provide *selective* "gating" functions, including the gating of sensory information processing from MRF via the thalamus to the cortex. Following *visual* stimulation, ACh is released *selectively* in the visual cortex, but not in other cortical regions (Marczynski, 1978).

MTFCS In addition to MRF, the mediothalamic fronto-cortical system (*MTFCS*) influences the inhibitory activity of R, however, in a manner which differs functionally from that of the MRF. While MRF induces sustained (tonic) positive shifts in R, but negative shifts in the prefrontal cortex (PFC), MTFCS induces phasic negative shifts in R and positive shifts in the cortex. Furthermore, MRF stimulation inhibits R units and evokes EEG desynchronization. But EEG desynchronization can also be seen following blockade of the excitation of R units by the frontal cortex (cf. Skinner, 1978). Steriade & Wyzinski (1972; cited after Skinner, 1978) observed short-latency responses in R units after stimulation of the frontal cortex, consistent with the assumption of monosynaptic pathways from frontal cortex to R, running through the inferior thalamic peduncle (ITP; cf. Skinner, 1978).

The outlined processes have been related to the generation of ERP components such as the CNV. Skinner (1978) suggests the two components of the CNV (see *paragraph 3.2*) to reflect two successive stages of R regulation by MRF and the frontal cortex. Following the WS an orienting response ("O-wave") occurs due to a strong excitation of MRF; with the decrease of the MRF output a selective preparatory excitation of R cells by the frontal cortex provides a closing of all thalamic gates except for the relevant one ("E-wave"). This assumption is empirically supported by the result of abolished early and late CNV components following blockade of the inferior thalamic peduncle (ITP), which includes pathways between the frontal cortex and R (Skinner, 1978). Since ITP blockade abolishes evoked responses in both frontal cortex and R following conditioned stimuli, conditioned slow potentials in R seem to depend on the response of the frontal cortex (Skinner, 1978). In line with this reasoning, Marczynski (1978) assumes the orbito-frontal cortex to determine the spatio-temporal patterns of activity in R under the indication of a goal. A prospective or preparatory variant of this PFC influence, plausibly enabling attentional focusing, is the excitation of those parts of R which gate the sensory afferents that shall *not* be attended to. Selective attention, including its preparatory variant, therefore amounts to the active inhibition of irrelevant channels, sparing the thalamo-cortical information transmission in the relevant channels. The observed enhancement of ERP to irrelevant stimuli following ITP blockade supports this concept (Skinner, 1978).

To summarize: While the regulation of cortical potentials by the MRF is nonspecific and related to arousal processes, the MTFCS system regulates sensory processes selectively and is assumed to be involved in selective attentional processes. Therefore, thalamocortical activity is considered extremely flexible and capable of analyzing frequently changing environmental and proprioceptive stimuli. In any event, the pattern of R activity, determined by the joined influence of MTFCS and MRF, determines the pattern of sensory input to the primary sensory areas.

Marczynski (1978) proposed to consider *limbic* structures relevant for ERP generation. Limbic descending pathways are capable of exciting or inhibiting the brainstem reticular formation (BSRF), which is part of the ARAS (BSRF-ARAS). According to Marczynski (1978) P300 waves, including those following the WS in a CNV task, may reflect a transient strong inhibition of the BSRF-ARAS. By the same mechanism the skilled performance positivity (Papakostopoulos, 1978) and the accompanying 9-11 Hz oscillations can be explained (Marczynski, 1978).

The neurophysiological ERP model of Skinner & Yingling (1977) has been formulated for the ERP generation during processes of sensory information processing. Brunia (1993) suggested an extension of this model to processes of motor preparation. As is the case for sensory thalamic nuclei, motor thalamic nuclei are inhibited by R, while R is controlled by the frontal cortex. This inhibitory influence enables R to "gate" motor commands from basal ganglia and cerebellum to the thalamic motor nuclei[169]. For these reasons, motor preparation can be understood as a process of selective attention directed towards an output

[169] The results of Ikeda et al. (1994; see *paragraph 3.2*) suggest that cerebellar commands are necessary during RP but dispensable during the tCNV observed prior to motor responses.

channel (Brunia, 1993). However, the underlying effects appear somewhat complementary to those described by Skinner & Yingling (1977). Results on the differential modulation of the Achilles Tendon-Reflex reported by Brunia (1993) in muscles which prepare for movements, compared to those which do not prepare for movements, suggest that a selective non-activation of the involved muscle during motor preparation, possibly combined with a selective pre-synaptic inhibition of the respective motoneurons, takes place within an environment of increased motoneuron excitability. "Thus it appears that a selective lack of activation is present in the middle of a selective generalized facilitation of disinhibition" (Brunia, 1993, p. 333).

7.1.2 Cortical Facilitation and Dysfacilitation

The *functional significance* of cortical potentials considering the neurophysiological evidence has been outlined by Rockstroh & Elbert (1990; Elbert, 1993; Rockstroh et al., 1989). According to this model the joined action of MTFCS and MRF on R provides a mechanism for the regulation of the thresholds of excitability of cortical neuronal networks. The term "excitability" is related to the modulation of depolarization in dendritic trees of cortical pyramidal neurons, causing an efflux of negative charges into extracellular space which shows up as surface negative potential shifts. This concept is intimately related to the concepts of cell assemblies (Hebb, 1949, cited after Rockstroh et al., 1989, p. 152) and feedback control of thresholds of cortical activity (Braitenberg, 1978).

The term "cell assembly" (Hebb, 1949, cited after Rockstroh et al., 1989) refers to a functional grouping of neurons which is defined by the reciprocal excitation of its constituent neuronal elements and allows for the temporal maintenance of excitation. Such neuronal reverberatory circuits have been considered to provide flexible mechanisms for memory functions (Elbert, 1993; Hebb, 1949, cited after Rockstroh et al., 1989). The increase in strength of the connection between two simultaneously activated neurons, together with the observation of unaffected strengths of the connections with a third, not simultaneously activated neuron (Gustafson et al., 1987), supports the cell assembly concept and has been referred to as "synaptic plasticity" (cf. Elbert, 1993). Synaptic plasticity is assumed to be realized by NMDA-receptors (Gustafsson et al., 1987). Since cortical neurons constitute dense networks with predominantly excitatory interconnections, the selective formation of cell assemblies must be flanked by the inhibition of neurons irrelevant to the "concept" represented by the activated cell assembly. Threshold (feedback) control mechanisms have been suggested to prevent an uncontrolled spreading of activation by raising firing thresholds in and, hence, providing inhibition of the parts of the cortical network *not* included in the current cell assembly (Braitenberg, 1984; Elbert, 1993; Rockstroh & Elbert, 1990). This widespread "dysfacilitation" is presumed to be seen in positive waves (Elbert, 1993;

Rockstroh & Elbert, 1990). Lowering of the firing threshold, reflected by negative waves, facilitates the mutual activation of cell assemblies.

Nonspecific thalamocortical pathways, synapsing with apical dendrites are assumed to regulate cortical excitability (Elbert, 1993). The aforementioned tuning of the thalamo-cortical afferents must take into account the current activity of the cortical areas that shall be regulated. This function is possibly achieved by pathways from the respective cortical regions to the thalamus via the *basal ganglia*, which are known to receive input from every cortical region but project primarily via thalamus to the frontal cortex (not: *pre*frontal cortex, see *paragraph 2.2.6.*; Wise, 1991)[170].Within the basal ganglia the striatum is assumed to inhibit the pallidum, which projects to the thalamus, "preventing an explosive ignition of the threshold regulatory circuit between thalamus, cortex, striatum, and pallidum" (Rockstroh et al., 1989, p. 154). This raising of excitability thresholds enables the focusing of cortical input to relevant events, suppressing irrelevant stimuli. The pathways from pallidum to thalamus are assumed to be regulated by the MTFCS, thus enabling prospective regulation of cortical excitability. ERP such as the P300 or the CNV can be interpreted in terms of the cortical excitability model. Following relevant stimuli (which activate the MTFCS via pathways from brainstem to R, releasing fronto-cortical activity) the reduction of cortical activity in order to store the relevant stimulus in reverberating networks can be achieved by increasing firing thresholds. This increase in firing thresholds is reflected by P300-*b* like waves.

Similarly, stimuli signaling the occurence of a behaviorally relevant event may activate the MTFCS and elicit threshold (down-) regulation in those networks associated with the processing of the behaviorally relevant event, as presumably reflected by the CNV components.

Several results support the interpretation of surface negativity as increased cortical excitability. Reduced reaction times have been observed when the motor cortex contralateral to the responding hand showed biofeedback-induced negativities (Rockstroh et al., 1990). Similarly, response speed and error rates were dependent upon the presentation of a certain task during spontaneous negative *versus* positive DC-shifts (Bauer, 1984; Stamm, 1984). Certain drugs reduce both cortical excitability *and* surface negativity (Rockstroh et al., 1991).

Another attempt to investigate the inhibitory versus facilitatory nature of cortical potential shifts is the presentation of "probe-"stimuli. Such stimuli are presented during a task which elicits an evoked potential, for instance, a P300 or a CNV. If the "probed" ERP is facilitatory in nature, the amplitudes of the probe-evoked responses should be augmented compared to a control condition (e.g. probe EP of baseline probes). If, however, the "probed" ERP has an inhibitory character, probe EP should have reduced amplitudes. The potentials which may

[170] This function is not specific to the basal ganglia. The PFC, too, receives informations from all primary and associational cortical areas (cf. Fuster, 1989; Skinner, 1978). Furthermore, projections from all cortical associational areas converge at the entorhinal cortex, which projects to the hippocampus (Marczynski, 1978).

be expected to occur after probe-stimuli are the vertex potential and, if the probe is task-relevant, the P300.

Applying the probe-technique during an "Oddball" task, Rockstroh et al. (1992) observed larger probe EP following standard than target stimuli. As in regular "oddball-"tasks, target stimuli were associated with larger P300 amplitudes. This result was interpreted to indicate the dysfacilitatory nature of the P300. In another experiment of Rockstroh et al. (1994) evidence concerning the facilitatory character of the CNV, especially its late component was obtained. The N1/P3 peak-to-peak amplitude was larger for probes presented between WS and IS (in this case: WS-offset) than for probes presented during the pretrial baseline. Furthermore, the amplitudes of the sequence of probe EP paralleled the development of the CNV.

7.1.3 Research Questions

On this theoretical and experimental background the question arises as to the nature of the PINV seen in schizophrenic patients. As yet, only a few studies have addressed this question, yielding partially conflicting results. On the one hand, Timsit-Berthier et al. (1971) observed reduced probe-induced N100 during the PINV (compared to the tCNV) interval in schizophrenic patients, suggesting that PINV reflects an inhibitory state of the brain. On the other hand, in the experiment of Rockstroh et al. (1994) N1/P3 probe EP were larger during CNV and PINV in schizophrenic patients and healthy control subjects, when compared to N1/P3 in response to probes delivered during the pretrial baseline. Also, reaction times following probe-stimuli during the PINV interval were shorter than reaction times following pretrial baseline probes. These results favor a facilitatory interpretation of the PINV which is seemingly at odds with the interpretation of Timsit-Berthier et al. (1971). Nonetheless, Rockstroh et al. (1994) replicated the main result of Timsit-Berthier et al. (1971), in that in patients the N100 amplitude following PINV probes (and late tCNV probes) was smaller than the N100 amplitude following pretrial baseline probes. Although both studies have a common result, namely reduced N100 amplitudes following PINV probes, they are not entirely comparable since in the study of Rockstroh et al. (1994) probe stimuli were associated with a task and, hence, a P300.

Since in the present experiments probe-stimuli were task-irrelevant (and introduced as such to the subjects, see *paragraph 4.1*) only the vertex potential, comprising the N100 and P200 subcomponents (Rockstroh et al., 1989), but no P300 as in the experiment of Rockstroh et al. (1994) are to be expected.

The N100 component is among the earliest *endogenous* EP, since it has a scalp topography which is independent of the physical characteristics of the eliciting stimulus (cf. Rockstroh et al., 1989). Research on N100 using a variety of recording techniques (topographical analysis, intracranial recordings, dipole model analysis, and neuromagnetic measurements) suggests this component to be composed of several, functionally different components (cf. Giard et al., 1994).

The first and second N100 subcomponents are the modality specific fronto-central N100 after about 100 ms, which is associated with passive attentional processes, and the "T complex", with maximal negative peaks after 150 ms at midtemporal leads preceded by a small positive peak at about 100 m (cf. Rockstroh et al., 1989).

The N100 is followed by a positive deflection after about 170-200 ms, the P200, which is supposed to indicate a compensatory threshold increase following the enhancement of cortical activity (Rockstroh & Elbert, 1990). The N1/P2-*complex* forms the *third* N100 subcomponent, which is modality non-specific and has a topographical maximum over the vertex ("vertex potential"). This component is regarded as one of the "arousal components" and is the only subcomponent which is related to psychological and state variables (cf. Rockstroh et al., 1989). It has been shown to decrease in amplitude with stimulus repetition, recovering after stimulus change (Simons et al., 1987). The sensitivity of the N1/P2 complex to stimulus onset, offset, and change suggests that it is related to an initial orienting response (cf. Rockstroh et al., 1989). The P200 has been associated with selective attention, novelty, arousal, habituation, and stimulus encoding (cf. Faux et al., 1987). The N1/P2 component (cf. Giard et al., 1994) and the N100 (Javitt et al., 1993) are especially sensitive to refractory phenomena (amplitude diminution) following rapid stimulation (see below).

Current Source Density (CSD) analyses indicate a sink and a source over temporo-parietal areas as well as bilateral frontal currents between 65 and 120 ms after the onset of auditory stimuli presented at a rate of 1 Hz (Giard et al., 1994). N100 peak amplitudes of slightly less than -5µV and N1/P2 peak-to-peak amplitudes of about 8 µV were achieved at Cz with monaurally presented 70 dB tones in the study of Giard et al. (1994).

In schizophrenic patients reduced amplitudes of the N100 and the P200 components during an "Oddball" task have been found to be independent of the clinical state and the neuroleptic medication (Eikmeier et al., 1991; Faux et al., 1987)[171]. Larger right (T4) than left-temporal (T3) P200 amplitudes were found in schizophrenic patients than in healthy control subjects. This asymmetry was independent of the left-temporal reductions of the P300 frequently observed in this group (Faux et al., 1987).

Another clinically interesting aspect is the recovery of the vertex potential and its refractory period. In general, an inverse relation between stimulation rate and the amplitude of the vertex potential has been found (Roth et al., 1980). For example, the N1/P2 peak-to-peak amplitude in response to stimulation with an interstimulus interval (ISI) of 1 s is about one-sixth of the amplitude during stimulation with 6 s ISI (cf. Roth & Cannon, 1972). The amplitude recovery during slow stimulation is attenuated in schizophrenic patients (reviewed by

---

[171] For an extensive review of the literature, see Cohen (1990).

Shagass et al., 1978) with visual and somatosensory stimulation (cf. Roth & Cannon, 1972)[172, 173].

7.2 Hypotheses

The ANOVA designs presented in *paragraphs 7.8* and *7.9* include a leapfrog strategy with confirmatory and exploratory parts. This distinction is legitimate but should be made explicit (see *paragraph 4.7.4*).

Confirmatory Part - Experiments A and B[174]

Hypothesis 1: Schizophrenic patients exhibit diminished vertex potential amplitudes. *Reason*: Reports on reduced N100/P200 amplitudes in schizophrenia, literature cited in *paragraph 7.3*. *Statistical form*: Group main effect.

Hypothesis 2a[A]: Vertex potentials elicited during the PINV have larger amplitudes than vertex potentials elicited during the prestimulus baseline. *Reason*: PINV is assumed to reflect a state of cortical excitability; literature reported in *paragraph 7.2*. *Statistical form:* Condition main effect and significant apriori contrast for level baseline versus level PINV of the condition factor.

Hypothesis 2b: Vertex potentials elicited during the PINV have the same amplitudes like vertex potentials elicited during the tCNV. *Reason*: Since both CNV and PINV are assumed to reflect cortical excitability, vertex potentials elicited during CNV and PINV should not differ; literature cited in *paragraph 7.2*. *Statistical form*: insignificant apriori contrast for the level tCNV versus level PINV of the condition factor[175].

Hypothesis 3: The vertex potential will be larger at central compared to frontal leads ("vertex potential"). *Reason*: Known topography of this component; literature reported in *paragraph 7.3*. *Statistical form*: Electrode main effect.

---

[172] The reduced temporal recovery of EP in schizophrenic patients has been attributed tentatively to the presence of auditory hallucinations as "a source of stimulation that might be functionally equivalent to exogenous stimuli" (Roth et al., 1980). But see *paragraph 2.1.2* for recent "output" theories on auditory hallucinations in schizophrenia which conflict with this "input"-theoretical approach to auditory hallucinations.

[173] The refractory effect turned out to be present in experiment B and will be referred to again during result description.

[174] All hypotheses stated pertain to the N1/P2, N100, and P200, which will be refered to as "vertex potential" for the sake of conciseness. If a hypothesis is stated only for one of the two experiments, it will be labeled with the index [A] or [B], depending on the experiment to which it pertains.

[175] It should be noted that this statistical hypothesis does *not* strictly correspond to the research hypothesis, which states *"no difference ..."* . Accepting the H_0 is not equal to rejecting the H_1 ! However, the classical Fisherian statistical approach is only designed to reject the H_0. Furthermore, the apriori definition of the β-error was impossible, since there had been no rationale to precisely assume condition-specific means.

Hypothesis 4^B: An enhancement of the PINV amplitudes under the ambiguous condition assumed: The vertex potential will be larger under the ambiguous than under the clear condition. *Reason*: An augmented PINV is assumed to reflect augmented cortical excitability, resulting in larger vertex potential; literature reported in *paragraph 7.2*. *Statistical form*: Condition main effect[176].

Exploratory Part

Other questions can be put at the probe results of experiments A and B. For example, is it possible that the vertex potential of schizophrenic patients is differently sensitive to the length of the interprobe interval than the vertex potential of the control group ? Is it possible that the assumed amplitude diminution in schizophrenic patients is differently pronounced at different leads?

7.3 Experimental Data Analyzed in Chapter 7

Because the focus of this chapter is on probe EP, the methods of acoustical probe presentation will be described only. The composition of the experimental samples, the task structure, the general data analysis, etc. are described in detail in *chapters 4, 5,* and *6*.

7.4 Presentation of Probe Stimuli in Experiments A and B

Throughout the experimental period (including the intertrial intervals) of experiment A acoustical probe stimuli were presented at regular intervals of 1.0 s, starting 0.5 s after the WS onset. The probe stimuli were clicks with in intensity of 50 dB[177] and a duration of 20 ms and were presented via headphones during practice and experimental trials. Subjects were told to ignore the probe stimuli since they were task irrelevant. A continual presentation was chosen in order to avoid a dual task situation.

The results of experiment A, however, suggested a change in this mode of probe presentation. High-frequency probe presentation was associated with diminished EP amplitudes, possibly because of refractory effects. Therefore, only two probe stimuli per trial were presented in experiment B. The stimuli of 20 ms duration and 80 dB intensity were delivered during the pretrial baseline (1.0 s before WS onset), during iCNV (1.5 s after WS onset), tCNV (0.5 s before IS onset), and PINV (1.5 s after IS onset) development (see **figure 7.2**). The

---

[176] Recall that there was no differences between the groups for the ambiguity effect along the sagittal midline (see *chapter 6.9.3.2*).

[177] A probe intensity of 80 dB had been provided. However, an intensity diminution of about 30 dB compared to the programmed intensity values occurs in the NEUROSTIM package if acoustical stimuli are not presented with earphones but with headphones. This was unfortunately detected too late.

resulting 6 combinations of 2 probe stimuli were presented with equal probability during each condition (comprising 2 x 48 (clear), 60 (nogo), and 48 (ambiguous) trials).

7.5 Data Reduction and Statistical Analysis of the Probe EP

Experiment A Following the artifact correction procedures described in *paragraph 4.7.3*, the single subject averages of experiment A were inspected. As the grand averages of the patient and the control group revealed low amplitude probe EP, especially concerning the N100 component, only the data of subjects exhibiting discernible probe EP were included in the statistical analyses. Nine control subjects and eight patients out of 19 subjects in each group met this criterion. For these subjects parameters of the N100 and the P200 components were determined.

Prior to the parametrization, the data were 3-7 Hz bandpass-filtered[178]. After bandpass-filtering of the single subject averages, parameters were computed for the different components by means of the KN program MINMLATS. The N100 was defined as peak negativity between 80 and 140 ms, while the P200 was defined as peak positivity between 140 and 240 ms, each following the probe onset and relative to the pretrial baseline (500-0 ms prior to WS). The vertex potential (N100-P200) was defined as the peak to peak amplitude difference between the N100 and the P200[179]. For a subset of channels from which slow potentials were obtained, the N1/P2 as well as the N1 and P2 separately were compared between groups and probe delays (background activity). The selection of frontal and central channels seemed reasonable since the "vertex potential" —according to its label— predominantes over the vertex. Previous probe-studies frequently investigated the vertex potential only at Cz. In addition, due to the scalp distribution of the slow potentials (in particular the frontally predominant PINV) described in *chapters 5* and *6* a possible topographical variation of the probe-EP amplitudes should be explored by including frontal leads. Finally, while in experiment A the analysis of probe EP at fronto-lateral (F7, F8) and temporal (T3, T4) leads would have been reasonable

[178] This filter procedure takes into account that the EP evoked by the different probes are superposed on positive- or negative-going slow potential shifts, depending on the "probed" background activity (e.g. tCNV vs PINV). Referring the N100 or P200 components to a prestimulus baseline would over- or underestimate the N100 or P200 peak amplitudes, depending on the component chosen and the direction of the slow amplitude shift of the background activity. This effect can, in principle, differ between all within and between subjects factors of a given experimental design and should increase error variance. Out-filtering the slow potential shifts should circumvent this problem. Since all slow negative shifts below 3 Hz (including the CNV and PINV) were filtered out, the pretrial interval (500-0 ms prior to WS-onset) was chosen as the baseline.

[179] The DOS version 6.04 of the Statistical Analysis System (SAS) was used for experiment B and the Mackintosh program SUPERANOVA for experiment A.

with respect to the literature on EP in schizophrenia (see *paragraph 7.1.3*), probe EP amplitudes at these sites were even less pronounced than at frontal and central leads.

For these reasons, two frontal (F3, F4) and two central (C3, C4) electrodes were chosen, due to the topography of the vertex potential following acoustic stimulation. Furthermore, only results from Session B (varying WS diamond arrangements) for the pooled clear and ambiguous conditions will be reported because slow potentials, in particular the PINV, were more pronounced in this session than in session A (see *chapter 5*). N100, P200, and N100-P200 data were separately analyzed by ANOVA procedures. The experimental factors GROUP (patients, controls), BACKGROUND ACTIVITY (pretrial baseline, tCNV, and PINV), HEMISPHERE (left, right), and ELECTRODE (frontal, central) were simultaneously considered. Correction of p-values for sphericity violations were accomplished according to Huynh & Feldt (1976).

Figure 7.2 Combinations of Probes Selected for the Statistical Analyses of Experiment B

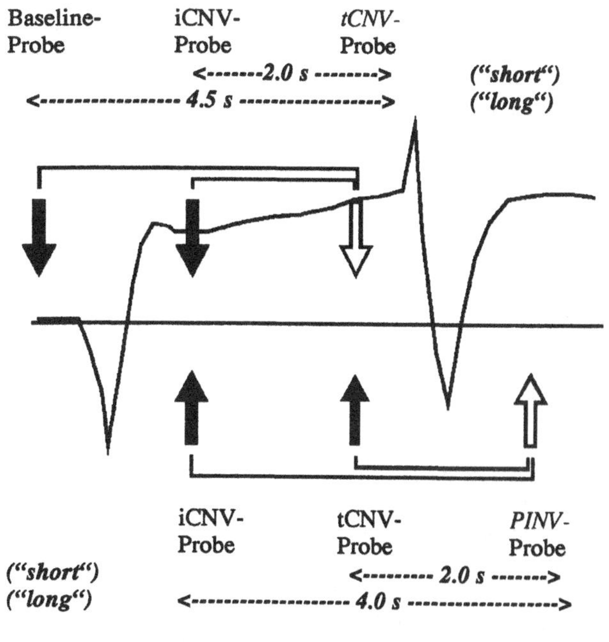

Experiment B Following the analysis and correction procedures described in *chapters 4* and *6*, trials with correct responses under the conditions clear and ambiguous (excluding the nogo condition because of the low-amplitude PINV) including (a) tCNV probes that were preceded either by an iCNV or by a baseline probe and (b) PINV probes that were preceded by tCNV or iCNV probes (see **figure 7.2**) were selected.

These data were 3-7 Hz bandpass-filtered. In this way 8 probe "conditions" were distinguished: two experimental conditions (clear, ambiguous), *by* two background activities (tCNV, PINV), and *by* two intervals separating the interesting probe and the preceding probe (short versus long previous interprobe interval, see **figure 7.2**). As described in the preceding paragraph, peak amplitudes for the N100 and the P200 components and N1/P2 peak-to-peak differences were analyzed. The Fz and Cz leads were chosen for statistical analyses. Each of the components (N100, P200, N1/P2) including 8 variables was submitted to an ANOVA comparing the experimental factors GROUP (patients, controls), BACKGROUND ACTIVITY (tCNV, PINV), CONDITION (clear, ambiguous), INTERPROBE INTERVAL ("short", "long"), and ELECTRODE (Fz, Cz) simultaneously. Since none of the experimental factors comprises more than 2 levels, sphericity assumptions are not pertinent.

7.6 Results of Experiment A

Probe-evoked potentials typical of the total patient and control groups were shown in **figures 5.3, and 5.4**. As can be seen from these figures, the N1/P2 complexes evoked by the probe stimuli, especially the N100 component, are rather small, since none of the subjects tested had a vertex potentials with sufficiently large amplitudes. **Figure 7.3** exemplifies the probe EP of 2 schizophrenic patients and 2 healthy controls for the pooled clear and ambiguous condition of session B (high memory load).

The subjects 1-3 of **figure 7.3** exhibit vertex potentials with larger amplitudes at the central than at the frontal leads. In subject 4 only small vertex potentials can be observed. Probe EP following the tCNV probe are larger than the EP induced by the other probes.

Table 7.1 summarizes the ANOVA results for the subset of subjects (8 patients, 9 controls) who showed sufficiently pronounced probe EP. A considerable similarity between the ANOVA results for the N1/P2, N100, and P200 parameters is evident. For this reason, only results on the N1/P2 complex will be described in detail.

Probe EP are generally larger at central ($-1.24 \pm 0.83 \mu V$) than at frontal (-1.62 ± 0.89 μV) sites, as indicated by the significant main effect **Electrode**. The **Background Activity by Electrode** interaction rises from amplitude differences between the frontal and central leads, depending on the type of background activity (pretrial baseline, tCNV, or PINV) and is again significant for the composite N1/P2 and its subcomponents, N100 and P200 (see **table 7.2**).

While the amplitudes of the EP elicited by probes during the PINV are similar to those elicited during the pretrial baseline, EP to probes presented during the tCNV have larger amplitudes primarily at central leads.

Table 7.1 ANOVA Results on N100-P200, N100, and P200

		N1-P2		N100		P200	
Effect	df1,df2	F	p	F	p	F	p
B	2,30	2.03	.17	3.35	.07	0.89	.37
B * G	2,30	1.91	.19	1.85	.19	1.62	.22
H	1,15	0.34	.57	0.56	.47	1.79	.20
H * G	1,15	0.28	.60	0.21	.65	1.17	.30
E	1,15	10.79	.005	8.63	.01	10.45	.006
E * G	1,15	0.87	.37	0.57	.46	0.98	.34
B * H	2,30	2.20	.14	2.80	.10	0.13	.75
B * H * G	2,30	0.24	.76	0.23	.72	0.04	.86
B * E	2,30	13.71	.0002	7.26	.003	12.81	.002
B * E * G	2,30	0.29	.71	0.76	.48	0.01	.94
H * E	1,15	0.17	.68	0.53	.48	0.00	.98
H * E * G	1,15	6.11	.03	3.97	.06	5.29	.04
B * H * E	2,30	0.28	.73	0.54	.59	0.63	.45
B*H*E*G	2,30	0.04	.95	0.08	.92	0.00	.99
Group	1,15	0.01	.95	0.00	.97	0.01	.93

B=Background Activity Factor (Baseline, tCNV or PINV); H=Hemisphere Factor (left, right); E=Electrode Factor (frontal (F3,F4), central (C3,C4)); Results of repeated measurement ANOVA with Huynh-Feldt corrected p-values; average mastoids reference

Table 7.2 Background Activity by Electrode Interaction on N1/P2

	Frontal	Central
Baseline	-1.20 (0.8)	-1.47 (0.6)
tCNV	-1.34 (0.9)	-2.00 (1.2)
PINV	-1.20 (0.9)	-1.40 (0.6)

unit is µV; SD in parentheses

Finally, the interaction of **Hemisphere by Electrode by Group** for the N1/P2 and the P200 components is significant, while the N100 component just falls short of significant (p=.06).

Compared to controls, schizophrenic patients exhibit slightly larger N1/P2 amplitudes at the frontal leads, identical amplitudes at C4, and diminished amplitudes at C3 (see **table 7.3**).

Figure 7.3 Probe-EP in 2 Patients and 2 Controls

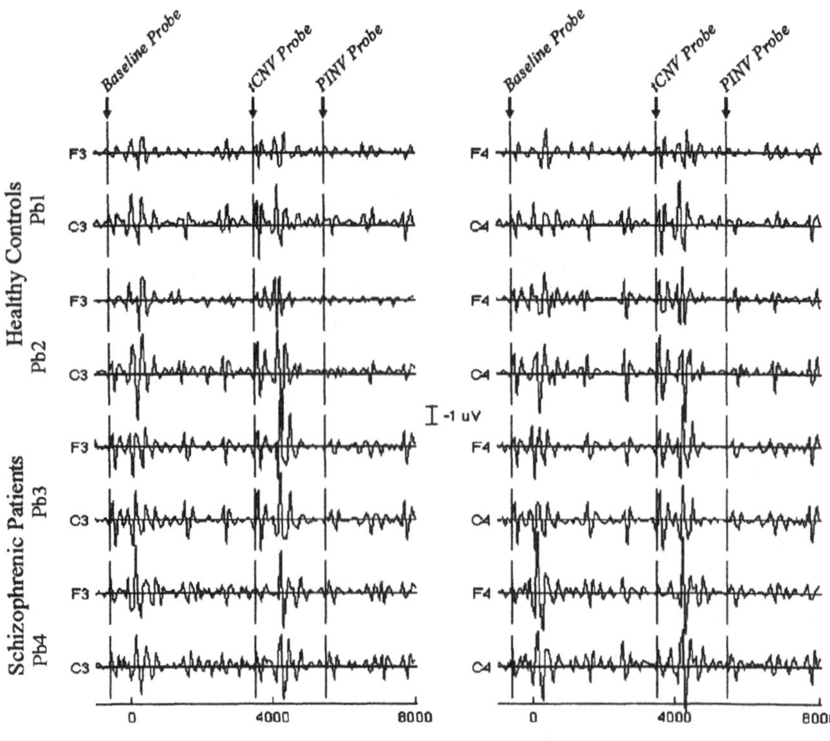

WS onset at 0 ms, IS onset at 4000 ms; the onset of the relevant probes is depicted by vertical bars and arrows

Table 7.3 Hemisphere By Electrode By Group Interaction on N1/P2

	Left Hemisphere		Right Hemisphere	
	Frontal	Central	Frontal	Central
Controls	-1.23 (0.7)	-1.77 (0.9)	-1.17 (0.8)	-1.59 (0.7)
Patients	-1.33 (1.0)	-1.52 (0.9)	-1.25 (0.8)	-1.59 (1.1)

unit is µV; SD in parentheses

To summarize: Except for slightly diminished (0.25 µV) N1/P2 amplitudes at C3, schizophrenic patients do not differ from healthy controls in the probe modulation. In both groups probe EP are larger at central than at frontal leads. The probe EP during tCNV is larger than the probe EP during baseline and PINV, the latter diminutuin occuring more at central than at frontal sites.

7.7 Results of Experiment B

Figures 7.4 and 7.5 depict the probe EP following auditory probes delivered during tCNV (**figure 7.4**) and PINV (**figure 7.5**), separated for the experimental conditions (clear, ambiguous) and for the duration of the previous interprobe interval (upper 4 rows: "long", lower 4 rows: "short"). The single-subject averages of patients and controls (N_{total}=22), which do not differ significantly overall (see **table 7.4**), are superposed in order to demonstrate the astonishing homogeneity of the probe responses within the *total* sample. Narrow peaks of the N100 and the P200 indicate little latency jitter within the total group. Larger Cz than Fz EP are as easily discernible as the amplitude diminishing effect of a short interprobe interval.

Larger central than frontal probe EP components for the N100, P200, and N1/P2 are documented by the significant **Electrode** main effect (see **table 7.5**). As indicated by the significant **Background Activity by Electrode** interaction, the centro-frontal amplitude gradient is steeper for tCNV compared to PINV probes (see **table 7.5**). This interaction falls short of significance for the P200 component.

The **Interval** main effect is significant for the N1/P2 and P200 components only, and is due to larger probe EP following long (-8.36±4.7 µV) compared to short (-7.46±3.9 µV) previous interprobe intervals (IPI).

For the N100, but not for the P200 or the N1/P2, a group specific ambiguity effect is indicated by the significant **Electrode by Condition by Group** interaction. While in schizophrenic patients N100 amplitudes are enhanced under the ambiguous (compared to the clear) condition both at Fz and Cz, slight reductions at Fz and no effect at Cz can be seen in controls (see **table 7.6**).

Table 7.6 reveals also, that group differences are larger under the clear than under the ambiguous condition, especially at Fz.

Furthermore, the significant **Background Activity by Interval by Condition by Group** suggests a complex differentiation of condition- and group-specific effects on the modulation of the probe EP (see **table 7.7**).

Generally, there is a diminution of probe EP with short as compared to long previous interprobe intervals. While for the tCNV probe in controls, this effect is larger under the clear than under the ambiguous condition, the interval effect on the PINV probe is larger under the ambiguous as compared to the clear condition.

Figure 7.4 Probe EP during tCNV

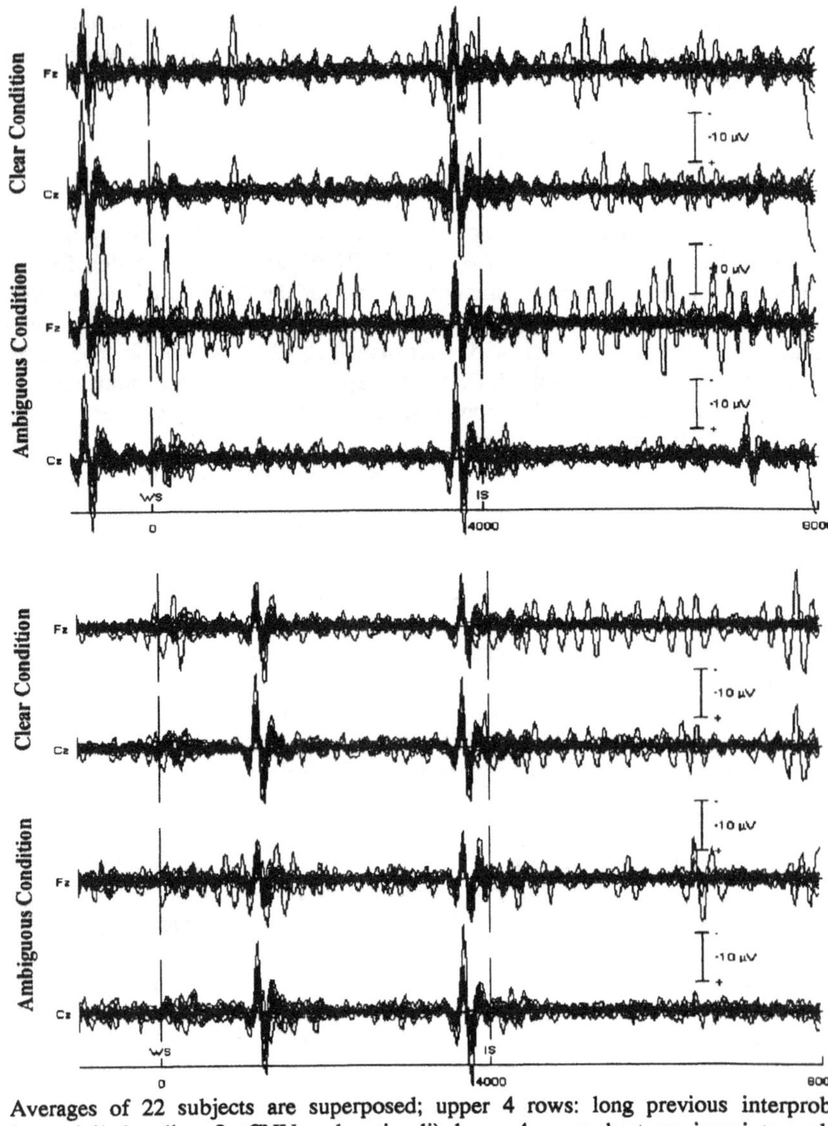

Averages of 22 subjects are superposed; upper 4 rows: long previous interprobe interval (1. baseline, 2. tCNV probe stimuli), lower 4 rows: short previous interprobe interval (1. iCNV, 2. tCNV probe stimulus); 3-7 Hz bandpass-filtered data; average mastoids reference

Figure 7.5 Probe EP during PINV

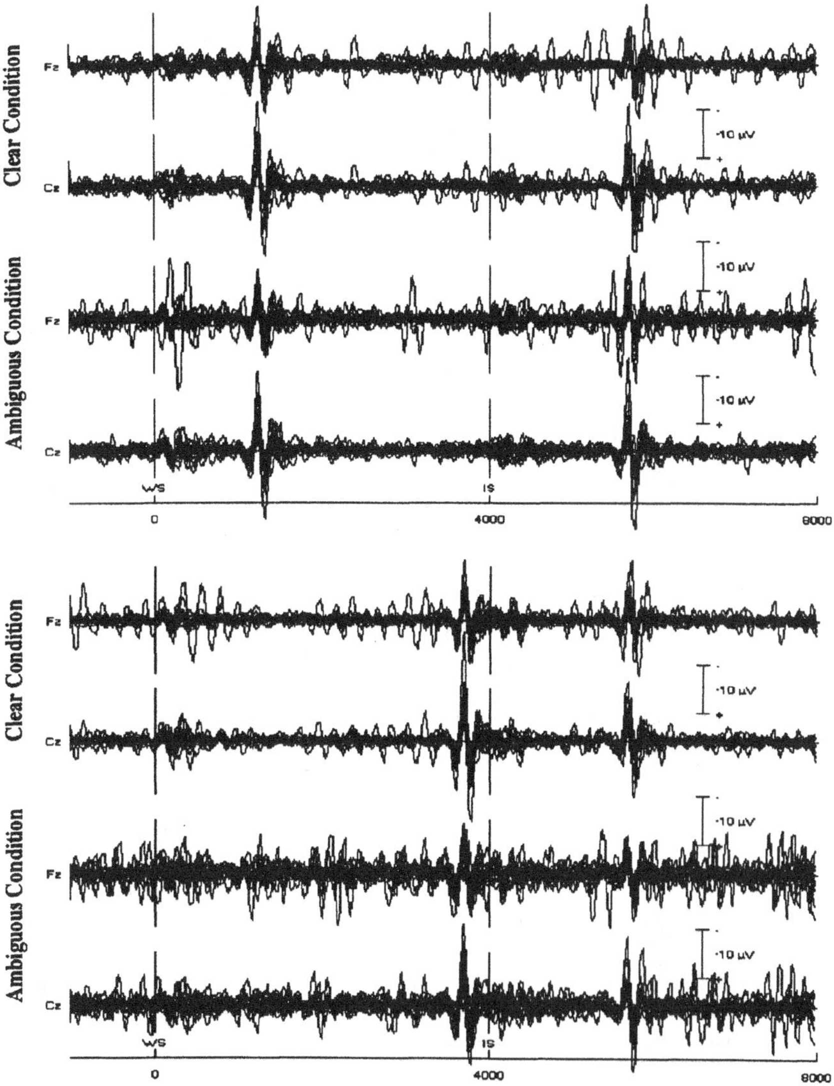

Averages of 22 subjects are superposed; upper 4 rows: long previous interprobe interval (1. iCNV, 2. PINV probe stimulus), lower 4 rows: short previous interprobe interval (1. tCNV, 2. PINV probe stimulus); 3-7 Hz bandpass-filtered data; average mastoids reference

Table 7.4 ANOVA Effects for the N100-P200, N100, and P200 Amplitudes

		N1-P2		N100		P200	
Effect	**df1,df2**	**F**	**p**	**F**	**p**	**F**	**p**
E	2,20	19.35	.0003	18.07	.0004	19.37	.0003
E * G	2,20	0.00	.88	0.02	.88	0.01	.94
B	2,40	0.93	.35	1.41	.25	0.23	.64
B * G	2,40	0.38	.54	3.45	.07	0.46	.51
I	1,20	5.35	.03	1.26	.28	10.61	.004
I * G	1,20	0.12	.74	0.07	.79	0.14	.71
C	1,20	1.45	.13	1.32	.26	2.84	.11
C * G	1,20	0.53	.48	1.42	.25	0.12	.74
E * B	2,40	5.90	.02	6.13	.02	3.67	.07
E * B * G	2,40	1.78	.20	0.09	.77	0.30	.59
E * I	2,40	0.01	.93	1.92	.18	1.31	.27
E * I * G	2,40	0.64	.43	0.02	.88	0.04	.84
B * I	1,20	1.57	.22	1.05	.32	1.64	.22
B * I * G	1,20	0.02	.88	0.00	.96	0.05	.82
E * C	2,40	0.00	.99	0.13	.72	.03	.87
E * C * G	2,40	3.34	.08	5.50	.03	1.07	.31
B * C	1,20	0.06	.81	0.01	.92	0.30	.59
B * C * G	1,20	0.68	.42	0.02	.89	2.02	.17
I * C	1,20	0.38	.55	0.40	.54	0.29	.60
I * C * G	1,20	0.36	.55	0.20	.66	2.13	.16
E* B * I	2,40	0.37	.55	0.60	.45	0.05	.82
E *B *I * G	2,40	0.07	.80	1.96	.18	0.81	.38
E * B * C	2,40	0.17	.32	0.93	.35	0.05	.82
E* B* C *G	2,40	1,10	.31	2.12	.16	0.43	.52
E * I * C	2,40	3.28	.09	3.28	.09	2.68	.12
E * I*C * G	2,40	1.28	.27	0.65	.43	1.55	.23
B * I * C	1,20	0.28	.60	1.63	.22	0.09	.77
B* I * C *G	1,20	2.74	.11	10.30	.004	0.04	.84
E* B * I *C	2,40	0.42	.52	0.01	.94	0.88	.36
E*B*I*C*G	2,40	0.49	.49	0.26	.62	2.13	.16
Group	1,20	0.81	.38	0.65	.43	0.83	.37

E=Electrode factor (Fz, Cz), I= Previous Interprobe-Interval (short vs long); B=Background Activity factor (tCNV vs PINV), C=Condition factor (clear, ambiguous), G=Group factor (patients vs controls); average mastoids reference

Schizophrenic patients exhibit the reverse pattern, i.e., larger interval effects on probe evoked N100 under the ambiguous than under the clear condition during tCNV, and larger interval effects under the clear condition as compared to the ambiguous condition during the PINV.

Table 7.5 Background Activity by Electrode Interaction on N1/P2

	Frontal	Central
tCNV	-6.37 (3.2)	-9.73 (5.5)
PINV	-6.31 (3.5)	-9.21 (5.6)

unit is µV; SD in parentheses

Table 7.6 Electrode by Condition by Group Interaction on N100

	Fz		Cz	
	Clear	Ambiguous	Clear	Ambiguous
Patients	-1.83 (1.1)	-2.31 (1.2)	-3.29 (1.8)	-3.64 (2.0)
Controls	-2.82 (1.5)	-2.72 (1.7)	-4.02 (2.8)	-4.11 (2.9)

unit is µV; SD in parentheses

Table 7.7 Background Activity by Interval by Condition by Group Interaction on N100

	tCNV		PINV	
	Clear	Ambiguous	Clear	Ambiguous
Patients				
Long	-2.55 (1.6)	-3.14 (2.0)	-2.47 (1.5)	-2.43 (1.8)
Short	-2.71 (1.3)	-2.63 (1.5)	-1.66 (1.4)	-2.53 (2.2)
Controls				
Long	-3.29 (2.5)	-3.06 (2.4)	-3.36 (2.1)	-3.38 (2.6)
Short	-2.89 (2.0)	-3.30 (1.7)	-3.08 (2.1)	-3.11 (2.4)

unit is µV; SD in parentheses

To summarize: In both experiments probe EP proved to be larger at central compared to frontal leads. While there are no overall group differences in the modulation of probe EP between schizophrenic patients and healthy controls, the magnitude of the group difference varies with the topographical location. In experiment A still small (0.25 µV) group differences in N1/P2 amplitudes were obtained at the left-central position (C3). In experiment B diminished N100 amplitudes in schizophrenic patients were found for the Fz lead, however, only under the clear condition (0.99 µV). Similar influences of the background activity on the probe EP can be found in the two experiments: In both experiments the probe EP elicited during the tCNV is larger than its PINV counterpart (additionally in experiment A: larger than the pretrial baseline probe EP, which does not differ from the PINV probe EP). The effect of the background activity is largest at the central positions (C3,C4 in experiment A, Cz in experiment B). The duration of the interval separating the probes clearly affects the probe EP amplitude: Probe EP are larger following long compared to short previous interprobe intervals. This effect does not differ between groups, leads, background activities and electrodes in any simple manner. Only the tCVN probe EP under the clear condition of the patient group does not correspond to this effect.

7.8 Discussion

The unexpected result of experiment A was the small amplitude of the EP evoked by the probe stimuli which were much smaller than those reported by Rockstroh et al. (1994). Several procedural differences between experiment A and the study of Rockstroh et al. (1994) may account for the result divergency. *Firstly*, the intensity of the probe stimuli in the present task was less than that used by Rockstroh and colleagues (80 dB APL); *secondly*, while in the study of Rockstroh et al. (1994) only one probe stimulus per trial was provided, in experiment A one probe stimulus per second was delivered; *thirdly*, in the study of Rockstroh et al. (1994) the probe stimulus was associated with a certain task (to press a button), while in experiment A this stimulus was task irrelevant. All these distinguishing procedural features may account for the difference in probe amplitudes between the two experiments. In experiment B, clearly larger probe EP were obtained.

The investigation of the probe EP data of experiments A and B revealed no statistically significant overall group differences. This does not correspond with published results on N100 or P200 in schizophrenia. Referring to experiment A, it should be considered that only small EP were elicited by the probe stimuli. This may indicate a non-specific "floor" effect, but is certainly also related to procedural characteristics of probe delivery in experiment A as discussed in the preceding paragraph.

However, the magnitude of the group differences varied with the leads (in both experiments) and conditions (in experiment B only). But even with this differentiation, group differences remained small (0.25 µV at C3 in experiment A, 1.06 µV at Fz in experiment B), suggesting that the functional significance of slow negative potential shifts such as CNV and PINV is normal in schizophrenic patients. However, the topographical distribution of maximal group differences might stimulate some speculation, despite the weakness of these effects. Temporo-parietal and frontal sinks and sources have been found for the N1/P2 (Giard et al., 1994). Furthermore, *left*-temporal differences between schizophrenic patients and controls have been found for the P200 (Faux et al., 1987) and the mismatch negativity (Javitt et al., 1993). Hence, although the group differences are too weak to be confirmed statistically, the topographical *pattern* is compatible with some of the results of other studies. This might have suggested a re-analysis of the probe data of experiment A, since in this study the EEG was recorded from the T3 (T4) and F7 (F8) channels as well. However, the very small probe EP amplitudes at these sites render this possibility not promising.

In both experiments the EP amplitudes of the PINV probe were smaller than those of the tCNV probe, and in both experiments this effect was most pronounced at the central leads, where probe EP were generally largest. In experiment B this interaction was significant for the composite N1/P2 and the N100, but fell short of significance for the P200 component (p=.06). In experiment A for all components the Background Activity by Electrode interaction was significant, but the F-value (which is related to the effect *size*) was smallest for the N100 component. Altogether, these results suggest that the vertex potential, including *all* of its components (N100 *and* P200), is reduced during PINV when compared to tCNV. The reduction of the N100, respectively N1/P2 component of the probe EP in experiments A and B, confirms similar observations reported in the literature (Rockstroh et al., 1994; Timsit-Berthier et al., 1971). Hence, reduced orienting and attentional processes associated with the N1/P2 component have been observed repeatedly during the PINV relative to tCNV. Schizophrenic patients do not differ from controls in the reduction of these processes from tCNV to PINV, confirming similar observations by Timsit-Berthier et al. (1971). This result supports the dysfacilitatory or inhibitory interpretation of the PINV suggested by Timsit-Berthier et al. (1971), but is compatible with a facilitatory interpretation of the tCNV as well.

A third aspect of the comparative investigation of probe EP is the sensitivity to refractory phenomena. This effect could be compared for the tCNV and the PINV under the clear and ambiguous conditions in patients and controls. Unfortunately, the interval effect proved to be rather inconsistent. Amplitude diminutions were found for patients for the tCNV probe EP under the ambiguous condition and for the PINV probe EP under the clear condition. In controls only for the tCNV probe EP under the ambiguous condition was this effect *not* found. Some of the refractory effects found here correspond to those reported by Roth et al. (1976). Refractory effects on the N100 were found at the Fz and Cz leads (where N100 was maximum) but not at Pz, with similar effects for the P200, which was maximum at Cz.

Altogether, these results remain somewhat inconlcusive concerning the functional significance of the PINV. While comparisons between tCNV and PINV probe EP support the inhibitory/dysfacilitatory interpretation of the PINV relative to tCNV, comparisons between different levels of the background activity correspond to a facilitatory interpretation of the PINV in schizophrenic patients. *This* interpretation is confirmed by the P300 and RT data of the study of Rockstroh et al. (1994).

Divergent results obtained with different dependent variables (N100, P200, P300, and RT) clearly suggest that the phenomenon under scrutiny is not one-dimensional but possibly involves different aspects. It should be mentioned in this context that in the studies of Rockstroh et al. (1994) and Timsit-Berthier et al. (1971) as well as in the studies presented here, probe stimuli and stimuli associated with the CNV task had different sensory modalities (visual versus acoustic). Hence, procedural differences pertaining to the sameness or difference of the sensory modalities of probe and the other experimental stimuli can not account for the divergence of results. Furthermore, the divergence between different dependent variables could be found *within* one and the same study (Rockstroh et al., 1994).

A tentative conclusion may be that PINV reflects a psychophysiological state of facilitation concerning conscious and effortful information processing (reflected by the P300 and RT data), but dysfacilitation of automatic processing.

Aposteriori it appears questionable whether *intra*individual comparisons provide an adequate approach to the investigation of the comparative facilitatory or dysfacilitatory nature of background activities such as baseline, tCNV, and PINV. A number of psychological and physiological processes may be considered to influence the DC level at a given point in time, but not necessarily during all segments of a forewarned reaction time trial with the identical "composition". For instance, in experiments A and B preparation for stimulus evaluation certainly took place during the tCNV and, to some extent, also during the intertrial interval[180], but perhaps not during the first seconds following the IS since the experimental design provided intertrial intervals of *at least* 4 seconds, rendering the occurence of a further task-relevant stimulus (in addition to the IS) impossible during this period. Similarly, preparation for motor responding was certainly present during the tCNV but not during the intertrial interval and the PINV. Again other processes may underlie the PINV, e.g. the verification of the past response, associated with response uncertainty, but not the tCNV, and so on. The presence of such *qualitative* differences in the underlying psychological state may be considered to be supported by the different topography of the components to be compared, in particular tCNV and PINV. While a negative potential shift may be a *common* consequence of all these presumed processes, they may diverge with respect to the preparedness for the processing of the probe stimuli. For instance, preparation for the evaluation of visual stimuli (outward directed attention) may facilitate the processing of acoustical probe-stimuli *more* than "verification of the past response" (rather inward directed attention).

In conclusion, CNV and PINV may *share* "physiological excitability" by virtue of the state of depolarization of apical dendrites in those cortical areas that comprise their physiological sources; but these components may differ with respect to the preparedness for the processing of sensory stimulation.

Some consequences and predictions can be derived from these speculations. Firstly, probe EP following probes delivered during tCNV and PINV are not comparable, since they are due to qualitatively different states of "psychological preparedness". Secondly, if "physiological excitability" is associated with modality-specific selectivity of facilitation of sensory information processing (or response preparation), probe EP should be larger if probe and experimental stimuli are of the same sensory modality than if stimulus modalities

---

[180] Even with varying lengths of the intertrial interval it is not plausible to assume that subjects are not awaiting the task onset. This is the more plausible, since the range of the intertrial interval is of restricted size, e.g. 4 to 12 s.

differ (e.g. acoustic-acoustic *versus* acoustic-visual). This hypothesis can, however, only be tested for the tCNV, but not for the PINV, since the status of the PINV with respect to modality-specific facilitation is not known. Secondly, probe EP during CNV and PINV should be larger for task-relevant compared to task-irrelevant probe stimuli (blockwise varied), since only with the task-relevant probe stimuli preparation for their processing will be part of the set of "excitability generating" processes. A further aspect of this hypothesis is a change of the topography of the background activity between trial blocks with task-relevant compared to task-irrelevant probe stimuli. Thirdly, under identical experimental conditions (including the quasi-experimental condition "nosological group" in clinical studies) interindividual comparisons should reveal a positive correlation between background activity and probe EP amplitudes. Fourthly, effects of within-subject factors on the level (not: the topography) of the background activity should be paralleled by effects on the probe EP amplitudes.

Chapter 8 Topography of the PINV in Schizophrenic Patients and in Groups of Healthy Subjects

Summary of Chapter 8 In the final studies of this dissertation the topography of the PINV was examined in schizophrenic patients and healthy subjects. The chapter starts with a brief introduction into biophysical aspects underlying EEG measurement which should help to understand the presented scalp distribution (maps) of the PINV. All maps are based on average reference and spherical spline-interpolated data. These maps include scalp and cortex amplitude and scalp laplacian maps. Cortex amplitude and scalp laplacian maps, which are similar, reveal that two separate PINV maxima are underlying the "broad" scalp amplitude topography of this component. These two separate PINV maxima are located over lateral right and left frontal regions. In healthy controls and subjects with low scores on a German translation of the Schizotypal Personality Questionnaire (SPQ), an initial left-frontal PINV maximum is supplemented by a second, even larger, and rapidly increasing right-frontal maximum. In schizophrenic patients and subjects scoring high on the SPQ, the initial left-frontal PINV maximum is only weakly supplemented by the second right-frontal maximum, giving the impression of a more balanced and less asymmetric distribution of the PINV in these groups. The statistical analysis of the *cortical* amplitude values of integrated 500 ms intervals (starting 1.0 s after the IS) for a selection of 12 frontal and temporo-central leads, covering most of the area of post-imperative negativity, confirm the qualitatitve descriptions of the group maps. While 1.5 s after IS onset there is a clear right-lateralization of the PINV in healthy and low-scoring subjects, schizophrenic patients exhibit a slight left-hemisphere preponderance of the PINV but no significant lateralization throughout the interval; high-scoring subjects show an initial statistically significant left-sided predominance of the PINV that changes into a non-lateralized pattern 1.5 s after IS onset. Results of normalized data indicate different generator structures of the PINV in the experimental and control groups.

These results confirm the reported convergence of cortical amplitude and scalp laplacian maps and their higher spatial resolution compared to conventional scalp amplitude maps, and the power of high-resolution EEG techniques to delineate differences between clinical and control groups in the spatio-temporal modulation of ERP components such as the PINV.

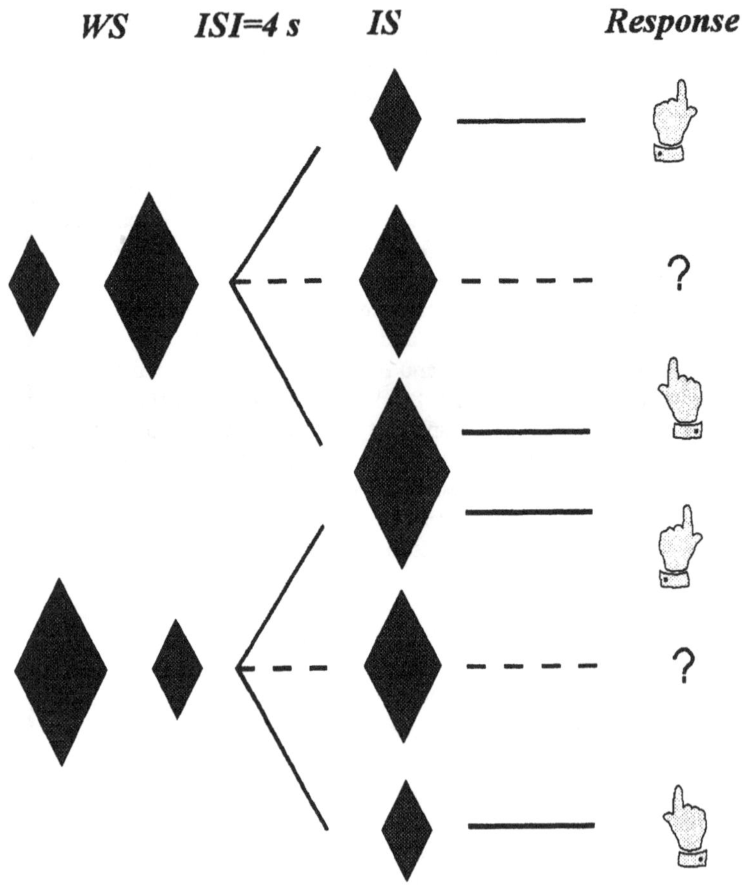

Figure 8.1

8.1 The Theoretical Background of Experiments C and D

PINV in schizophrenic patients has been observed to be pronounced especially over frontal (Rizzo et al., 1979, 1983) and central (Chouinard et al., 1977; Delaunoy et al., 1978; Rizzo et al., 1979, 1983) scalp regions, in studies using only 1 to 3 mid-sagittal electrodes. Although a few studies applied additional lateral (C3, C4; Van den Bosch, 1983, 1984) or even up to 11 electrodes (Eikmeier et al., 1992), results were reported mainly for Fz and Cz positions. Using Fz and Cz referenced to linked-mastoid, Eikmeier et al. (1992) observed a significant ANOVA "group by lead location" interaction. The difference score (Fz-Cz) distinguished schizophrenic patients from healthy controls. In the study of Van den Bosch et al. (1988) the Fz-Cz difference of the left ear-referenced data was significantly correlated with "social closeness" (r=-.59) and the "sequential matching memory task" (r=-.63) in schizophrenic patients. However, Van den Bosch (1983) found identical values of the Fz-Cz score in schizophrenic patients and healthy controls with the left-ear reference. Thus, with a simple anterior-posterior differentiation no consistent picture of a possible typical PINV topography of schizophrenic patients seems to emerge.

Investigating *lateralizational* patterns of the PINV, Eikmeier et al. (1992) observed neither a "group" main effect, nor a "group by lead location" interaction with the C3 and C4 electrodes. On the other hand, Werther et al. (1993) observed a significant left-hemisphere PINV predominance at central leads in schizophrenic patients compared to a healthy control group. Furthermore, Klein et al. (1993; see *chapter 5*) found a left-hemisphere PINV predominance in schizophrenic patients and a right-hemisphere predominance of the PINV in healthy control subjects. Both Werther et al. (1993) and Klein et al. (1993) used a linked mastoid reference. While the results of Werther et al. (1993) apparently conflict with those of Eickmeier et al. (1992), the results of Klein et al. (1993) suggest that group differences between schizophrenic patients and healthy subjects in the PINV topography may be revealed with a broader distributed spatial sampling of this component.

A topographical overlap of the terminal CNV component and the PINV can not be excluded. Hence, the topography of the preceding tCNV should at least be considered, if not corrected for, when assessing the PINV topography.

For the anterior-posterior axis some results on tCNV were briefly reviewed in *chapter 3*. These results indicate a less reduced frontal CNV (relative to parietal amplitudes) in schizophrenic patients (Klein et al., in press; Rockstroh et al., 1994; Wagner et al., 1993). Furthermore, differences between the CNV at Cz and Fz discriminated different psychiatric groups (Van den Bosch, 1984), and CATEGO$^+$ schizophrenic and healthy subjects were best discriminated by the CNV differences between Cz and Fz (Verhey et al., 1986).

A left-hemisphere predominance of the CNV can be observed frequently, also in schizophrenic patient groups (see also *chapter 5*). Furthermore, a left-hemisphere preponderance of the iCNV and tCNV was observed in a variety of psychiatric groups and healthy subjects in the study of Van den Bosch (1984), although this asymmetry was least expressed in schizophrenic patients. It is interesting to note that a left-hemisphere PINV predominance has been reported only in those studies that attempted to correct for the preceding tCNV. Werther et al. (1993) chose a regression analytic approach, by including the tCNV amplitudes as covariates in the ANOVA for PINV analysis. Klein et al. (1993; see also *chapters 5* and *6*) defined the PINV topography as *topography change* between tCNV and PINV by taking the tCNV as the baseline for PINV parametrization. The only study not reporting the left-hemisphere PINV preponderance in schizophrenic patients did not attempt to take the tCNV level into account (Eikmeier et al., 1992).

Although only few studies could be cited, these results indicate that topographical differences in the PINV between schizophrenic patients and controls may be preceded by topographical differences in the tCNV. Positive instances of significant amplitude correlations between tCNV and PINV (see *chapter 3*) underline that PINV variance may be partially explained by tCNV variance. A comparison between the simple PINV topography (referenced to the pretrial baseline) and the tCNV-PINV *topography change* may not only help separate the tCNV and the PINV topographies but provide a *partial* answer to the question: which aspects of the PINV topography in schizophrenic patients may be due to the tCNV[181].

The possibility of typical topographical PINV patterns in schizophrenic patients raises the question whether such topographical patterns can also be found in other subject groups. Lutzenberger et al. (1981, 1983) reported global PINV increases in subjects with high scores in physical anhedonia and body image distortion questionnaires and significant correlations between the left-central PINV amplitude (C3) and the central PINV lateralization (C3-C4) on the one hand and anhedonia scores on the other hand.

The present experiment aimed at exploring the topography of the PINV in schizophrenic patients and healthy controls by recording the ERP from 33 channels using the basic experimental design described in *paragraph 4.1*. In addition, the PINV topographies were compared in healthy subjects with low versus high scores in a German translation of the Schizotypal Personality Questionnaire by Raine (1991). Since a discussion of the schizotypy concept or the strengths and weaknesses of the different methods of the assessment of schizotypal features is beyond the scope of this chapter, the presentation of the

[181] The use of analysis of covariance procedures by Werther et al. (1993) is a flexible approach, in principle. In the case of complex repeated measurement designs, including a variety of electrode factor levels and other experimental conditions, statistics and statistical programs soon become disproportionately complex, including many possibilities for violating the crucial ANCOVA assumption of homogeneous regression slopes (Winer, 1971).

topographical results of this study is intended to add information about the PINV topography[182].

8.2 *Biophysical Principles of EEG Recording*

The topographical results of the present experiments will be illustrated by a variety of maps. In order to understand *what* is displayed by such maps some neurophysiological and biophysical issues will be dealt with, although a thorough methodological treatise of these issues is beyond the scope of this chapter. As long as other authors are *not* cited, the following arguments will follow the "classical" *Electric Fields of the Brain* by Nunez (1981), which is a still recommended monograph (Snyder, 1991).

8.2.1 Cortical Architecture

The human cortex, like all mammalian cortex, is arranged into columns (perpendicular to the cortical surface) and layers (parallel to the cortical surface) of neurons which can be subdivided into two groups: pyramidal and non-pyramidal (stellate) neurons (see also Elbert, 1993). About two-thirds to three-quarters of all cortical neurons belong to the former cell type, which occupies cylindrical volumes due to the arrangement of their axons and dendrites. Most of the neuronal cell surface stems from dendritic branches, only 10% from cell bodies. Cortical neurons are interconnected by intracortical (short-range) and corticocortical fibers terminating mainly on dendritic spines and shafts of pyramidal and stellate cells. Corticocortical fibers provide a large part of the subcortical white matter. Within the range of a few centimeters each cortical column can communicate with almost every other cortical column, making synchronous activity possible. The neuronal activity within such vertical slabs or minicolumns of 20-50 µm width is highly correlated. Most of the cortical input and output systems are excitatory in nature. This is especially true of the pyramidal cell axons, which terminate exclusively in excitatory synapses. Each pyramidal cell receives about 10^3 to 10^4 synaptic contacts from other cortical cells. In contrast to pyramidal cell axons, stellate cell axons terminating on pyramidal cells are inhibitory in nature. Although inhibitory cortical neurons are less frequent than excitatory, inhibition is rather effective since inhibitory synapses terminate on the cell bodies or proximate dendrites and not on distal dendrites.

---

182 The interested reader will find a description of the SPQ and its German translation in **Appendix C**.

8.2.2 Biophysical Processes in Homogeneous Media

Information transmission between neuronal cells includes the presynaptic release of neurotransmitters which cross the synaptic cleft by diffusion and induce an Excitatory or Inhibitory Post-Synaptic Potential (EPSP or IPSP) at the target cell. Although each EPSP brings the neuronal cell membrane closer to the threshold for the firing of an action potential (AP), about 50 EPSP are needed in order to induce an AP. This effect is reversed by IPSP.

EPSP are associated with a heightened permeability of the cell membrane, resulting in an inflow of positive ions. Therefore, even in the subthreshold state the cell subsynaptic membrane is a local current *sink* (positive charges are flowing in, extracellular space is relatively negative), with the current flowing along the cell membrane through the intracellular fluid, crossing the membrane into the extracellular space at remote sites (which are *sources*, extracellular space is relatively positive), and flowing back to the synapse, thereby forming a closed loop. Contrary to EPSP, IPSP cause local current *sources* at the synapse (outflow of K^+-ions) and sinks at distant locations. At sufficiently distant points the resulting potential behaves like a *dipole*, for a dipole is composed of a current source (I^+) and a current sink (I^-) which are separated by a distance d. The potential due to a dipole is the sum of the constituent (monopolar) source and sink[183]. Potentials produced by dipoles decrease with the inverse square of the distance r to the dipole ($1/r^2$) (only in an infinite homogeneous medium, i.e., a conductor of constant, not location dependent conductivity)[184]. In the case of neural generators oriented *in parallel* and exhibiting their activity *synchronously*, the time-averaged potential corresponds to the number m of generators close to the electrode ($|\Phi_{p(average)}| \approx m$). This is to a much lesser extent the case if the generators are *randomly arranged* and *non-synchronously* active ($|\Phi_{p(average)}| \approx \mathrm{sqrt}(m)$).

Since cortical neurons are oriented perpendicular to the cortical surface, *EPSP at apical (surface-near) dendrites* are associated with a current sink close to the cortical surface and a current source at deeper locations, with the resulting potential falling off as a dipole ($1/r^2$, see next footnote). In the case of *EPSP at synapses in the middle of the dendrite*, current (positive ions) can flow both in

---

[183] The potential due to a dipole is given by $4\pi\varepsilon_0\Phi(\vec{r})=qd\cos\Theta/r^2$ ($r>3d$ or $4d$) with ε_0 being a constant called permittivity of empty space and the spherical coordinates r (distance between field point and dipole) and Θ (angular location wiht respect to the dipole orientation).

[184] In principle, each source-sink configuration can be expressed as a *multipole expansion*, including a monopolar (one charge, r^{-1}), dipolar (a positive and a negative charge, r^{-2}), quadrupolar (two dipoles of opposed orientation, r^{-3}) etc. contribution. While in the present case the monopolar contribution is zero, the "higher" than dipolar contributions fall off faster than the dipolar with increasing distance from the synapse, rendering the dipolar contribution dominant at large distances. $\Phi \cong Id\cos\Theta/4\pi r^2$, for $r \gg d$, gives the approximation of the potential at distances of 3d or 4d. I is the electric current, i.e. the flow of charges, which is defined as $\vec{I}={}_i\Sigma\rho_i\vec{\upsilon}_i$, where ρ_i is the charge density of various positive and negative ions and $\vec{\upsilon}_i$ the average velocity of the ions in m/sec.

proximal (surface-far) and distal (surface-near) directions. In this case two dipoles of opposite orientation and, hence, a cancellation effect occurs, with the net potential falling off as a quadrupole with increasing distance r ($1/r^3$)[185,186]. *EPSP at proximal (soma-near) dendrites* are also associated with current flows in proximal and distal directions. However, due to the large area of the soma, source and sink are less distant for the proximal than for the distal dipole, resulting in a smaller proximal than distal dipole and resulting surface-positive and depth-negative potentials (see footnote on this page). In the case of *inhibitory synapses which terminate at proximal dendrites* the situation is similar, but of reversed "sign" (surface-negative and depth-positive). Since the two last mentioned EPSP are more quadrupole- than dipole-like, the resulting potentials fall off with $1/r^3$.

The aforementioned PSP cases make clear that although a diversity of inhibitory and especially excitatory processes can occur at a microscopic level, due to the respective source-sink combinations (and the resulting potential fall-off), only PSP at apical dendrites, which are excitatory, can play a major role in potential measurements at locations relatively distant to the synapses. This holds also for the EEG. The effects of inhomogeneity will be discussed below.

However, in the case of EEG measurement the activity of *dipole layers* including the activity of up to 10^5 neurons (Braun, 1991) and not of single dipoles is measured. This is anatomically possible due to the perpendicular orientation of about 65% to 75% of cortical neurons and their high interconnectivity. Accordingly, dipole layers include all *synchronously* active neurons which are *perpendicularly oriented* with respect to the cortical surface. The potential at any location amounts to the sum of the potential contribution of each single dipole. At vertical distances much larger than the radius b of the dipole layer, the potential behaves like a single dipole.

In addition to PSP, action potentials (AP) propagated along cylindrical axons are associated with certain source-sink configurations. While the depolarized axonal segment is a current sink, the adjacent axonal segments are current sources. In the case of AP the attenuation (r^{-n}) of the extracellular potential Φ_e depends on the radial distance r (in the direction perpendicular to the axon) at all axial locations along the axon, augmenting with increasing radial distance r. In other words, while the extracellular potential due to an AP falls off slowly at low radial distances, it falls off faster the larger this distance becomes. At distances exceeding the axial extent of the AP by a factor of 3 to 5, the

[185] The distances between the central sink and the two sources are d_1 and d_2, respectively. If the resulting potential is measured from relatively large distances (y) relative to d_1 and d_2, it can be expressed in the following way: $\Phi \cong (I/4\pi\sigma d)\ [(d_1^2-d_2^2)/2y^2+(d_1^3+d_2^3)/3y^3+(d_1^4+d_2^4)/4y^4+\ldots]$. The terms within brackets correspond to the dipolar, quadrupolar, octupolar etc. contributions. Only if $d_1=d_2$ (symmetry) is the dipole term zero and the potential falls off with r^{-3} or faster. For the previous and the following PSP examples $d_1 \neq d_2$ (asymmetry).

[186] Synchronously active dipoles located at opposed sides of fissures (quadrupole) cancel each other, with very low net surface potentials.

potential falls off like a quadrupole (r^{-3}). Furthermore, since d is rather small in the case of AP, $Id \approx 0$. Therefore, AP can not be considered to have a substantial impact on the EEG.

For these reasons, potentials measured at a macroscopic level, e.g. by scalp or cortical EEG techniques, are mainly determined by EPSP at apical dendrites if larger cortical regions exhibit the EPSP synchronously (see also Elbert, 1993). Cortical activity due to EPSP at non-apical dendrites, proximal IPSP, and AP will probably not be recorded by EEG techniques[187].

8.2.3 Biophysical Processes in Inhomogeneous Media

The previous considerations did not take into account the effects of location dependent conductivities (σ) which are present in brain and head, and amount to an inhomogeneity of the volume conducting medium[188]. Since currents follow the path of least resistance, potential distributions in homogeneous media can be complicated considerably by the introduction of inhomogeneity. Although these problems cannot be treated here in detail, some causes and effects of inhomogeneity within the head will be sketched, since they pertain to the comparisons of scalp and cortical maps presented later in this chapter.

An elementary case is given when a current source is embedded in a medium with conductivity σ_1 and located at a certain distance below a second medium with conductivity σ_2. While in a homogeneous medium ($\sigma_1=\sigma_2$) the potential in the upper medium is $\Phi \approx I/4\pi\sigma r$ (with r being the distance from the source), in inhomogeneous media ($\sigma_1 \neq \sigma_2$) this potential is $\Phi=I/2\pi(\sigma_1+\sigma_2)R$ (where R is the distance between source I and the field point). In other words, the potential in inhomogeneous media results from the potential computation for homogeneous media with the homogeneous conductivity (σ) being replaced by the *average conductivity* of the two media ($(\sigma_1+\sigma_2)/2$)[189]. If the $\sigma_2=0$, which means that all current flow is confined to the first medium, the same potential is *doubled* when compared to the homogeneous medium case. For the EEG, the effect of the non-conductive air ($\sigma_{air}=0$) surrounding the scalp is an *increase* of the potentials measured at the scalp surface. Immersing the head in saline solution (high conductivity) would reduce scalp EEG potential amplitudes. In general, the potential Φ in the upper medium of conductivity σ_2 due to sources in the lower medium of conductivity σ_1 is related to its value Φ_H of homogeneous medium conductivity σ_1 by the relation $\Phi/\Phi_H=2\sigma_1/\sigma_1+\sigma_2$. If the upper medium has a larger conductivity than the lower medium ($\sigma_2>\sigma_1$), the potential in the upper medium is reduced, while the reverse holds for $\sigma_1>\sigma_2$. This relationship applies to

---

[187] Possible influences of glial cells will not be discussed here (cf. Bauer et al., 1993; Rockstroh et al., 1989).

[188] Conductivity pertains to a mass of tissue and is $ohm^{-1}m^{-1}$.

[189] In the case of complicated inhomogenities this averaging is no longer possible.

comparisons between scalp and cortical surface potentials, since the conductivity of the scalp is much larger than that of the skull[190].

To summarize: the zero-conductivity of the surrounding air results in an increase, but the low conductivity of the skull results in a decrease of scalp amplitude potentials when compared to their homogeneous medium analog. The latter of these two opposite effects is predominant.

Since subcortical structures have been demonstrated to be involved in ERP generation (cf. Rebert 1993), it is interesting to compare the aforementioned effects for central versus surface-near dipoles. According to the three concentric sphere model (defined by the three layers brain, skull, and scalp), if the conductivity of the skull is assumed to relate to the conductivity of the brain by 1/80, the surface potential due to a *central* dipole is increased to three times its homogeneous value due to the confinement of all current flow within the spherical volume. The additional potential reducing effect of the low-conductivity skull amounts roughly to a net *doubling* of the potential compared to its homogeneous value, with smearing "effects" being absent in the case of *central* dipoles. By contrast with *central* dipoles, dipoles in the cortex are reduced to a small fraction of their homogeneous values at locations directly above the radial dipole (angular location $\Theta=0$). Starting directly above the radial dipole, the potential falls off to the sides (increasing angular location Θ) faster in the homogenous than in the heterogeneous case. This means, that at larger angular locations the potential on the inhomogeneous sphere is larger than that on the homogeneous sphere. Therefore, "smearing" effects are present in the case of surface potentials due to cortical dipoles in inhomogeneous spheres.

Although central dipoles are increased and cortical dipoles are decreased in inhomogeneous compared to homogeneous media, this does not mean that the former contribute mainly to the surface potentials. Within the three concentric sphere model, the potential between a certain surface point B above a cortical dipole layer and another surface point A is $\Phi_{BA}=V_0$ x Area of active Zone/Area of Outer Sphere x GAIN[191], with V_0 corresponding to the potential difference across the dipole layer. This formula can be used in order to obtain estimates of surface potentials due to cortical versus subcortical structures. Similar comparisons have been undertaken by Braun et al. (1990) using the formula V_{max} [in µV] = 2 x area of active region [cm^2] x voltage across structure [mV], with V_{max} being the scalp voltage maximum. Braun et al. (1990) suggest that ERP with scalp amplitudes ≥ 10 µV are mainly (over 80%) generated in the cortex (see also Elbert, 1993; Lutzenberger et al., 1987).

---

[190] Although, in principle, higher conductivity of the cortex and cerebrospinal fluid compared to the skull results in horizontal current flow, "smearing" the scalp potential, this effect of higher conductivity is negligible when compared to the effect of the highly resistive skull and will not be discussed further here.

[191] The nondimensional parameter GAIN is related to the total area of the active zone, the depth of the dipole layer, the conductivity ratios, and the sphere radii, and will not be explained further here.

An attenuation ratio (AR) has been suggested as an empirical measure of the comparative intracranial and scalp recording, and is defined as the ratio of maximal cortical surface amplitude and maximal scalp amplitude (Goff et al., 1978). The AR depends mainly on the depth and the extension of a given source, with focal cortical surface sources having high and extended depth sources having low AR (Goff et al., 1978).

The above mentioned arguments can be summarized in the following way. Scalp recordings are subject to 3 kinds of signal degrading. Firstly, an amplitude reduction; secondly, a reduced spatial resolution; thirdly, an increased spatial averaging (due to superposition, Goff et al., 1978).

8.2.4 Reference Dependence and EEG Recordings

In EEG recordings scalp potentials are not measured directly but as the potential difference between two electrodes, the latter being the reference electrode, the activity of which one is traditionally not interested in. The relative strength of the contribution of a given source i to the potential at electrode k may be expressed as: $c_{ki}=c(e_k; r_i; o_i; x_i)$, with e_k being the location of electrode k, r_i the location of source i, o_i the orientation of source i, and x_i the extension of source i (Scherg & Picton, 1991). The contributions of several sources sum up linearly, and the potential at time t at electrode k is Σ $(c_{ki}-c_{ri})$ $s_i(t)$, with r referring to the reference electrode, si(t) being the dipole source activities and Σ summing up from i=1 to the number of sources m (Scherg & Picton, 1991).

There is no principle difference between “recording“ and “reference“ electrode, and only electrode pairs instead of single electrodes can be considered sensitive or insensitive with respect to a certain generator location and orientation. The potential difference between two recording electrodes A and B due to a certain dipole source depends not only upon the location but also upon the orientation of the dipole (Katznelson, 1981, in Nunez, 1981). For these reasons, the choice of the reference electrode is crucial to the identification of neuronal generators and to the topographical identification of components, as the reference dependency of the P300 amplitude maximum illustrates (Gevins & Cutilo, 1986).

A reference which has frequently been used in EEG research is the *linked-ears reference.* Linking the two ears, however, can change the potential distribution over the whole scalp by providing a possibly least resistive path for the current flows (Katznelson, 1981, in Nunez, 1981).

The single or average *mastoid(s) reference* does not provide such additional current paths. However, it has been questioned whether the mastoids are indeed inactive[192]. As a consequence, potentials at the other scalp electrodes

[192] The mastoid reference can be assumed to be rather sensitive to distortions of potential distributions due to current flows through skull openings such as the ears (Katznelson, 1981; Nunez, 1981).

always contain mastoid reference *activity*, a fact which may be considered to induce some topographical ambiguity (Lehmann, 1986). Furthermore, cephalic skin potentials occur primarily at the mastoids (Kenemans et al., 1991), and may alter CNV and P300 amplitudes artificially (Picton & Hillyard, 1972). Finally, the mastoid reference, like any common reference, can not "be neutral with respect to sources having different locations and different orientations" (Hjorth, 1986).

Another reference which has been advocated especially for multi-channel recordings is the *average reference*, which "avoids privileging of one particular location as a no-change-over-time point in space" (Lehmann, 1986). The average reference is the mean of the potentials of all electrodes (referenced to a common reference) for each point in time, i.e. $\Phi_{i(\text{average reference})} = (\Phi_i - \Phi_r) - (1/N)\ \Sigma\ (\Phi_j - \Phi_r) = \Phi i - (1/N)\ \Sigma\Phi_j$, with $(\Phi_i - \Phi_r)$ being the potential difference between the ith electrode and the common reference, N being equal to the number of electrodes and Σ summing up from j=1 to N. Since the average reference includes information from all electrodes, the potential $\Phi_{i(\text{average reference})}$ for a given scalp location depends upon the locations of all other electrodes as well as their activity, rendering the average reference especially "sensitive" to artifacts. Although artifacts at a given electrode may be treated by excluding the electrode, this procedure reduces the desirable representativity of the average reference. Also, the dependency of the average reference upon the whole electrode montage imposes at least some constraints upon the comparability of ERP data from different studies[193]. An additional serious problem of the average reference arises, if the same potential shifts occur in cortical areas which are large relative to the scalp area covered by the total of recording electrodes. In this case the average *reference* will contain much of the signal itself, and the conclusion of, for instance, "a broadly distributed potential shift" may be substituted by the conclusion of "no spatial information". Furthermore, the principle property of most EEG recordings, namely that "the potential at any point is not a characteristic of that point but rather a characteristic of the path to that point" (Katznelson, 1981, in Nunez, 1981), is not circumvented by the average reference[194].

Accordingly, a method is needed "that characterizes a local property of the electric field which could be tied more closely with the underlying generators" (Katznelson, 1981, in Nunez, 1981).

As has been argued above, these generators may be identified with synchronous EPSP at apical dendrites in large cortical layers. Electric current is injected at certain locations from the cortex (sources) into the scalp through the skull. These currents must re-enter the brain through the skull at different locations (sinks). At other regions of the scalp, current will be neither accepted

---

[193] Despite this feature, the average reference is recommended by the International Pharmaco-EEG Group "for comparison purposes" (IPEG, 1989)

[194] Other arguments against the average reference in recordings of early somatosensory EP have been given by Desmedt & Tomberg (1990).

from nor injected into the skull but will be confined to the scalp, flowing parallel to the skull.

If the sum of the second spatial derivatives is applied to the potentials, the *Laplacian* of the electric potential results, which is the two-dimensional divergence of the scalp current density (CSD)[195], providing a quantity which is proportional to source or sink intensity, depending on the sign. The CSD is a reference-free quantity (Srebo & Purdy, 1990), which is independent of volume conductive models of the head (Nunez, 1989). This quantity can only be approximated with EEG measurements and must even be replaced by the derivation of the average potential gradient to a given electrode from the surrounding electrodes if inter-electrode distances are rather large (Katznelson, 1981, in Nunez, 1981). Since the activity of a given electrode is related to the weighted compound activity of its neighboring electrodes, the electrode is mainly sensitive to local sources, both tangential and in depth (Nunez et al., 1994). For this reason, the CSD method has been compared to a spatial high-pass filter (Wong, 1991). This feature amounts both to a relative insensitivity to remote sources and to an enhanced sensitivity to noise, rendering the assessment of the temporal stability of the CSD distribution necessary (Giard et al., 1994). Although due to their small distance currents in superficial cortical laminae are most influential on the surface EP, the whole intracortical pattern of current source densities is relevant. The weighting function taking into account the effects of varying distances is, however, non-linear (Kraut et al., 1986). Application of the source derivation results in an improved spatial resolution compared to, e.g., referential derivation, which is confirmed by a reduction of the inter-electrode signal correlations by a factor of 0.5 to 0.25 (Hjorth, 1986). There exist several versions of this procedure, which range from the use of some local electrodes (Hjorth, 1975; Nunez, 1981; Law et al., 1993) to global measures which are based on spherical spline interpolations (Perrin et al., 1989; Law et al., 1993). Similar spatial resolution and maps similar to the scalp Laplacian have been achieved by "cortical imaging" methods (Nunez et al., 1994).

In addition to the quality of the interpolation algorithm itself, the spatial sampling density is crucial for the correctness of subsequent current source density estimates (problem of spatial aliasing). CSD distributions are regarded as rather sensitive to spatial sampling (Giard et al., 1994). For dipoles of *extreme* eccentricity inter-electrode distances of 1 cm may be mandatory in order to grasp the topographical information completely (Braun, 1991). Srebo & Purdy (1990) suggest 2 cm sampling intervals to be sufficient. With computer simulations, Nunez et al. (1994) were able to resolve cortical sources separated by about 3 cm with 64 channel spatial sampling using the combined spline-Laplacian method.

---

[195] The scalp current density is the total current crossing a unit of cross-sectional area of a conductor (in ampère/cm^2).

8.3 Hypotheses

Research hypotheses as well as a number of research questions which do not have the quality of hypotheses but are exploratory in nature will be presented in the following discussion. When both hypotheses and questions are simultaneously treated statistically ("leapfrog strategy"), the distinction between the two kinds of tests should be made explicit. Therefore, hypotheses and questions will be stated separately in a confirmatory and an exploratory part. Results from experiment A and B will be taken into consideration unless they do not contradict results reported in the literature. Prior to the presentation of the research hypotheses and questions it shall be recalled that the primary intention of the experiments C and D is mainly to investigate the PINV topography in schizophrenic patients and groups of healthy subjects as revealed by a succession of PINV maps. These maps will clarify whether results and conclusions drawn from experiments A and B can be substantiated by a more extensive assessment of the scalp distribution, the distribution calculated for the cortical surface and the scalp Current Source Density maps. Exploiting these techniques can justify speculations about generator structures and biophysical sources which were briefly made in *chapters 5* and *6*. This chapter will focus on the PINV, rendering the analysis of CNV and LPC secondary. Both a qualitative description of the group maps and quantitative-statistiscal analyses are provided. Since identical hypotheses and questions pertain to the comparison of schizophrenic patients versus healthy controls in experiment C and high-scoring versus low-scoring subjects in experiment D, the following list can be applied to both experiments C and D. For the sake of conciseness patients and high-scoring subjects will be referred to as the "experimental group" and healthy controls and low-scoring subjects as the "control group".

Confirmatory Part of the Experiments C and D
PINV - Quantitative Analyses[196]

Hypothesis 1: PINV will be larger over frontal than over temporo-central leads in the control and experimental groups. *Reason:* Results of experiments A and B. *Statistical form:* Coronal Row (see *paragraph 8.15*) main effect.

Hypothesis 2a: Ambiguous imperative stimuli enlarge the PINV amplitudes in the experimental and the control groups. *Reason:* Literature and arguments presented in *paragraphs 3.5-3.7, 5.1* and results of experiments A and B. *Statistical form:* Ambiguity main effect.

Hypothesis 2b: Ambiguous imperative stimuli exert their strongest PINV increasing impact at frontal leads. *Reason:* Results of experiment A. *Statistical form:* Ambiguity by Coronal Row (see *paragraph 8.15*) interaction with a

[196] It should be noted that the following quantitative hypotheses pertain to the statistical analyses of cortex amplitude data, while experiments A and B analyzed scalp amplitude data. Cortical data were chosen since the signal quality is considered improved over the scalp data. The validity of the cortical projection is suggested by the similarity of cortex amplitude and scalp laplacian data (see below).

significant apriori contrast between the coronal rows (F9,F7,F3,F4,F8,F10) and (FT9, T3, C3,C4, T4, FT10)[197].

Hypothesis 3a: Experiment C: Schizophrenic patients and healthy controls reveal different PINV lateralizations. *Reason:* Results of experiment A and B. *Statistical form:* Hemisphere by Group interaction and significant hemisphere effects in the within-group ANOVA[198].

Hypothesis 3b: Experiment D: High-scoring and low-scoring subjects reveal reverse PINV asymmetries. *Reason:* Known association between questionnaire anhedonia and PINV lateralization; literature reported in paragraph 8.1. *Statistical form:* Hemisphere by Group interaction and significant hemisphere effects in the innergroup ANOVA.

PINV - Qualitative Analyses

Hypothesis 4: Cortex amplitude and scalp laplacian maps will reveal similar results. *Reason:* Theoretical relatedness of both techniques; literature present in *paragraph 8.2.4.*

PINV- Exploratory Part

The topographies of the PINV in schizophrenic patients and healthy controls as revealed by the different kinds of maps will be assessed and compared exploratively. Due to the almost unknown topography of this component, this is interesting in its own right. Questions referring to the topographical "broadness" of the PINV in patients and controls and to the (group specific ?) spatio-temporal pattern of its generation (much better seen in maps than in curves) will be addressed to the data.

---

[197] Since this effect was not found in experiment B, and in experiment A the ambiguity effects at (F7, F8) were only descriptively larger than the ambiguity effects at (C3, C4), and since a different set of electrodes will be entered into the statistical analyses (compared to experiment A), this hypothesis is lacking deductive stringency. It is formulated in this way, since the overlap of leads entered into the statistical analyses in experiments C and D is larger with experiment A (8 out 12 leads) than with experiment B (4 out of 12 leads) and since those leads obviously particularly relevant to the PINV (F7 and F8) are analyzed in experiment A, C, and D, but not B. Assuming a preponderance of the ambiguity effect at (F3,F4) would result in a statistical Ambiguity by Coronal Row by Sagittal Row hypotheses for experiments C and D, which will not be stated because of the —refuting— results of experiment B. The Ambiguity by Electrode by Group interaction will be skipped, because the stronger frontal PINV increase following ambiguous imperative stimuli in the control group of experiment A was confined to the (F3, F4) leads and was not replicated in experiment B.

[198] This hypothesis is also lacking deductive stringency, since different but not reverse lateralizational patterns were observed in experiment B. The hypothesis is stated none the less because of the larger overlap in investigated leads between experiments A and C than between experiments B and C (see previous footnote).

8.4 *The Laboratory*

Experiment C was conducted in the psychophysiological laboratory of the University of Konstanz in the State Hospital Reichenau. Experiment D was conducted in the psychophysiological laboratory of the University of Konstanz in the University. Both laboratories have been described in detail in *paragraph 5.2*.

8.5 *Recruitment of Subjects and Assessment of Psychopathological Variables in Schizophrenic Patients*

8.5.1 Subject Groups in Experiment C

Schizophrenic patients and healthy controls were recruited for the present experiment as described in *paragraph 4.3*. As clinical variables the Brief Psychiatric Rating Scale (BPRS) and the medication of the day of testing were assessed (described in *paragraph 4.4*). Twelve out of 20 participating psychotic patients from the State Hospital Reichenau and 11 out of 14 participating healthy control subjects could be analyzed statistically. Data of the other subjects had to be omitted from further analysis due to artifact contamination. In **table 8.1** the socio-demographic features of both experimental groups are presented.

Table 8.1 Socio-Demographic Features of the Patients and Controls Participating in Experiment C

	Patients	Controls
N	12	11
Age (Range)	30.0 (23-42)	28.5 (21-35)
Male : Female	11 : 1	8 : 3
Smoker : Non-Smoker	8 : 4	8 : 3
Education (Years)		
8	6	4
10	4	3
13	1	3
>13	1	1
Right-Handedness	11	9

Eight out of 12 schizophrenic patients of the final sample were inpatients from the research ward of the State Hospital Reichenau. Three former patients of the research ward lived in the community, two of whom were currently taking part in a rehabilitation program. Another patient was recruited from a ward for long-term chronically ill patients. All patients met a DSM-III-R diagnosis of a chronic schizophrenic disorder. All schizophrenic patients were receiving neuroleptic medication during the time of testing (mean=190.9 mg/day CPZ equivalent, range=116-435). The mean psychopathological rating as assessed by the BPRS was 42.8 (SD=12.1, range=26-64).

In addition to the 12 schizophrenic patients described above, 11 healthy control subjects participated in experiment C. Prior to testing they were questioned for psychiatric and neurologic diseases and the intake of drugs. All questions were answered in the negative by all subjects.

8.5.2 Subject Groups in Experiment D

Forty-four subjects selected from students of the University of Konstanz by means of a German translation of the Schizotypal Personality Questionnaire by Raine (1991) participated in experiment D.

Table 8.2 Socio-Demographic Features of the Subjects Participating in Experiment D

	High-Scoring Group	Low-Scoring Group
N	19	18
Age (Range)	22.1 (20-30)	23.1 (19-32)
Male : Female	11 : 8	12 : 6
Smoker : Non-Smoker	4 : 15	3 : 15
Right-Handedness	17	18
$SPQ_{total\ score}$ (Range)	41.8 (38-58)	15.2 (2-37)

Groups are defined by median-split (SPQ_{total}=38) of the total sample.

The 74-item Schizotypal Personality Questionnaire (SPQ) was filled in by the subjects together with an abbreviated, 50-item version of the Eysenck Personality Questionnaire (EPQ-KR; Ruch, in preparation) and 3 scales of 12-items each of the Hamburg Personality Questionnaire (HPF) by Andresen (in preparation). Selection criterion for experiment D was the total SPQ score. From a total of 665 subjects who completely filled in the compound 160-item questionnaire, 47 subjects covering a SPQ total score range from 2 to 58 were selected for experiment D.

Due to electrode and eye movement artifacts data of 10 out of 47 subjects could not be analyzed. Groups were defined by a median-split of the sample. The median in the total sample of participating subjects is SPQ_{total}=38. Socio-demographic features of the subject groups of experiment D are displayed in **table 8.2**.

8.6 The Testing Procedure

The testing procedure applied in the experiments C and D was identical for both and is described in *paragraph 4.5*. With each subject the time of his/her experimental session was arranged after he/she had agreed to participate in the experiment. Before starting with electrode applications the laboratory and its equipment were shown to each subject. After explaining the goal and meaning of the experiment, written consent was required. Voluntary participation was paid with DM 35.- in experiment C and with DM 50,- in experiment D. (A financial bonus of DM 10,- for fast and correct responding was included in each case.) None of the subjects in experiments C and D wished to terminate the session prematurely. However, due to insufficient understanding of the instructions two patient sessions had to be terminated prematurely by the experimenter.

After the preparation for physiological recordings, described in *paragraph 4.5.1*, each experimental session began with the execution of calibration eye movements for the off-line eye artifact correction. The calibration task is described in *paragraph 4.5.2*.

Prior to the 205 trials of the main block 20 practice trials without ambiguous IS were presented in order get the subjects acquainted with the experimental task.

After termination of the experimental task, electrode cables were removed from the pre-amplifier and the subject was led into the experimenter room with all electrodes still affixed to the head. The 3-dimensional electrode positions were recorded with a digitizer (POLHEMUS). The electrode positions were used for eye artifact correction and for mapping (see below and *paragraph 4.7.3)*.

8.7 The Task Presentation

Figure 8.1 (title-page of *chapter 8*) illustrates the experimental design of experiments C and D. The experimental design is identical with the basic design described in *paragraph 4.1* (experiments A and B without acoustical probe stimuli and nogo IS). From a total of 205 trials of the main block, 140 trials (68%) pertained to the clear condition and 65 trials (32%) to the ambiguous condition.

8.8 Apparatus and Physiological Recordings

The technical equipment of the laboratories of the experiments C and D are the same and were described in *paragraph 4.2*. Amplifier settings are presented in **table 4.1**.

The experiments C and D were both conducted with a 32-channel electrode configuration. The positions of the EEG electrodes were determined with the aid of an "electrode cap" (MEDILOG), providing no electrodes but 5-6 mm large holes at the appropriate positions instead. Electrode positions not defined by the positions of the electrode cap holes are F9, F10 (corresponding to the horizontal EOG channels at the outer canthi), VE1 and VE2 (the two EOG channels 2-3 cm vertically below the eyes), Nz (corresponding to the nasion), FT9 and FT10 (EEG channels located about 1 cm anteriorly to the preauricular points), TP9 and TP10 (the left and the right mastoids), and CB1 and CB2 which are positions over the cerebellar region, defined as one-third and two-third segments between the left and the right mastoid, respectively. The electrode at position AFz served as the common recording reference. AN earth electrode was placed on the forehead.

Figure 8.2 Electrode Montage in Experiments C and D

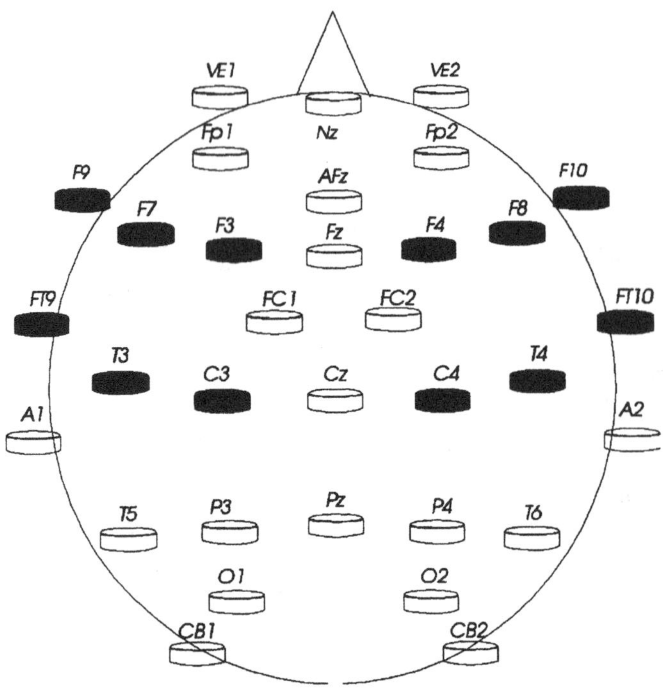

The cortical ERP data of the shaded electrode positions were used for the statistical analyses.

Figure 8.2 illustrates the electrode positions of the experiments C and D. These electrode positions and their nomenclature are defined according to recent guidelines of the *American Electroencephalographic Society* (1991). In contrast to the international 10-20 system this is a 10-10 system, since all "10% positions" are defined herein.

8.9 Data Reduction and Data Analysis

The EEG data recorded by the DCMES system were at first corrected for DC shifts with the program DCAUS (MES, Munich; see *paragraph 4.2*). For this DC correction a fourth-order polynom is fitted to the mean DC levels of a 1.0 s pretrial baseline interval in order to compute each trial as a difference from the statistically fitted DC level.

The subsequent steps in data analysis were performed with the analysis programs described in detail in *paragraph 4.2*. From most of the following programs applied in that order 386 PC versions were taken. The first steps in data analysis are identical with those described in *paragraphs 5.* and *6.* Transformation of data format from the DCMES to the Konstanz raw data format; categorization of each trial according to its combination of stimulus and response event marker and computation of RT; initial baseline subtraction; re-referencing of raw data on an average reference, re-referencing of average reference data on an average mastoids reference, artifact identification, computation of regression coefficients for blinks, eye activity correction and condition-wise computation of single subject averages, averaging over predefined conditions (see below), and computation of mean amplitudes over predefined intervals for statistical analyses (see below).

Differences to the analysis steps of experiments A and B arose from the different eye activity correction, using the programs DOPCAKN (PCA on electrode channels in order to get the loading of each channel on the eye movement factors), and 3EOGCE (EOG artifact correction with the MSEC method). The quality of the eye activity correction was checked by visually inspecting each trial, outselecting those trials that still contained eye activity in the EEG channels. After that the following steps were done: interpolation from measured to standardized electrode positions with spherical spline interpolation, condition-wise computation of single subject averages, averaging over predefined conditions (see below), computation of mean amplitudes over predefined intervals for statistical analyses (see below), computation of scalp maps with spline interpolation, and projection from scalp to cortical maps. Single-subject averages were computed for artifact-free (after eye artifact correction) trials only. **Table 8.3** summarizes the mean relative frequencies of correct, false, or missing reactions, and of artifact-free EEG trials, differentiated for conditions and groups.

The eye activity corrected single-subject averages were interpolated from the empirically determined to standardized electrode positions with the program

ELINTERP, using a spherical spline-interpolation algorithm (Perrin et al., 1989, see below). In order to generate scalp maps, grand averages were computed from the single-subject averages. Subsequently the cortical surface distribution was calculated from the scalp distribution using a method based on that of Gonzales-Andino et al. (1992). This method assumes an "isotropic volume conductor of arbitrary geometry" (which is composed of a number of homogeneous layers, each with a different conductivity) and linear relations between the electric potential of the scalp and the potentials on the inner conductivity interfaces. Matrices for the transformation of the measured data into equivalent infinite homogeneous conductor measurements are computed for a given set of measurement locations and a given volume conductor model. In this case, the 33 standardized electrode positions served as measurement locations and the three concentric sphere model was used. Eccentricity of the dipole layers was set to 0.85 of the head radius.

Table 8.3 Speed and Correctness of Reactions and Artifact-Free Trials

(a) Relative Frequencies of correct, false or missing responses

Condition	Clear				Ambiguous		
Response	Correct	False	No	RT	Correct	No	RT
Patients	86	13	1	649±147	98	2	817±143
Controls	87	13	0	479±493	99	1	588±1031

SD in parentheses

(b) Artifact-Free EEG Trials (Mean, (SD), (%))

Condition	Clear		Ambiguous	
	Mean±SD	%	Mean±SD	%
Patients	98.4±17.6	70.3	52.2±7.3	80.0
Controls	110.3±12.6	78.8	52.4±7.0	80.6
High-Scoring Group	109.9±24.2	78.5	52.9±12.5	81.4
Low-Scoring Group	105.3±21.7	75.2	50.8±10.8	78.2

Patients versus Controls: Effects: Condition: $F(1,20)=7.37$, $p=.01$; Condition by Group: $F(1,20)=2.48$, $p=.13$; Group: $F(1,20)=1.27$, $p=.27$ (ANOVA); *Low-Scoring versus High-Scoring Group:* Effects: Condition: $F(1,36)=3.56$, $p=.07$; Condition by Group: $F(1,36)=0.0001$, $p=.99$; Group: $F(1,36)=0.38$, $p=.54$ (ANOVA)

Spherical spline interpolation was accomplished following the method proposed by Perrin et al. (1989). This interpolation method includes three separate steps: Firstly, the measured data are projected to a sphere, placing Cz on the top and Fpz and Oz, respectively T3 and T4, at diametrically opposite positions on the equatorial line. The second step comprises the interpolation of the data using spherical spline surfaces. Finally, the sphere is projected onto the plane. For the maps presented here, this "plane" comprises interpolated potentials from regions with angular locations up to 110° relative to the vertical head axis.

Finally, the *Current Source Density (CSD)* was estimated with the method of Perrin et al. (1989). With this method, the CSD is computed as the second derivative of the spherical spline interpolation of the potential distribution.

8.10 Results of Experiment C: PINV Topography in Schizophrenic Patients and Healthy Control Subjects

The following presentation of results consists of 3 sections: (a) grand average curves of the patient and control groups; (b) different types of maps (scalp and cortical amplitude maps as well as scalp and cortical laplacian maps); (c) statistical analyses of a subset of selected electrodes. For reasons outlined in *paragraph 8.14*, the statistical analyses will use the average mastoid reference $((A_1+A_2)/2)$. By contrast, the average reference makes optimal use of the restricted possibilities of a graphical differentiation of areas of positive versus negative potential shifts through white and gray areas, which is provided by the mapping program used here. Furthermore, the information from 2 additional electrodes (the mastoids) can be included to compute the spherical splines. For this reason the maps display average reference data. Since for a given time point the topography is reference-independent, this procedure introduces no inconsistencies between maps and statistics. For the grand average curves the average reference will be used in order to make curves and maps comparable. The maps of the PINV are a succession of 8 averages over adjacent 250 ms intervals, beginning with the map for the 1.0-1.25 s interval, and terminating with the map for the 2.75-3.0 s interval, in each case relative to IS onset. Except for the tCNV-referenced maps in *pararaph 8.x*, each map and each grand average curve is referenced to the pretrial baseline (1000 ms before WS onset).

Before describing the grand average curves and maps it should be recalled that the topography for a given point in time is reference-independent. In contrast, the location of the regions in which a polarity inversion occurs for a given ERP component *is* reference-dependent. For this reason no attention will be focused on the description of regions of polarity inversion.

8.10.1 Grand Averages

Figure 8.10.1a Grand Averages: Schizophrenic Patients

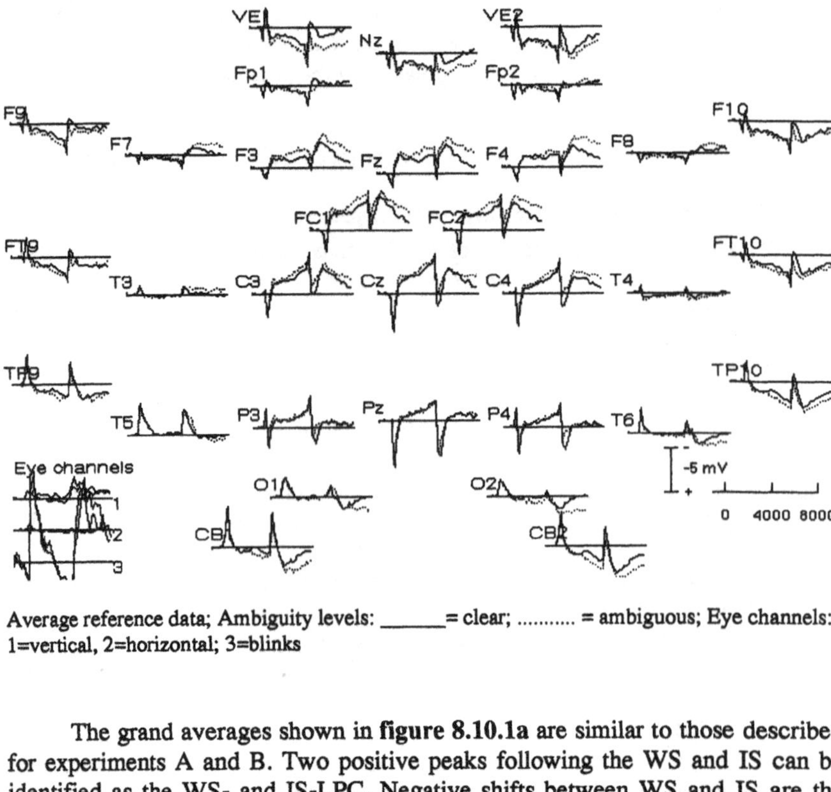

Average reference data; Ambiguity levels: ______= clear; .... = ambiguous; Eye channels: 1=vertical, 2=horizontal; 3=blinks

The grand averages shown in **figure 8.10.1a** are similar to those described for experiments A and B. Two positive peaks following the WS and IS can be identified as the WS- and IS-LPC. Negative shifts between WS and IS are the two CNV components: a fronto-central iCNV and a central-to-precentral tCNV. The PINV can be seen only at fronto-central leads. The multi-channel grand averages reveal a broader and more parietally-oriented topography of the CNV compared to the PINV. Notable is also the diminution or absence of the PINV at lateral to inferior frontal positions (F7, F8, F9, F10).

Figure 8.10.1b shows the multi-channel grand averages of the healthy control group from experiment C. Although, by and large, control group and patient group averages are rather similar, some distinct differences become evident in the multi-channel data of the control group, both compared to the multi-channel data of the schizophrenic patients and the grand averages of the control groups of experiments A and B (see **figures 5.3, 5.4, and 6.3 a-c**). First, the positive peaks of the LPC have reduced amplitudes in controls especially over central regions, as compared to patients LPC. By contrast with these diminution, CNV amplitudes in the control group are larger than those in the patient group, with maximal amplitudes of the iCNV at positions FC1 and FC2 and of the tCNV at Cz. Following the IS a PINV is clearly discernible in the control group under all conditions, but with larger amplitudes under the ambiguous than under the clear condition. By contrast with the patient group, exhibiting an almost uniform PINV shape, at least two distinguishable PINV wave forms are present in the control group. The first waveform can be seen most typically at those leads where CNV is largest and IS-LPC smallest, the FC1 and FC2 positions. At these positions PINV is similar to that observed under the loss of control conditions (cf. Rockstroh et al., 1989, p. 107) and has the waveform of a rapidly declining negativity. The second waveform can be seen at the right-frontal positions F4, F8, and F10, where PINV rises moderately (rather than decreasing) following a well-developed IS-LPC.

Figure 8.10.1b Grand Average: Healthy Controls

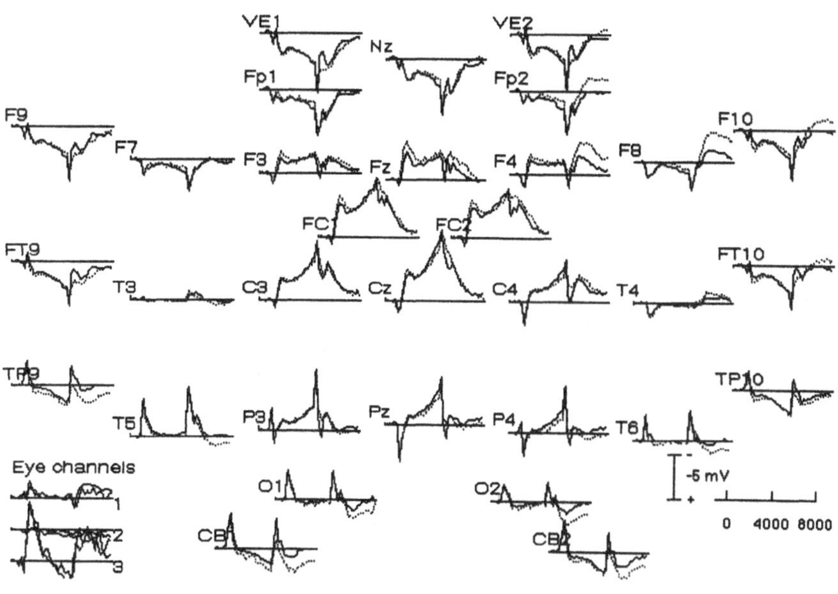

Average reference data; Ambiguity levels: ______= clear; .... = ambiguous; Eye channels: 1=vertical, 2=horizontal; 3=blinks

The waveform of the patient group is similar to the second waveform of the control group. However, the amplitude decline following shortly the first second after the IS is faster in patients than in controls.

Furthermore, as already mentioned, this waveform is more broadly distributed in patients than in controls. **Figures 8.10.1a,b** include the mean eye activity (bottom left). While there is some blink activity following the experimental stimuli, only minor vertical and virtually no horizontal eye activity can be seen.

8.10.2 Scalp Amplitude Maps

Figure 8.10.2a presents scalp amplitude maps of the healthy control group of experiment C. In this map, as will be the case in all further maps, areas with negative potential shifts are hatched, while white areas represent regions with positive going potential shifts, relative to a 1000 ms pretrial baseline. Lines represent isopotential or contour lines. From line to line there is an increase of the strength of the potential field. This increase is 0.75 µV for scalp amplitude, 1.0 µV/cm/cm[199] for CSD maps, and 1.25 µV for cortex amplitude maps, starting with zero at the border line between gray (negative) and white (positive) areas[200]. The left double-column of maps shows the PINV under the clear condition, the right double-column of maps shows the PINV under the ambiguous condition. On the right side potential shifts are displayed for selected channels (mainly F7, F3, F4, F8). For each of the four rows of maps and ERP the following holds: Map onset of the *left* map *within* each double-column (condition) corresponds to the time point indicated by the *left* vertical bar in the average ERP (right); and map onset of the *right* map *within* each double-column corresponds to the time point indicated by the *right* vertical bar in the average ERP. In addition, for each map the onset in seconds relative to the WS onset is documented with the number closest to the map.

Figure 8.10.2a presents the PINV in healthy controls under the ***clear condition***. It is broadly distributed over frontal to centro-parietal areas. A slight "diagonal" orientation with a more frontal extension over the right hemisphere and a more parietal extension over the left hemisphere is apparent. Although negativity extends into centro-parietal areas, the PINV maximum is always anterior to the central coronal line. Under the clear condition the PINV starts with a slight left-sided central maximum (5.0, 5.25, and 5.5 s). At 5.75 s (map 4) a second right-frontal maximum becomes apparent. The epoch terminates almost symmetrically with only a minimal rightward shift.

Under ***ambiguous conditions*** an initial lateral symmetry can be observed, with slight anterior maximum shifts over the right hemisphere and slight posterior

---

[199] This is only the theoretical unit of measurement.

[200] Different isocontour line-scalings have been tested. The presented maps provide scalings which appear optimally adjusted to the different kinds of maps.

maximum shifts over the left hemisphere again, giving the impression of a "diagonal" orientation. However, in contrast with the clear condition, a right frontal maximum shift develops soon (5.5 s), leading within the first 500 ms to a clear right-frontal predominance of the PINV in healthy controls. This right-frontal predominance, with substantial extensions into left central regions, remains stable for the rest of the interval, even during the decrease of the PINV towards the end of the trial. Referring to the grand average curves presented in **figure 8.10.1a**, the different course of the potential shifts at right-frontal (especially F4 and F8) leads compared to the other leads is confirmed by the scalp amplitude maps with its fast-rising and relatively sustained right-frontal predominance within an environment of slowly diminishing PINV amplitudes.

In **figure 8.10.2b** scalp amplitude maps of the schizophrenic patient group are presented. Again, PINV in schizophrenic patients under ***clear conditions*** shows a broad distribution, extending from bilateral fronto-central into left anterior frontal and left parietal regions. A left-sided shift of the region of maximal PINV amplitudes is stable.

Under the ***ambiguous condition***, when compared to the clear condition, only slight changes concerning the PINV topography occur in schizophrenic patients. With generally higher amplitude levels, there is a slight left-sided PINV preponderance which slowly changes into a more symmetric distribution with a fronto-central predominance. At 6.0 s the region of maximal amplitudes is somewhat larger over the right than over the left hemisphere.

Figure 8.10.2a ***Grand Averages: Healthy Controls***

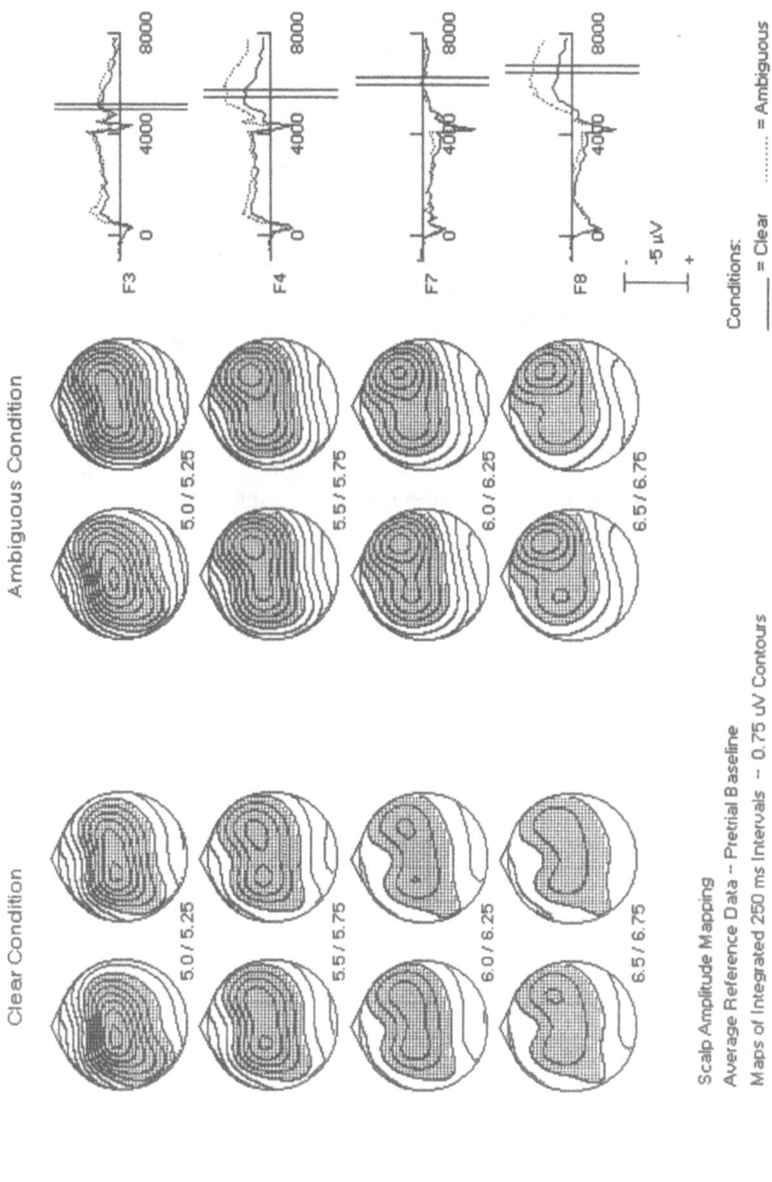

Figure 8.10.2b ***Grand Averages: Schizophrenic Patients***

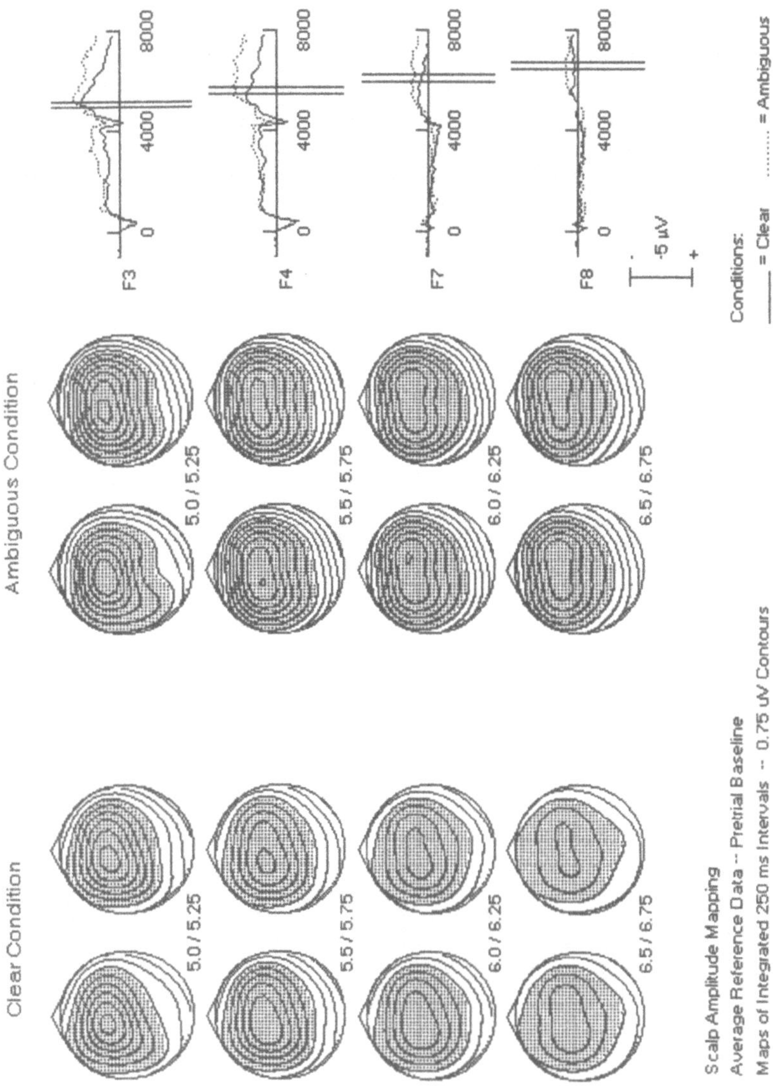

In sum: the maps of the multi-channel PINV data reveal spatio-temporal features which go undetected with fewer electrodes.

8.10.3 Cortex Amplitude Maps

Figure 8.10.3a presents cortex amplitude maps of the control group. The initial "cortical" PINV in healthy controls under the ***clear condition*** has a more restricted topography than its scalp amplitude counterpart. Extensions into parietal regions are largely reduced, except for a short-lasting small left temporo-parietal hill. The topographical "core" of the PINV distribution is a relatively small fronto-central band which has an initial left fronto-central maximum (map 1). At 5.25 s the cortical amplitude maps reveal *2* centers of maximum negativity. One of the two maxima is over the left hemisphere, slightly anterior to the central coronal electrode row, adjacent to the mid-sagittal axis. The other maximum is over the right hemisphere and is clearly oriented to anterior and lateral areas. For the remaining intervals, both maxima are by and large of equal strength and remain topographically "stationary", decreasing in amplitude already at 5.5 s (map 3).

Under the ***ambiguous condition*** the initial map has a similar topography like that of the initial map of the clear condition. However, a larger right-frontal extension and a *smaller* amplitude of the initial left-hemisphere maximum can be seen. The rising of *2* fronto-central PINV maxima becomes apparent at 5.25 s. Both maxima remain topographically stable for the rest of the presented intervals. The amplitude differences between the right-frontal area and the the left fronto-central area and the absolute amplitudes are larger under the ambiguous when compared to the clear condition at 5.5 s (map 3). For the rest of the period a clear right hemisphere PINV asymmetry is apparent. While the smaller left fronto-central hill starts to become flatter at about 5.75 s (map 4), the larger right-frontal hill shows no amplitude decrease until 6.75 s.

Figure 8.10.3a ***Grand Average: Healthy Controls***

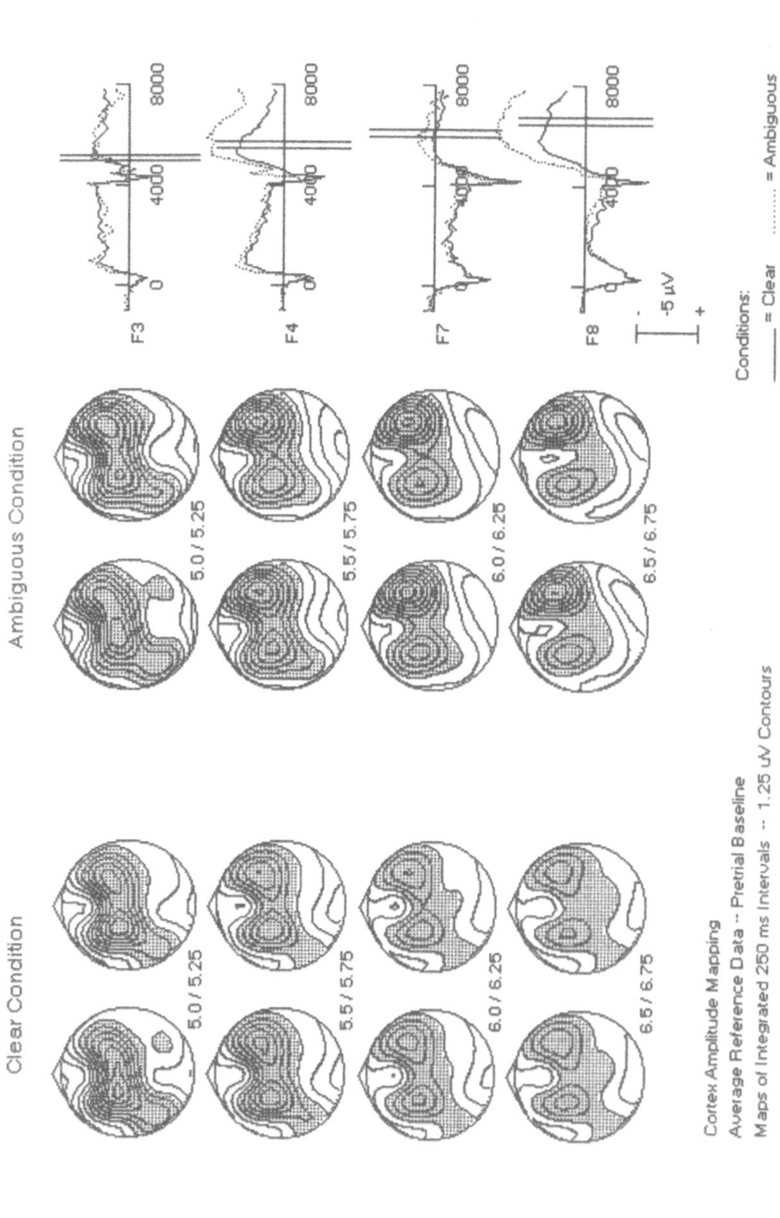

Figure 8.10.3b ***Grand Average: Schizophrenic Patients***

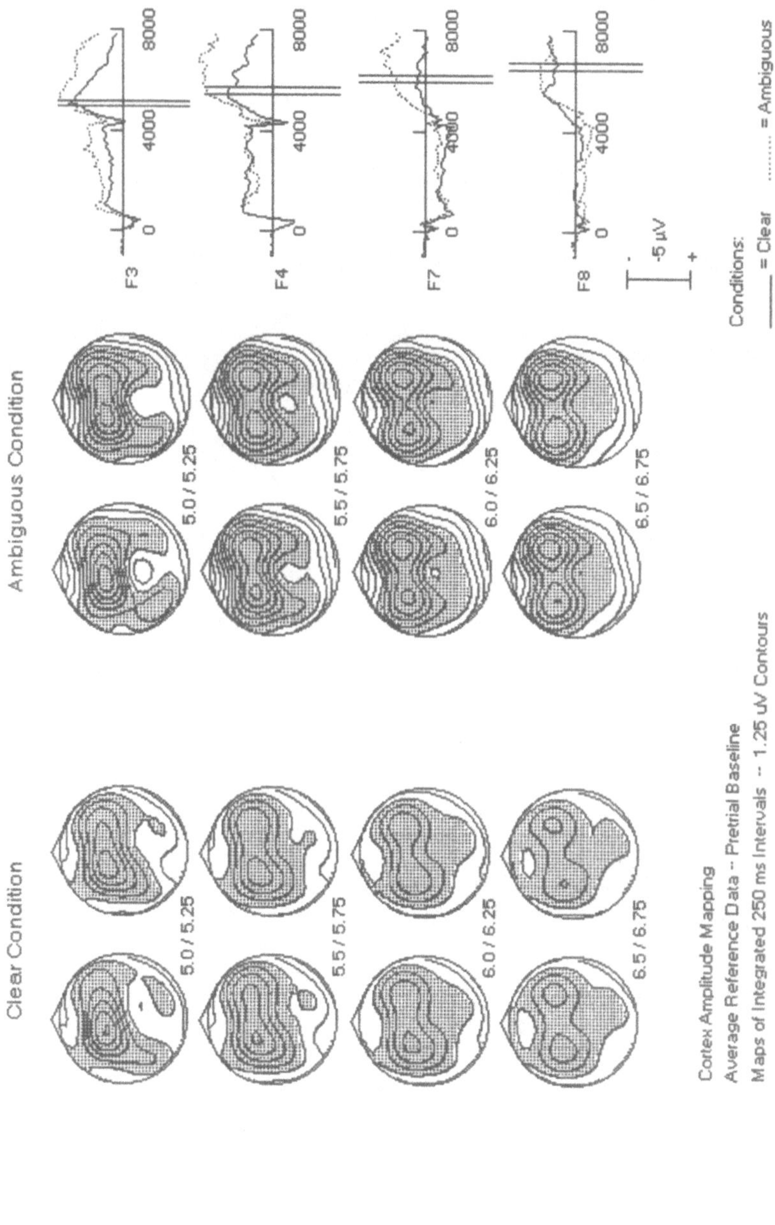

In **figure 8.10.3b** cortex amplitude maps of the patient group are presented. As was the case for the scalp-cortex comparison in controls, cortical maps of patients exhibit somewhat "sharper" contours than the scalp maps. Under the ***clear condition***, a relatively small bilateral negative band with a maximum over the left precentral region extends both into left temporo-parietal and right fronto-central regions. Until 6.0 s (map 5) a left-hemisphere preponderance of the PINV is preserved, which is followed by a terminal, almost symmetric distribution.

Under the ***ambiguous condition*** PINV topography is even broader than under the clear condition. This holds especially for its posterior extension, which slightly favors the left hemisphere and is absent in controls. A small positive trough over the anterior parietal region is the remainder of the diminishing IS-LPC (see **figure 8.10.1a**). As is true of all maps of the patient group, cortex amplitude maps reveal an initial left-hemisphere fronto-central PINV predominance. While this left fronto-central maximum remains topographically stationary for the rest of the intervals, a second right-hemisphere and slightly more anteriorly shifted maximum becomes apparent at 5.5 s (map 3). Eventually, two rather distinct maxima can be seen at 6.5 s (map 7). Again, as is the case for *all* maps (those of patients and controls), the right-hemisphere PINV hill is more anteriorly shifted than the left-hemisphere hill.

8.10.4 Scalp Laplacian Maps

Figure 8.10.4a ***Grand Average: Healthy Controls***

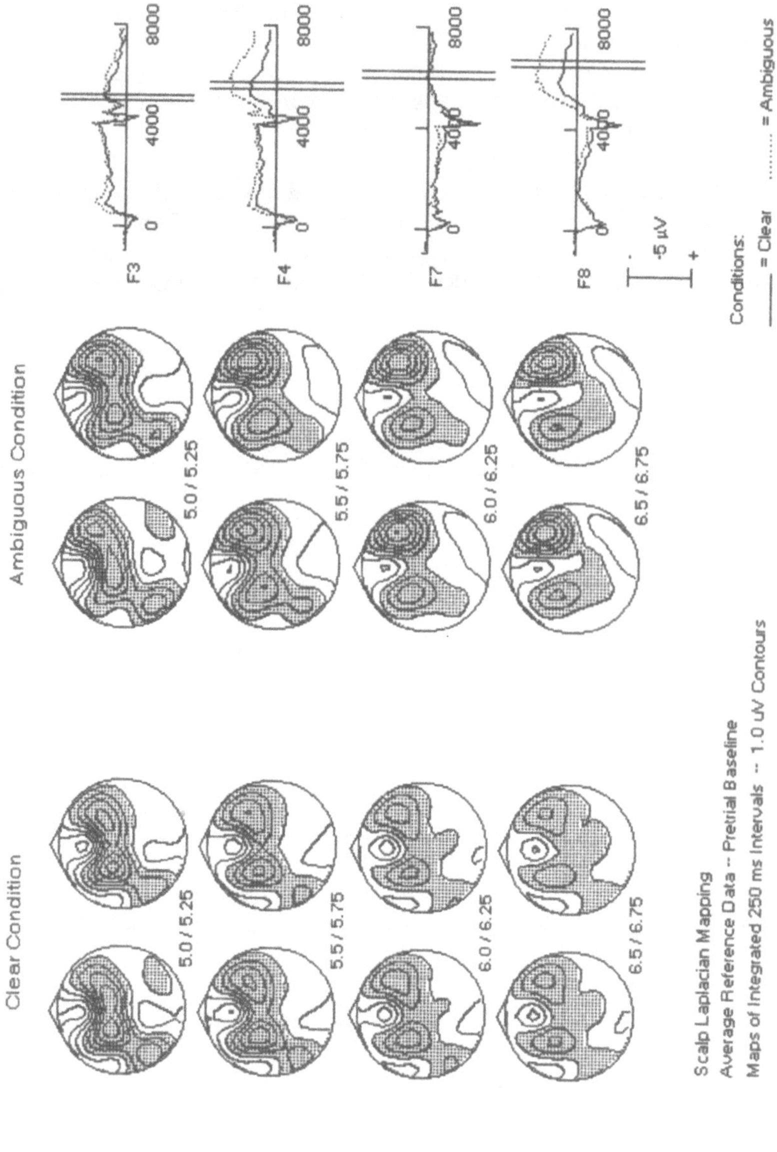

Figure 8.10.4b ***Grand Average: Schizophrenic Patients***

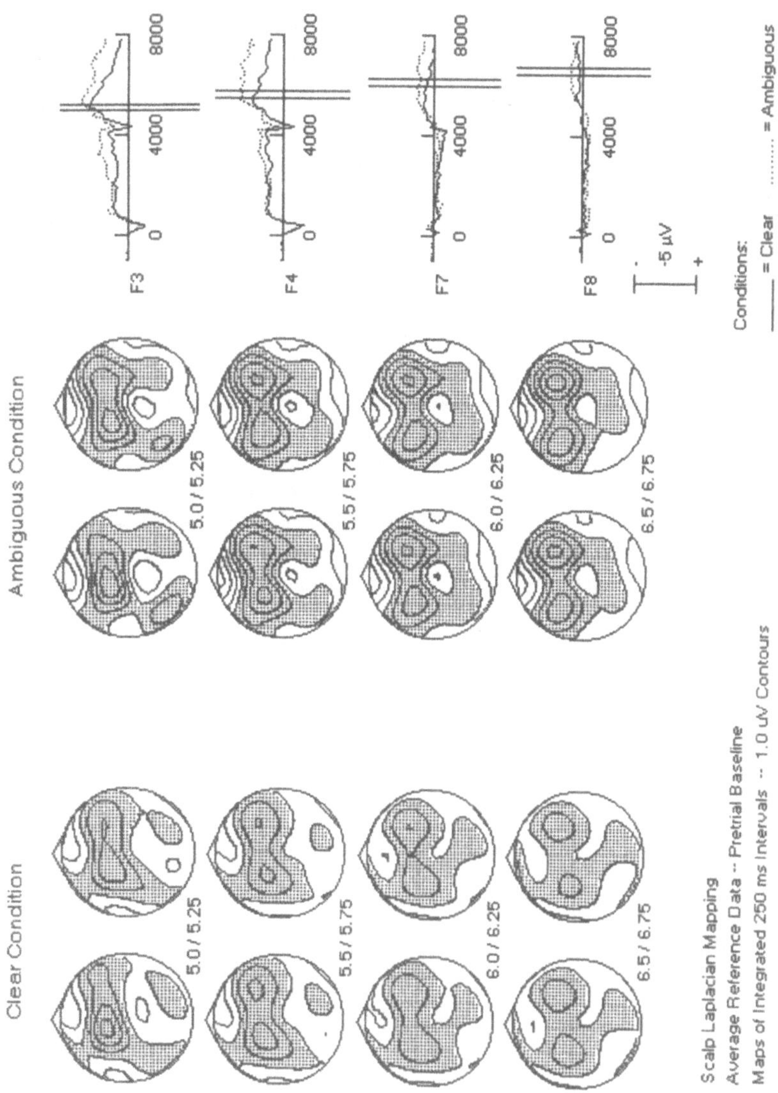

Figures 8.10.4a and 8.10.4b show the scalp laplacian maps of the healthy control and the patient groups. The scalp laplacian maps reveal hardly any *topographical* difference to the cortex amplitude maps. The only possible exception is the midline-frontal positive trough in the maps of the control group, which becomes more clearly apparent in the scalp laplacian maps than in the cortex amplitude maps. Furthermore, the left temporo-parietal negativity is separated more clearly in the laplacian maps (concerning the first four maps of each condition). Similar impressions arise from the maps of the patient group. However, some minor differences between scalp laplacian and cortex amplitude maps can be detected. Firstly, the cortical maps reveal a left-hemisphere preponderance in maps 2 to 5 of the clear condition which is not, or not as clearly, detectable in the scalp laplacian counterparts. Secondly, a more "shrunken" impression due to the sharper separation of the negative areas is given by the scalp laplacian maps. It should be recalled, however, that amplitude differences between the isocontour lines are not the same in the two kinds of maps, rendering such fine-grained comparisons somewhat ambiguous.

In sum: by projecting to the cortical surface and by applying the Laplacian to the scalp data, the negative and positive areas of the ERP distribution, which are "smeared" in the scalp amplitude maps, become clearer. This indicates the improved spatial resolution of the two methods when compared to scalp amplitude maps. Healthy controls exhibit a right-frontal PINV predominance, whereas the PINV in schizophrenic patients is less asymmetrically distributed. This group difference appears to result from a different spatio-temporal modulation of the PINV. In both groups a left-frontal PINV maximum is present throughout the interval. Also, in both groups a second right-frontal maximum develops. The crucial difference is that the development of this right-frontal maximum occurs not only faster but in particular much more pronounced in controls than in patients.

8.11 *Results of Experiment D: PINV Topography in Subjects Scoring High Versus Low on a Schizotypal Personality Dimension*

8.11.1 Grand Averages

Figure 8.11.1a Grand Average: Low-Scoring Subjects

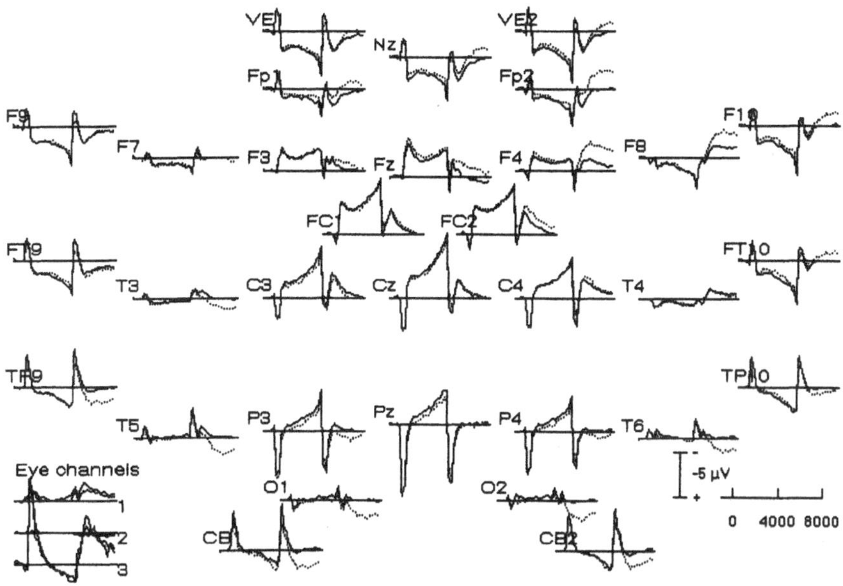

Average reference data; Ambiguity levels: ______= clear; .... = ambiguous; Eye channels: 1=vertical, 2=horizontal; 3=blinks

Figure 8.11.1a presents the grand average curves of the low-scoring group of experiment D. The grand averages of the low-scoring subjects are similar to those of the two groups of experiment C. In contrast to the control group of experiment C, however, this group exhibits pronounced IS-LPC amplitudes at the central (C3, Cz, C4) and precentral (FC1, FC2) leads. The development of the PINV at the inner (F4), middle (F8) and outer (F10) leads of the right-frontal region is similar in healthy controls of experiment C and the low-scoring subjects. Furthermore, in both groups there is no increase in negativity over the level of the pretrial baseline at the middle (F7) and outer (F9) positions of the left-frontal region. An augmentation of the PINV amplitudes from the clear to the ambiguous condition is most obvious at several mid-sagittal and right-sided frontal positions

(Fz, Fp2, F4, F8, F10) with minor enlargements over the left hemisphere (e.g., Fp1, F3).

Figure 8.11.1b presents the grand average curves of the high-scoring group of experiment D. The ERP of the high-scoring groups are similar to those of the low-scoring group.

Figure 8.11.1b Grand Average: High-Scoring Subjects

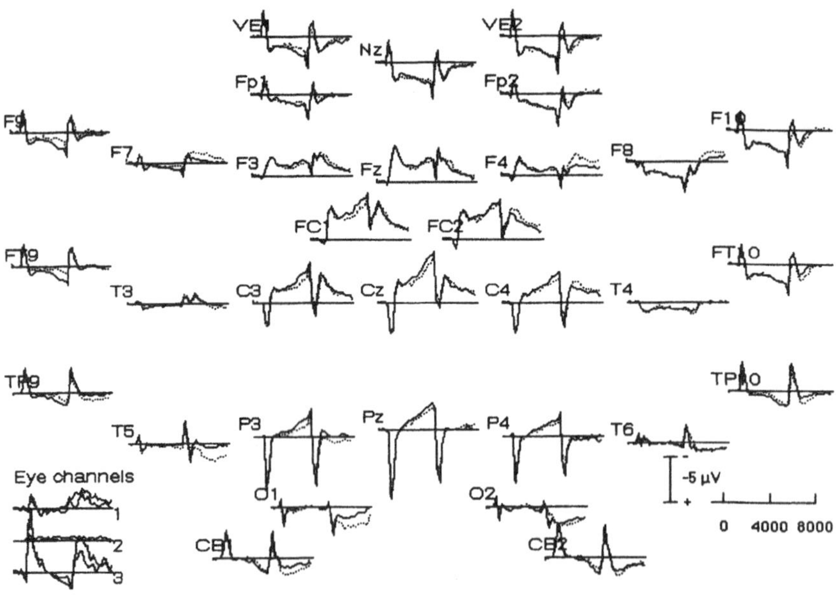

Average reference data; Ambiguity levels: ______= clear; .... = ambiguous; Eye channels: 1=vertical, 2=horizontal; 3=blinks

Despite this "first-sight" similarity, some differences can be detected. Firstly, tCNV amplitudes are smaller in this group, especially over the vertex (Cz) and at pre-central leads (FC1, FC2), and IS-LPC amplitudes are smaller at these positions. Secondly, PINV amplitudes are comparatively reduced at right-frontal positions, but larger in amplitude at midline-to-left central (C3, Cz), left pre-central (FC1), and left fronto-lateral (F7) positions. Thirdly, the amplitude augmentations due to ambiguous compared to clear IS are smaller in the high-scoring than in the low-scoring group, except for the F7 lead. The eye activity of the two groups (bottom left of **figures 8.11.1a,b**) reveals blink activity after the experimental stimuli, minor vertical but no horizontal eye activity.

8.11.2 Scalp Amplitudes Maps

Figure 8.11.2a Grand Averages: Low-Scoring Subjects

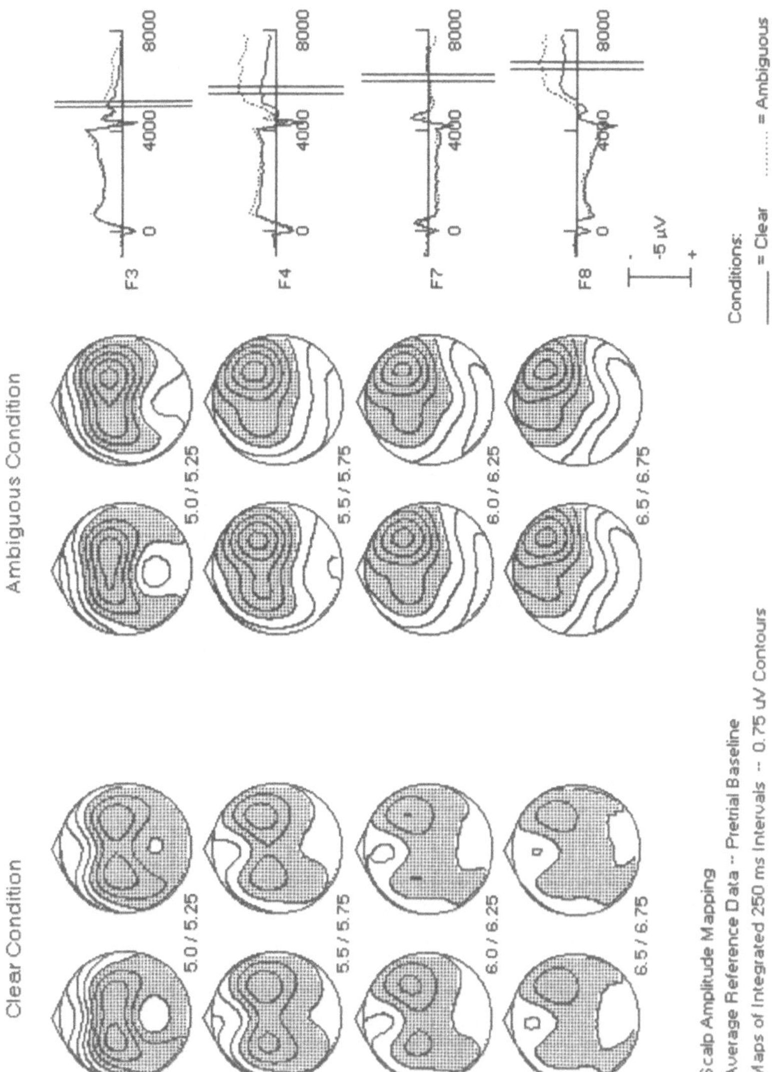

Figure 8.11.2 Grand Averages: High-Scoring Subjects

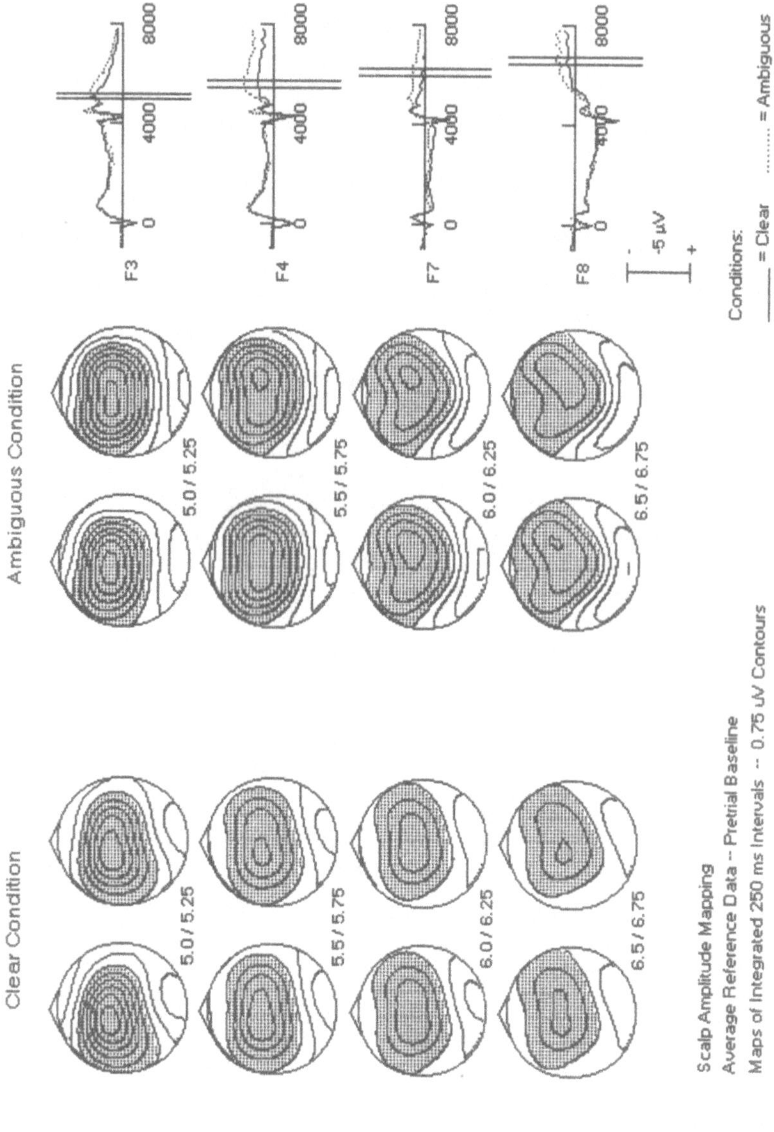

Figure 8.11.2a presents the scalp amplitude maps of the group of low-scoring subjects from experiment D.

The initial map of the ***clear condition*** shows the PINV as a flat negative coronal band over precentral regions, with a slight left-hemisphere maximum and a stronger left than right hemisphere extension into temporo-parietal regions. The second map (5.25 s) reveals the development of a right hemisphere pre-central hill which is slightly more anteriorly located than the left hemisphere hill. Both maxima decrease with the third map. While during the further course the left hemisphere maximum decreases, the maximum over the right hemisphere remains stable for 1 to 2 maps and decreases towards the end of the interval.

Under the ***ambiguous condition*** the right precentral predominance of the PINV becomes apparent with the initial map (at 5.0 s) and remains stable for the rest of the interval. Only its extensions into left-hemisphere regions start to decrease with the third map (5.5 s). An amplitude augmentation from the clear to the ambiguous condition is most apparent in right-frontal regions.

Figure 8.11.2b presents the scalp amplitude maps of the high-scoring group of experiment D. The initial map of the ***clear condition*** shows that the PINV is predominant over the left precentral region, extending anteriorly and posteriorly over the left hemisphere, but less into right fronto-lateral regions. This topographical maximum remains stationary for the rest of the interval.

Under the ***ambiguous condition*** the PINV maximum is situated over mid-sagittal central to precentral regions until 5.5 s (maps 1-3), extending more into left-sided than right-sided areas. A left frontal extension of the PINV with moderate amplitudes is obvious throughout the interval. At 5.75 s a right-hemisphere precentral maximum develops. Towards the end of the interval and during a general amplitude decrease this area broadens. The succession of maps using 0.75 µV isocontour lines reveal the amplitude enlargement with ambiguous compared to clear IS only with the third map. The ambiguity effect is weaker in high-scoring than in low-scoring subjects, but favors the right-sided areas likewise.

In sum: the maps give a vivid impression of the spatial aspects of the PINV development in both groups, which is not conveyed by the grand average ERP, let alone by montages using fewer electrodes. The two groups exhibit a different topography of the PINV, which is predominant over the right hemisphere, particularly for the ambiguous condition, in the low-scoring group, and broader distributed, extending into left fronto-lateral areas, in the high-scoring group.

8.11.3 Cortex Amplitude Maps

Figure 8.11.3a Grand Average: Low-Scoring Subjects

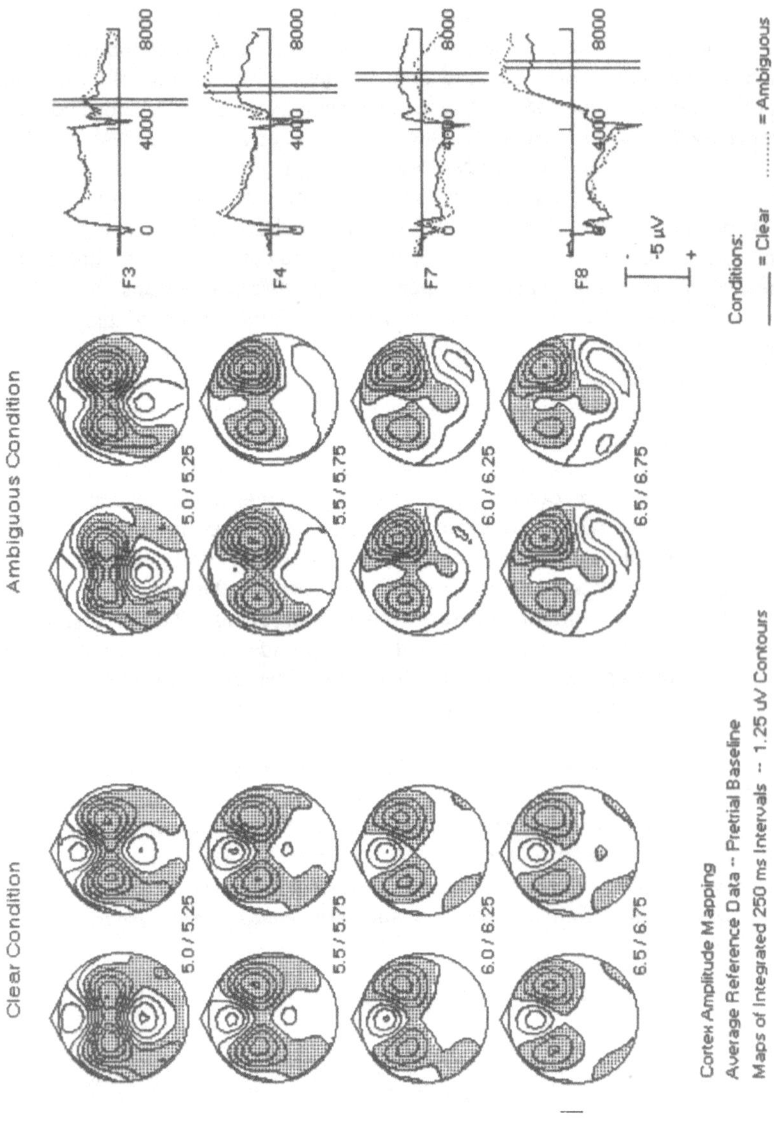

Figure 8.11.3b Grand Average: High-Scoring Subjects

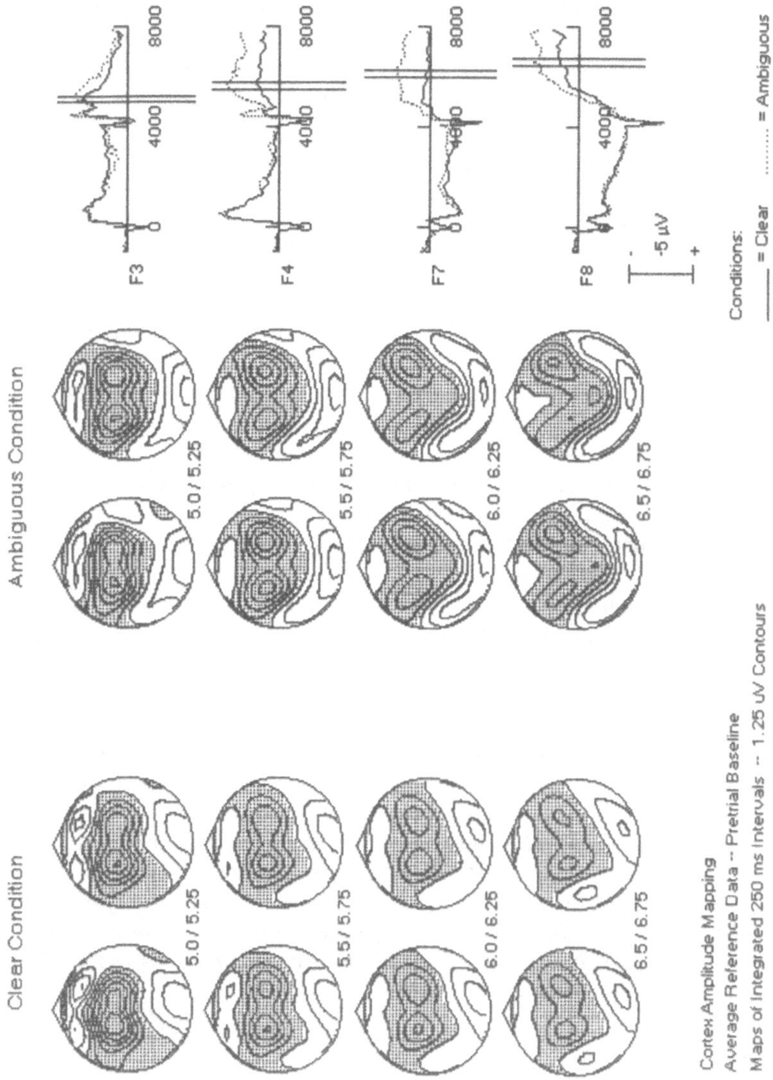

Figure 8.11.3a presents the cortex amplitude maps of the low-scoring group from experiment D. The initial map of the ***clear condition*** reveals two pre-central PINV maxima near the sagittal midline, which are not separated and extend over both hemispheres into fronto-lateral regions. Additional low-amplitude hills can be seen over the left posterior-temporal cortex and over right-temporal regions, separated by a positive trough, the IS-LPC. Already with map 2 the two pre-central hills exhibit slightly decreased amplitudes. With map 2 the mid-sagittal posterior trough is decreased. With map 3 a midline anterior-frontal positivity begins to arise. For the rest of the interval the right-frontal PINV maximum remains rather stable, while the left-frontal hill starts to decrease with map 3.

Under the ***ambiguous condition*** compared to the clear condition PINV amplitudes are enlarged. This holds especially for the right-frontal maximum, which has peak amplitudes from map 2 to map 7. The additional left-frontal negative area is again less pronounced and less augmented by ambiguous compared to clear IS than the right-frontal area. As was the case for the clear condition, a small frontal mid-sagittal trough can be seen during the whole interval, however, with smaller amplitudes when compared to the clear condition.

Figure 8.11.3b shows the cortex amplitude maps of the high-scoring group of experiment D. The initial map of the ***clear condition*** reveals the PINV as a narrow precentral negative band, with maximum amplitudes over the left-hemisphere. A second negative hill becomes apparent over the right precentral area with map 3. Until 6.0 s (map 5) this left-precentral PINV predominance remains present.

Under the ***ambiguous condition***, the initial map shows a bilateral precentral negative area with peak amplitudes over the left hemisphere, extending into left fronto-lateral regions. During the following 750 ms (maps 2 to 4) two precentral hills of similar amplitudes arise. Furthermore, a small anterior-frontal trough can be seen during the whole period. At 6.0 s (map 5) the negative area extends into mid-sagittal parietal regions.

In sum: The cortex amplitude data confirm the different spatio-temporal patterns of PINV regulation in low-scoring and high-scoring subjects. With the cortex amplitude maps it becomes apparent that the broadly distributed PINV, with its rapidly increasing preponderance over the right-frontal area in low-scoring subjects and its less asymmetric topography in high-scoring subjects, is based on two separate centers of peak negativity in both groups. Among these two local maxima the right-frontal is clearly more pronounced than the left precentral only in the low-scoring group. The higher spatial resolution of the cortical maps is capable of revealing local maxima that remain undetected by scalp amplitude maps.

8.11.4 Scalp Laplacian Maps

Figure 8.11.4a Grand Averages: Low-Scoring Subjects

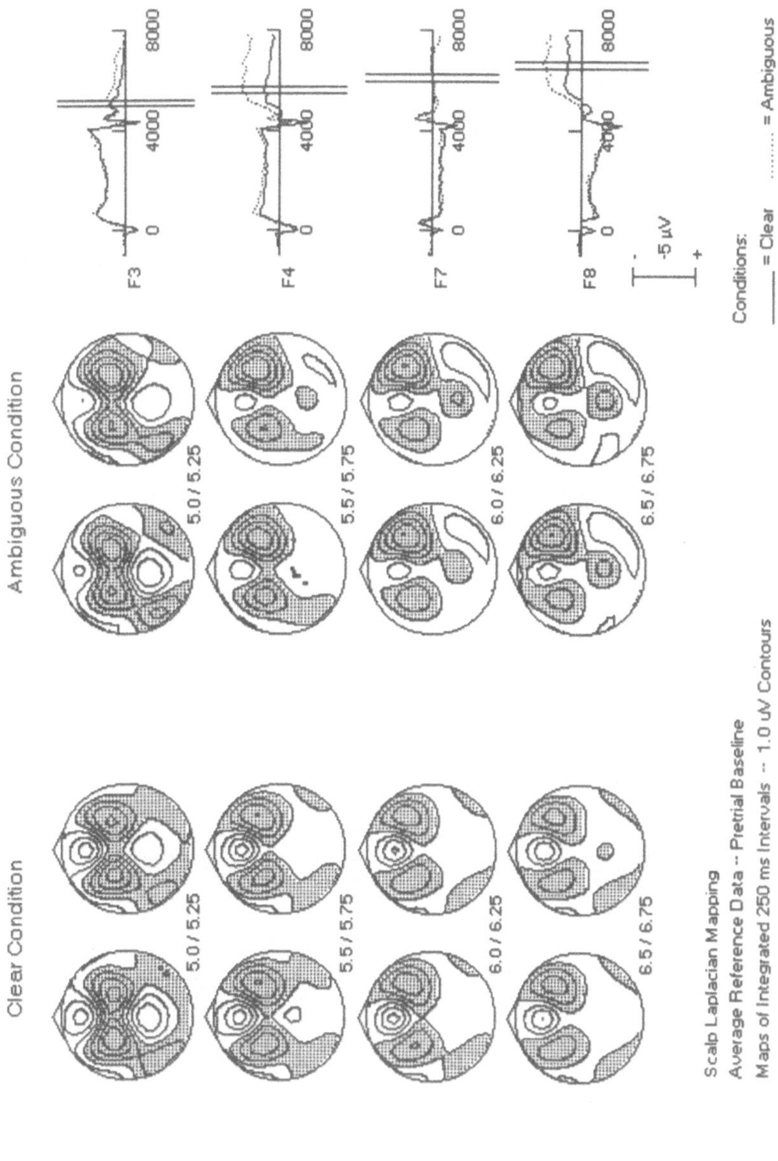

Figure 8.11.4b Grand Averages: High-Scoring Subjects

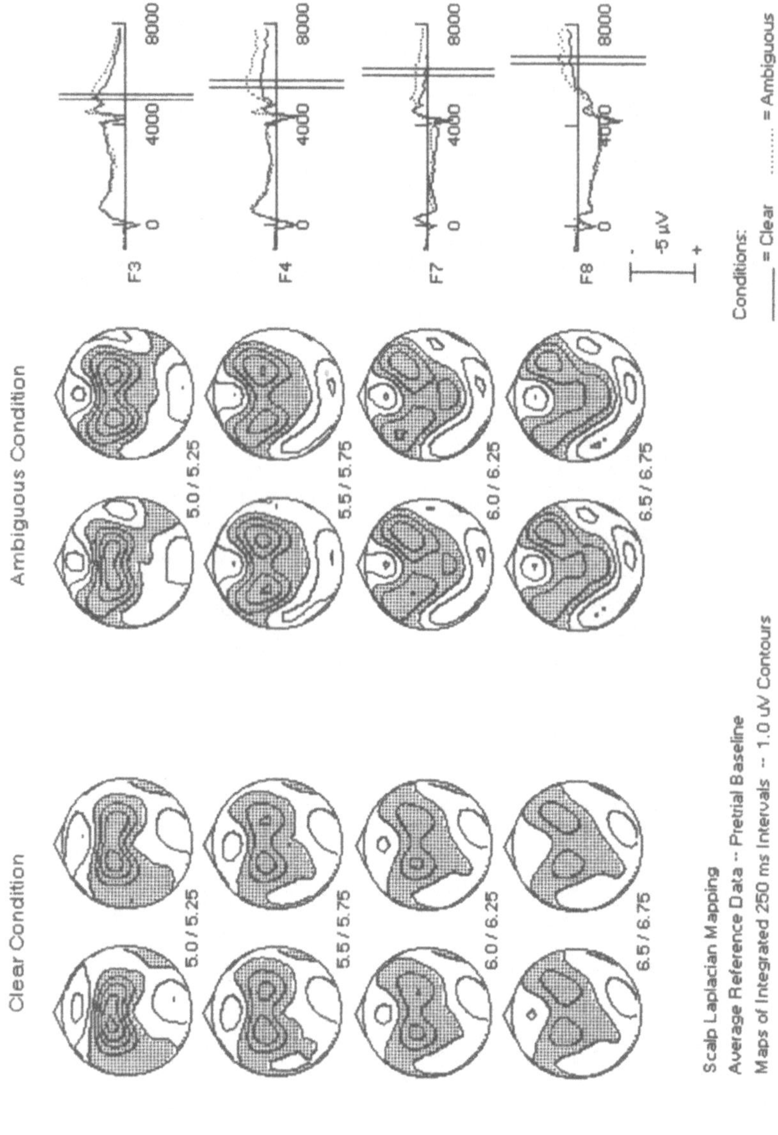

Figures 8.11.4a and **8.11.4b** present the scalp laplacian maps of the low-scoring and high-scoring subjects of experiment D. These maps are very similar to the cortical maps presented in the preceding paragraph. For this reason they will not be described further.

In sum: Projecting to the cortical surface and applying the Laplacian to scalp amplitude data reveal similar topographies of the PINV. Both are superior to scalp amplitude maps in that both provide a better spatial resolution, separating local maxima that are "smeared" in the scalp amplitude data. Hence, a more detailed description of the spatio-temporal regulation of the PINV as well as a more detailed description of differences between schizophrenic patients and healthy controls, respectively low- and high-scoring subjects, is enabled by exploiting these "high-resolution" EEG techniques. The importance of this result convergence can be expected on a theoretical basis and is discussed by Nunez & Westdorp (1994).

8.12 Contingent Negative Variation and Post-Imperative Negative Variation

In this paragraph topographies of iCNV, tCNV, and PINV shall be compared. For the sake of conciseness only scalp amplitude maps will be presented. Cortex amplitude and scalp laplacian maps can be found in **appendix B**. After this, PINV will be discussed as topographical change with respect to the tCNV. For that purpose, scalp laplacian maps will be presented. Scalp and cortex amplitude maps of the tCNV-referenced PINV can also be found in **appendix B.**

8.12.1 Comparative Topography of iCNV, tCNV, and PINV in Schizophrenic Patients and Healthy Controls

Figure 8.12.1a reveals that the iCNV has a topographical maximum over precentral mid-sagittal areas in healthy controls, extending into left temporo-parietal regions under both conditions. Compared to the iCNV, the tCNV maximum is slightly posteriorly shifted and extends with a broader topography into parietal regions. The topography of the PINV, which has in detail been described before, overlaps with the tCNV topography especially over central, pre-central, and left temporo-parietal regions.

Figure 8.12.1b shows the iCNV, tCNV, and PINV topography in schizophrenic patients. Lower iCNV amplitudes in patients when compared to controls can be seen. The iCNV of the patients extends into left temporo-parietal regions. The topographical maximum of the tCNV is more posteriorly located than that of the iCNV, but, in contrast to controls, clearly shifted towards the left central region. In patients, the tCNV extends also into the parietal area. The PINV overlaps topographically with the tCNV over precentral-to-parietal areas.

Figure 8.12.1a Topography of iCNV, tCNV, and PINV in Healthy Controls

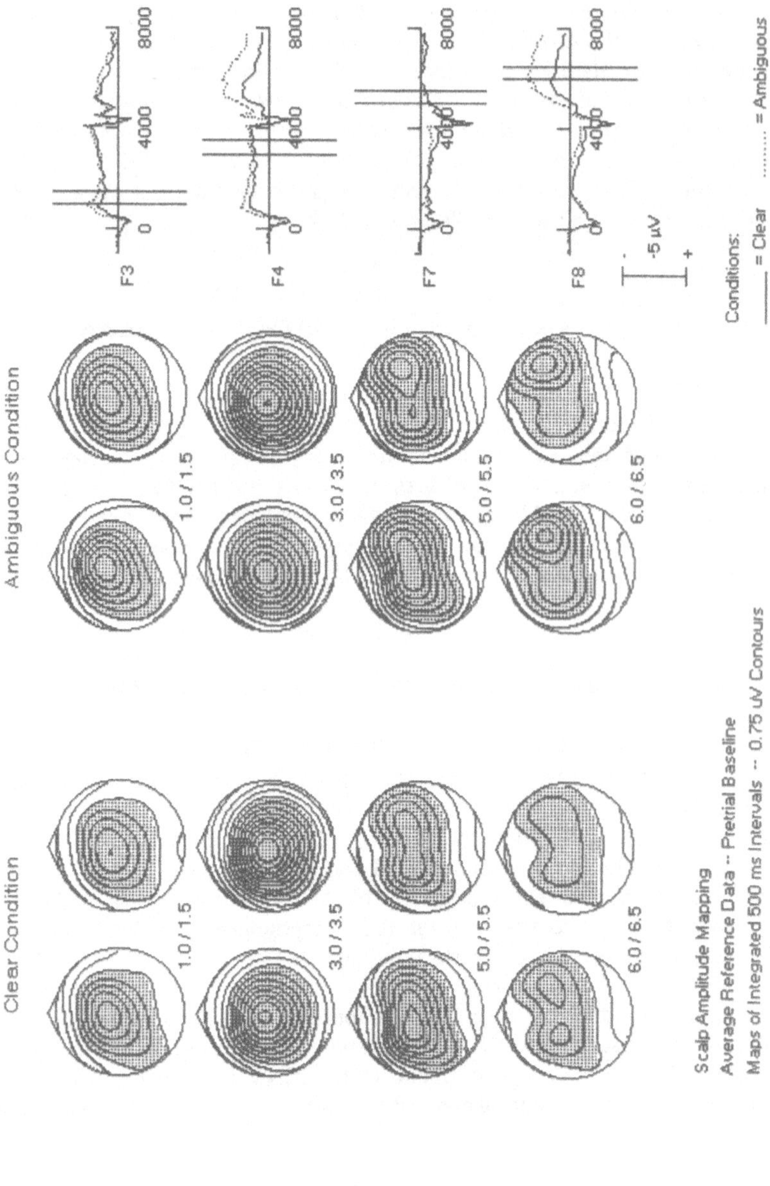

Figure 8.12.1b Topography of iCNV, tCNV, and PINV in Schizophrenic Patients

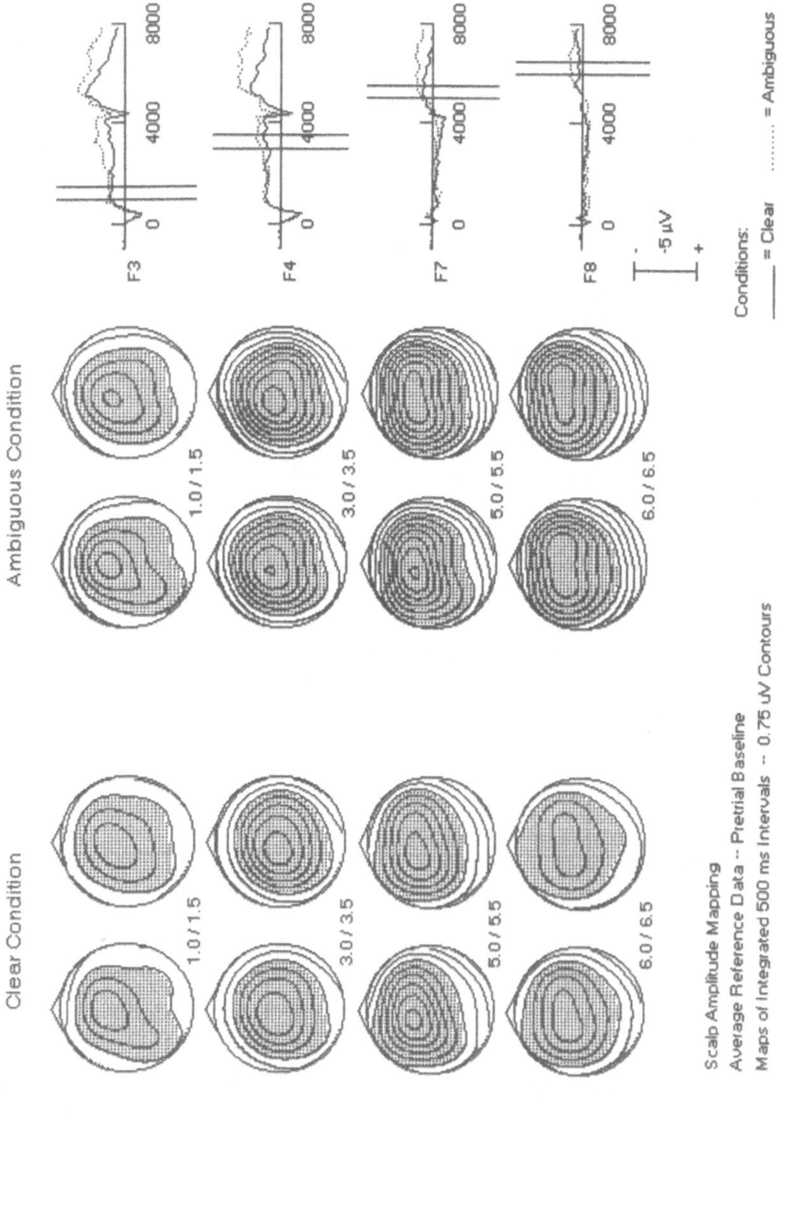

8.12.2 Topographical Changes After IS Presentation—PINV with the tCNV Baseline[201]

With the description of the topographical overlap between tCNV and PINV of the preceding paragraph the question arises, what topographical *change* takes place between tCNV and PINV. **Figures 8.10.1a/b** revealed that at least two *different* PINV wave forms can be distinguished. This distinction between amplitude decrease versus increase after the tCNV should become especially apparent when using the tCNV as baseline for the PINV.

Figure 8.12.2a shows the tCNV-referenced PINV of the healthy control group from experiment C. Except for initial left posterior-temporal negative shifts which can be seen in the first 3 maps of the clear condition, topographical changes from tCNV to PINV occur mainly over both hemispheres in fronto-lateral regions, especially over the right hemisphere. During the first 2 to 4 maps (depending on the experimental condition) the two fronto-lateral hills are separated by a large trough which is most pronounced over the vertex and extends into anterior areas. After 1.5 s (condition clear) or 1.75 s (condition ambiguous), an anterior-frontal negative band starts to close the "gap" between the two fronto-lateral hills.

Figure 8.12.2b presents the cortex amplitude maps of the tCNV-referenced PINV of the patient group. These maps show that in patients, as well, the broadest and largest increases in negativity following the IS can be seen bilaterally over fronto-lateral regions. The most apparent differences between patients and controls are over frontal regions. Firstly, the fronto-lateral maxima have lower amplitudes than those of the control group. Secondly, under the clear condition the fronto-lateral maxima of the patient group are not separated by the positive trough throughout the interval, while the negative band "connecting" the fronto-lateral negative areas develops in controls only with third map (5.5 s). Thirdly, under the ambiguous condition the fronto-lateral negative maxima are initially connected in patients and are separated during the second half of the interval, while in controls the reverse development takes place (at first separated, then connected). Finally, the right fronto-lateral negative maximum is smaller in patients than in controls, while the left fronto-lateral negative maximum of the patients is as large as or larger than the corresponding maximum in controls, especially under the ambiguous condition.

---

[201] The topographical changes occuring between tCNV and PINV described for schizophrenic patients and healthy controls have been corroborated by secondary statistical analyses which will be published elsewhere (Klein et al., submitted).

Figure 8.12.2a Topographical Changes from tCNV to PINV in Healthy Controls

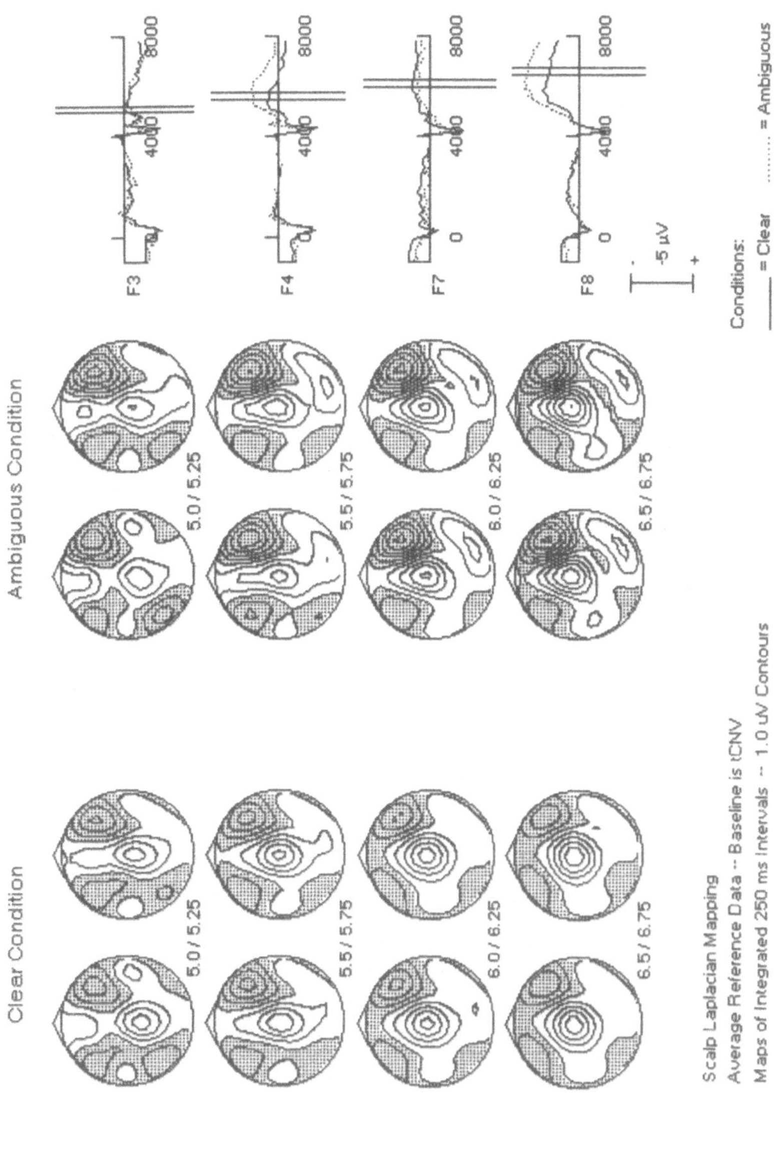

Figure 8.12.2b Topographical Changes from tCNV to PINV in Schizophrenic Patients

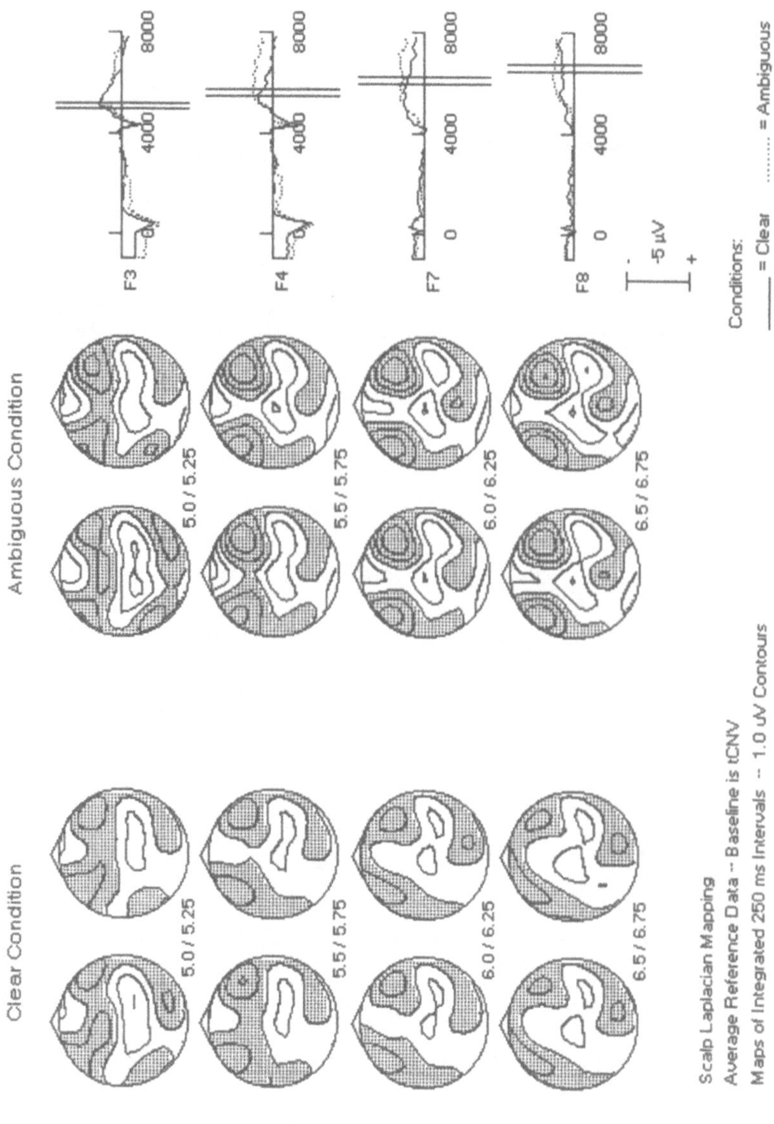

In sum: Referencing the PINV to the tCNV results in a more restricted PINV topography. While the central and precentral mid-sagittal near proportions "disappear", the fronto-lateral proportions are emphasized. Referring to the grand average curves presented in ***figures 8.10.1 a,b****, it is interesting to note that different "slopes" (or wave forms) could be observed at different leads. Negative slopes (decreasing amplitudes) were observed at central and precentral leads, while positive slopes (increasing amplitudes) were observed at fronto-lateral positions. This might suggest that topographically distinguishable proportions of the PINV show a different temporal course: Decreasing at central leads, increasing at fronto-lateral leads. The spatio-temporal development of the pretrial baseline-referenced PINV (cortex amplitude maps) corresponds to this interpretation, since in these maps the fronto-lateral PINV maxima evolve with a delay. Referencing the PINV to the tCNV seems to bring out this distinction more clearly.*

Figure 8.12.2c shows the "cortical" PINV topography with the tCNV baseline in the low-scoring group of experiment D. Under both conditions, two fronto-lateral PINV maxima, separated initially by a sagittal-midline trough are obvious. The right fronto-lateral maximum is clearly more pronounced than the left fronto-lateral for the entire interval. Under both experimental conditions bilateral posterior-temporal extensions of the PINV can be seen initially. Stable troughs can be seen over the vertex extending into mid-sagittal frontal and parietal regions.

A similar PINV pattern can be observed in the high-scoring subjects of experiment D, as illustrated in **figure 8.12.2d**. Two fronto-lateral PINV maxima are visible; between them the right-sided is larger than the left-sided. However, the amplitude difference between the two is smaller in the high-scoring compared to the low-scoring group.

Figure 8.12.2c Topographical Changes from tCNV to PINV in Low-Scoring Subjects

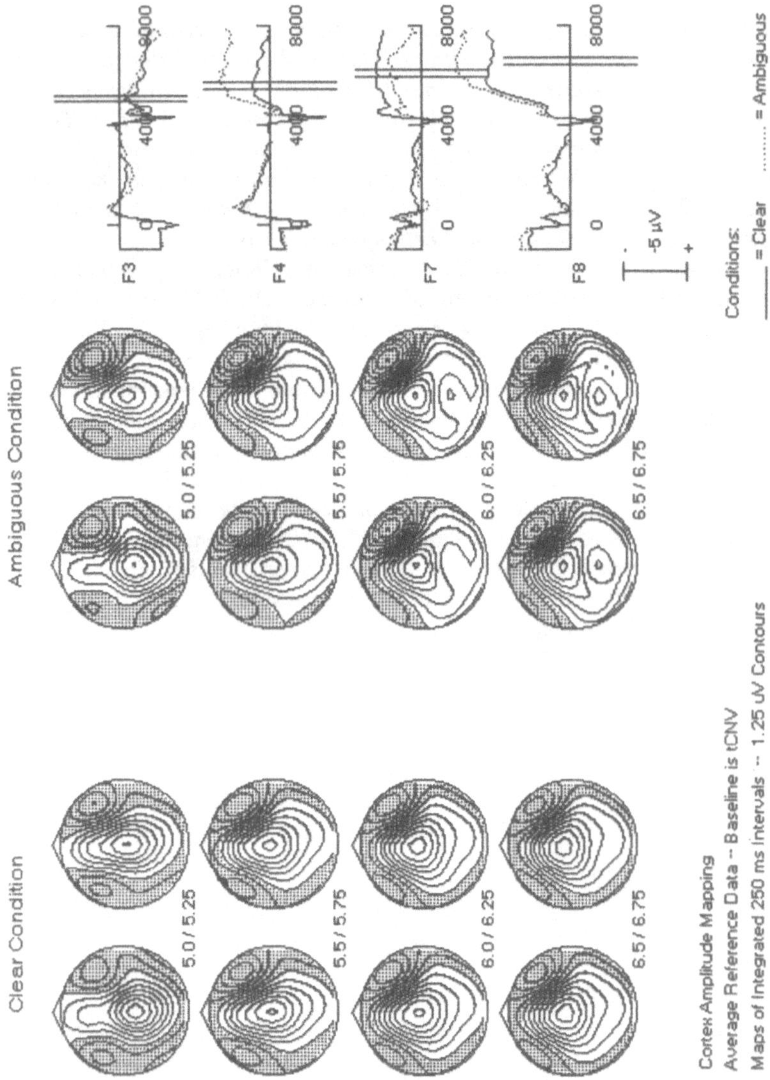

Figure 8.12.2d Topographical Changes from tCNV to PINV in High-Scoring Subjects

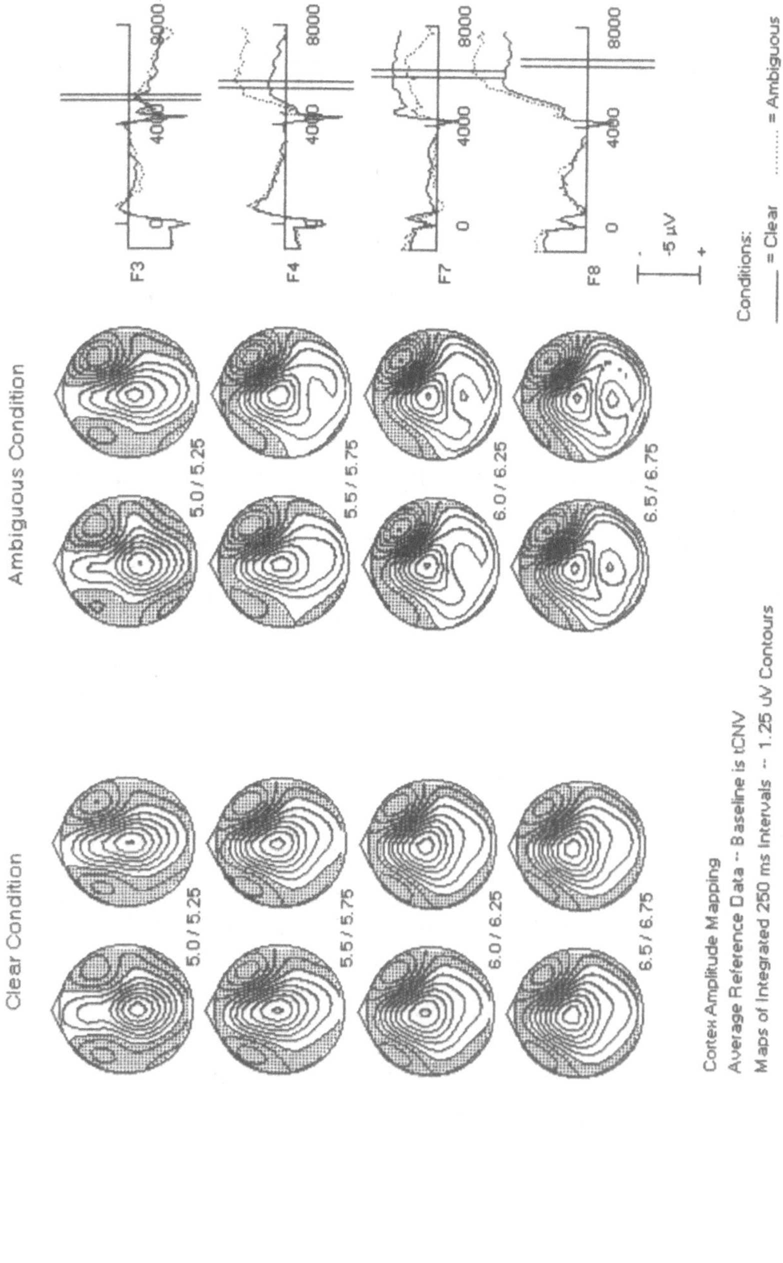

This is due to larger PINV amplitudes over left-frontal regions and smaller amplitudes over the right frontal region, as compared to the low-scoring group.

In sum: Referencing the PINV to the tCNV simplifies the PINV topography in that the negativity increases during the PINV interval are restricted to fronto-lateral areas over both hemispheres. Group differences in the PINV topography between low- and high-scoring subjects remain visible in the tCNV-referenced maps. The right fronto-lateral PINV is less pronounced in high-scoring than in low-scoring subjects, while the reverse holds for the left fronto-lateral PINV.

8.13 Control of Eye Movement Artifacts

The PINV maps presented in the preceding paragraphs revealed a fronto-central bilateral maximum of this component. Since eye movement artifacts are more distorting at anterior than at posterior leads, such topographies call for a discussion of such possible artifacts. For several reasons, an eye movement source of the PINV appears implausible, if not impossible. Firstly, subjects were continually instructed not to move their eyes during the recording periods and this was monitored throughout the sessions. Secondly, the Multiple Source Eye Correction (MSEC) method by Berg & Scherg, (1994) not only provides a correction of eye movements and blink activity, it is in addition capable of detecting systematic eye movements in the averages that may remain undetected by visual inspection (correction of vertical, horizonatal, and blink activity; identification of the movement type). Thirdly, following the application of this method, each trial was visually checked for the goodness of the eye movement correction and questionable trials were excluded from further analyses. During these final trial-by-trial checks, trials containing detectable *uncorrected* vertical or horizontal eye movements or blinks were excluded, while trials containing eye movements which were corrected by the MSEC remained in the sample. The eye channels (not: EOG channels) in **Figures 8.10.1 a,b** and **8.11.1 a,b** reveal blinks and presumably blink-related minor vertical eye movements following the IS, but no horizontal eye movements. This means that PINV asymmetries cannot be explained by eye movements. While some *symmetrical* bias of the EEG topography is possible, it can be expected to be less than after conventional eye activity correction (Berg & Scherg, 1994). The release of blink activity (blink disinhibition) after the termination of an epoch of allocation of visual attention is a known phenomemon (e.g., Drew, 1951) and can hardly be suppressed completely during an experimental session. However, blinks should have been corrected in the present data.

It is possible to assess the effects of *uncorrected* vertical eye movement and blink activity (which are very similar) by simulating the topographical distribution of such eye movement activities. In case of uncorrected eye movements in the ERP data, the "contaminated" ERP topography should be compatible with the eye movement topography. The following simulation of blink activity was done with the BESA program (Scherg, 1990).

As can be seen from **figure 8.13.1** blink activity induces a positive peak between the two fronto-polar positions Fp1 and Fp2, which spreads laterally and posteriorly, decreasing in amplitude. For vertical eye movements directed upwards a similar pattern can be expected, while downward eye movements would induce the same distribution, but of reverse sign (negative). This pattern is incompatible with the *bilateral fronto-lateral* distribution of the PINV.

At most the anterior-frontal positive troughs with slightly varying lateral locations (depending on groups and experimental conditions) which could be seen in the maps are topographically compatible with residual *un*corrected blink-related eye movement activity. *If* this were true, the mid-sagittal oriented flanks of the PINV areas should have been somewhat dampened.

Figure 8.13.1 Simulation of Blink Activity

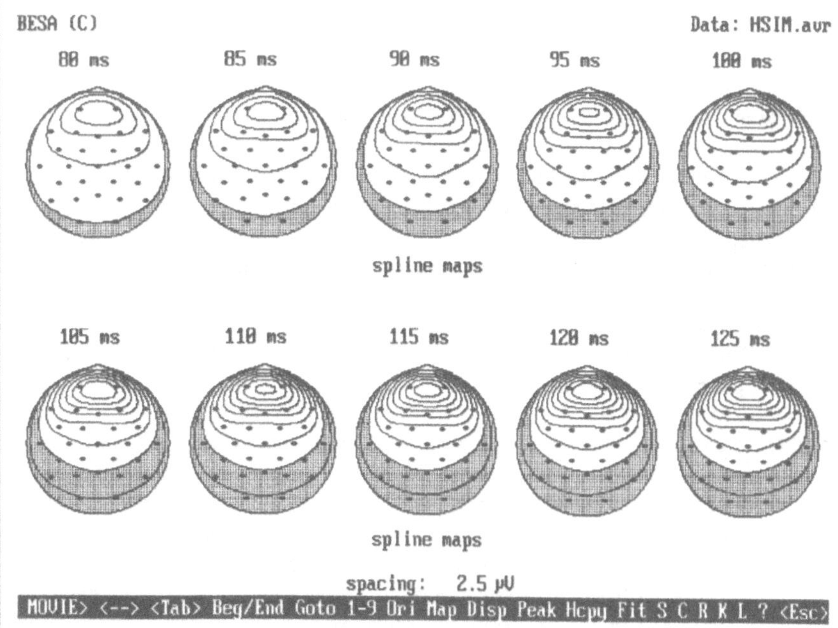

8.14 Statistical Analysis of the Map Data

In order to substantiate the results described so far statistically, mean amplitude values of the *cortical* data[202], using the average mastoid reference ($A_1+A_2/2$), for each subject, condition, and channel were determined by the program AREANOTX for one tCNV, and 4 PINV intervals. Mean amplitudes for the interval from 3.5 to 4.0 s following WS onset served as tCNV parameter. Four PINV measures (PINV-1 to PINV-4) were defined as adjacent, non-overlapping mean amplitude values of 500 ms segments for the intervals 1.0-1.5 s (PINV-1) to 2.5-3.0 s (PINV-4) after the IS onset. Twelve leads were selected for statistical analyses. The selection of leads was based primarily on the inspection of the cortical maps. The maps revealed a predominance of the PINV and of the experimental effects on PINV over fronto-central areas, extending far into lateral regions. The ERP parameters of these leads (see **figure 8.2**) were statistically analyzed with the GLM procedure of SAS. *Univariate* analyses are reported (see *paragraph 4.7.4*). Within a mixed-model repeated measures ANOVA, the between-subject factor "GROUP" (patients versus controls) and the within-subject factors "CONDITION" (referring to the IS types clear and ambiguous), "CORONAL ROW" (distinguishing frontal (F9, F7, F3, F4, F8, F10) and temporo-central (FT9, T3, C3, C4, T4, FT10) leads), "HEMISPHERE" (distinguishing left (F3, F7, F9, C3, T3, FT9) and the right (F4, F8, F10, C4, T4, FT10) leads), and "SAGITTAL ROW" (distinguishing the 3 sagittal rows over each hemisphere, i.e. (F3,C3; F4, C4), (F7, T3; F8, T4), and (F9, FT9; F10, FT10)) were analyzed. The 4 PINV segments were analyzed separately due to the low ratio of subjects to dependent variables. The statistical design is identical for experiments C *and* D. Following the ANOVA planned contrasts were accomplished in order to specify the origin of significant effects including the sagittal row factor (the only factor with 3 levels). In these cases the sagittal row levels "middle" (F7, T3; F8, C4) and "outer" (F9, FT9; F10, FT10) are contrasted with the level "inner" (F3, C3; F4, C4). Non-significant contrasts will not be reported or described. Results of the range normalization of the data following McCarthy & Wood (1985) will also be reported. A significant condition by topography effect which *remains* significant after the normalization of the data suggests the rejection of the hypothesis that a simple amplitude effect can account for this interaction effect, resulting in the assumption of a change in the generator structure of the ERP component. Such differences between generators may, however, result from amplitude differences of 2 generators that may be the same in two (or more) groups or conditions. As a consequence, the interpretation of the results of normalized data should be read with reservations (see *paragraph 4.7.4*).

---

[202] The CSD data were analyzed statistically in the same way as the cortical data. Almost identical results were obtained by these analyses, which will not be reported here for that reason (see Klein et al., submitted).

8.14.1 Schizophrenic Patients Versus Healthy Controls

For all PINV segments, PINV amplitudes differ between the sagittal rows, resulting in a significant **Sagittal Row** main effect. This main effect is due to significant contrasts between the middle and the inner sagittal row[203] and the outer and the inner sagittal row[204].

For all PINV segments amplitudes are largest along the inner and smallest along the outer sagittal row.

Table 8.14.1.1 ANOVA Effects on PINV in Experiment C

		PINV-1		PINV-2		PINV-3		PINV-4	
Effect	df1, df2	F	p	F	p	F	p	F	p
C	1,21	7.67	.01	19.05	.0003	18.68	.0003	18.20	.0003
C * G	1,21	1.89	.18	2.27	.15	1.64	.21	0.06	.82
Cr	1,21	2.72	.11	8.30	.009	17.83	.0004	21.37	.0001
Cr * G	1,21	1.55	.23	0.44	.51	0.00	.99	0.14	.771
H	1,21	0.09	.77	2.38	.14	4.53	.05	6.13	.02
H * G	1,21	4.26	[A].05[1.1]	9.02	[G].007[1.2]	10.68	[G].004[1.3]	11.78	[G].003[14]
Sr	2,42	29.91	.0001	31.11	.0001	23.82	.0001	16.64	.0001
Sr * G	2,42	0.28	.74	0.21	.81	0.72	.49	0.84	.44
C * Cr	1,21	5.13	[A].03[2.1]	7.40	[G].01[2.2]	15.41	[G].0008[2.3]	14.64	[G].001[2.4]
C * Cr * G	1,21	1.08	.31	0.00	.99	0.11	.75	0.01	.92
C * H	1,21	0.67	.42	0.95	.34	1.55	.23	0.87	.36
C * H * G	1,21	2.22	.15	3.05	.10	3.49	.08	4.43	[A].05[3.4]
Cr * H	1,21	6.57	.02	7.16	.01	7.86	.01	7.50	.01
Cr * H * G	1,21	2.06	.17	0.63	.44	0.66	.42	0.42	.52
C * Sr	2,42	1.10	.34	2.83	.09	2.65	.08	1.13	.33
C * Sr * G	2,42	2.55	.10	2.30	.12	0.78	.47	1.55	.22
Cr * Sr	2,42	3.31	.05	5.51	.01	7.00	.005	6.50	.007
Cr * Sr * G	2,42	1.73	.19	0.69	.50	0.35	.66	0.22	.76
H * Sr	2,42	1.14	.32	1.53	.23	1.56	.22	1.88	.17
H * Sr * G	2,42	0.76	.44	1.06	.35	2.53	.09	4.59	[G].02[4.4]
C * Cr * H	1,21	0.67	.42	1.60	.22	2.25	.15	3.31	.08

203 PINV-1: F(1,21)=17.74, p=.0004; PINV-2: F(1,21)=23.76, p=.0001; PINV-3: F(1,21)= 17.56, p=.0004; PINV-4: F(1,21)=15.10, p=.0009

204 PINV-1: F(1,21)= 41.23, p=.0001; PINV-2: F(1,21)= 50.33, p=.0001; PINV-3: F(1,21)= 41.37, p=.0001; PINV-4: F(1,21)= 28.45, p=.0001

Table 8.14.1.1 continued

		PINV-1		PINV-2		PINV-3		PINV-4	
Effect	**df1, df2**	**F**	**p**	**F**	**p**	**F**	**p**	**F**	**p**
C * Cr * Sr	2,42	0.52	.60	1.61	.21	4.52	[A].02[5.3]	6.49	[G].004[5.4]
C * Cr * Sr * G	2,42	0.60	.55	0.06	.94	0.05	.95	0.77	.47
C * H * Sr	2,42	1.87	.17	1.35	.27	0.75	.48	0.69	.51
C * H * Sr * G	2,42	1.41	.26	0.42	.66	0.32	.72	0.62	.54
Cr * H * Sr	2,42	2.80	.07	3.01	.06	3.10	.06	2.08	.14
Cr * H * Sr * G	2,42	0.40	.67	0.17	.84	0.02	.98	0.54	.59
C * Cr * H * Sr	2,42	0.18	.80	0.08	.91	0.07	.92	0.01	.98
C * Cr * H * Sr * G	2,42	0.47	.60	0.62	.53	0.82	.44	0.89	.41
Group	1,21	0.80	.38	1.00	.33	1.01	.33	1.53	.23

C*=Condition factor,* ***Cr****=Coronal row factor,* ***H****=Hemisphere factor,* ***Sr****=Sagittal row factor,* ***G****=Group factor, univariate ANOVA, Huynh-Feldt corrected p-values;* [G(G)]=Generator effect, [A(A)]=Amplitude effect indicated by results of normalization; [1.1]=F(1,21)=3.18; p=.09; [1.2]=F(2,21)=8.35, p=.009; [1.3]=F(1,21)=9.10, p=.007; [1.4]=F(1,21)=10.03, p=.005; [2.1]=F(1,21)= 2.19; p=.15; [2.2]=F(1,21)=4.98, p=.04; [2.3]=F(1,21)=13.46, p=.001; [2.4]=F(1,21)=12.65, p=.002; [3.4]=F(1,21)=3.07, p=.09; [4.4]=F(2,42)=3.68, p=.03; [5.3]=F(2,42)=2.93, p=.06; [5.4]=F(2,42)=5.07, p=.01

Except for PINV-1, the PINV is more pronounced along the frontal than along the temporo-central coronal row, as the significant **Coronal Row** effect indicates. A further differentiation of the Coronal Row and the Sagittal row main effects is indicated by the significant **Coronal Row by Sagittal Row** interaction. For all 4 PINV segments this interaction rises from a significant difference in the inner versus middle sagittal row concerning the difference between the frontal and the temporo-central coronal row[205]. The difference between frontal and temporo-central amplitudes is larger along the middle (F7, T3; F8, T4) than along the inner (F3, C3; F4, C4) sagittal row (see **table 8.14.1.2**).

Due to the significant interaction of **Coronal Row by Hemisphere** another differentiation of the Coronal Row effect is indicated. For PINV-2 to PINV-4 the frontal (versus temporo-central) preponderance is larger over the right than over the left hemisphere. For PINV-1 left-frontal amplitudes are slightly smaller than the left-central, while the right-frontal amplitudes are larger than the right-central (see **table 8.14.1.3**).

[205] PINV-1: F(1,21)=5.53, p=.03; PINV-2: F(1,21)=10.50, p=.004; PINV-3: F(1,21)= 14.28, p=.001; PINV-4: F(2,21)= 12.71, p=.002

Table 8.14.1.2 Coronal Row by Sagittal Row Interaction on PINV

	Outer	Middle	Inner
PINV-1			
Frontal	0.11 (3.5)	-3.28 (4.6)	-5.55 (4.3)
Central	0.10 (2.8)	-1.53 (4.6)	-5.52 (5.3)
PINV-2			
Frontal	-0.33 (3.1)	-4.21 (4.2)	-5.73 (3.7)
Central	-0.35 (2.3)	-2.09 (4.3)	-5.51 (4.1)
PINV-3			
Frontal	-0.37 (2.5)	-3.92 (3.9)	-4.99 (3.6)
Central	-0.27 (2.0)	-1.67 (4.1)	-4.50 (3.6)
PINV-4			
Frontal	-0.22 (2.5)	-3.46 (3.6)	-4.21 (3.4)
Central	-0.20 (1.9)	-1.31 (3.9)	-3.61 (3.3)

unit is µV; SD in parentheses

Table 8.14.1.3 Coronal Row by Hemisphere Effect on PINV

	Frontal		Central	
	Left	Right	Left	Right
PINV-2	-2.82 (4.1)	-4.24 (3.1)	-2.76 (3.2)	-2.54 (2.9)
PINV-3	-2.45 (3.8)	-3.93 (2.7)	-2.21 (2.9)	-2.09 (2.6)
PINV-4	-2.04 (3.5)	-3.44 (2.5)	-1.72 (2.8)	-1.70 (2.2)

unit is µV; SD in parentheses

There is *no* "overall" group difference in PINV amplitudes. While the **Hemisphere** main effect is significant only for the 3rd and 4th PINV segment (pointing to larger right than left hemisphere amplitudes), the interaction of **Hemisphere by Group** is significant throughout the series of PINV measures. Descriptively, in control subjects there is a stable PINV predominance at the right-sided locations; in contrast, patients exhibit a consistent left-hemisphere maximum which becomes gradually smaller during the course of the series. This descriptive hemisphere difference within the patient group, however, does not become statistically significant in the patient within-group ANOVA ($Fs \leq 1.38$). In contrast, the within-group ANOVA of the control group produce significant hemisphere effects for almost all PINV segments[206]. This difference is documented in **table 8.14.1.4**. The presence of different generator structures in the two groups may explain this interaction, as suggested by the significant Hemisphere by Group interaction with the normalized ERP data (for PINV-2 to PINV-4).

[206] PINV-1: $F(1,10)=3.36$, $p=.10$; PINV-2: $F(1,10)=11.39$, $p=.007$; PINV-3: $F(1,10)=15.69$, $p=.003$; PINV-4: $F(1,10)=18.02$, $p=.002$

Table 8.14.1.4 Hemisphere by Group Interaction on PINV

	Controls		Patients	
	Left	Right	Left	Right
PINV-1	-1.31 (3.1)	-2.62 (2.9)	-3.76 (4.5)	-2.78 (4.1)
PINV-2	-1.47 (2.6)	-3.40 (2.4)	-4.02 (3.9)	-3.38 (3.3)
PINV-3	-1.15 (2.4)	-2.99 (2.0)	-3.41 (3.6)	-3.02 (3.0)
PINV-4	-0.68 (2.3)	-2.43 (1.7)	-2.97 (3.3)	-2.69 (2.6)

unit is µV; SD in parentheses

Ambiguous compared to clear IS increase PINV amplitudes, which is documented by the significant **Condition** main effect. It is interesting to note the augmentation of F-values of the Condition effect from the 1st to the 2nd PINV segment, suggesting that the maximum impact of ambiguous IS evolves about 1.5 s after IS onset. The PINV enlargement with ambiguous IS is larger along the frontal than along the temporo-parietal coronal row, giving rise to the significant **Condition by Coronal Row** interaction. This effect is documented in **table 8.14.1.5**. Except for PINV-1, the effect is compatible with a change in the generator structure of the PINV from the clear to the ambiguous condition, since this interaction effect is also significant for the normalized data.

Table 8.14.1.5 Condition by Coronal Row Interaction on PINV

	Frontal	Central
PINV-1		
Clear	-2.26 (3.8)	-1.98 (3.7)
Ambiguous	-3.69 (3.9)	-2.65 (3.6)
PINV-2		
Clear	-2.53 (3.2)	-2.11 (2.9)
Ambiguous	-4.52 (3.6)	-3.20 (3.1)
PINV-3		
Clear	-1.99 (2.8)	-1.55 (2.5)
Ambiguous	-4.39 (3.6)	-2.74 (3.0)
PINV-4		
Clear	-1.46 (2.7)	-1.13 (2.3)
Ambiguous	-4.01 (3.3)	-2.28 (2.8)

unit is µV; SD in parentheses

At the end of the interval the lateralization of the Condition effect begins to differ between groups, as the significant **Condition by Hemisphere by Group** interaction indicates. In patients the PINV augmentation is larger over the left than over the right hemisphere, in controls this effect is larger over the right than over the left hemisphere (see **table 8.14.1.6**). While the interaction reaches conventional significance levels only with the 4th PINV segment, trends are present already for PINV-2 and PINV-3. An amplitude effect is suggested by the non-significant interaction in the normalized ERP data.

Table 8.14.1.6 Condition by Hemisphere by Group Interaction on PINV-4

	Patients		Controls	
	Left	Right	Left	Right
PINV-4				
Clear	-1.97 (2.9)	-1.94 (2.8)	-0.03 (2.3)	-1.13 (1.3)
Ambiguous	-3.98 (4.0)	-3.44 (2.7)	-1.33 (3.0)	-3.74 (2.7)

unit is µV; SD in parentheses

A final differentiation of the topography of the Condition effect rises from the significant **Condition by Coronal Row by Sagittal Row** interaction. The frontal versus temporo-central difference in the ambiguity effect differs between the inner ((F3,F4,C3,C4)) and the middle ((F7,F8,T3,T4)) sagittal row for both PINV-3 and PINV-4 [207] and between the inner and the outer ((F9, FT9, F10, FT10)) sagittal row for PINV-4 only [208].

Table 8.14.1.7 Condition by Coronal Row by Sagittal Row Interaction on PINV-3 and PINV-4

	Outer	Middle	Inner
PINV-3			
Frontal			
Clear	0.02 (2.8)	-2.52 (3.7)	-3.48 (3.2)
Ambiguous	-1.35 (2.6)	-5.33 (4.7)	-6.49 (4.6)
Central			
Clear	0.23 (2.4)	-1.21 (3.9)	-3.68 (3.3)
Ambiguous	-0.77 (2.1)	-2.14 (4.6)	-5.32 (4.1)
PINV-4			
Frontal			
Clear	0.25 (2.8)	-2.00 (3.6)	-2.63 (2.7)
Ambiguous	-1.33 (2.4)	-4.93 (4.3)	-5.79 (4.6)
Central			
Clear	0.44 (2.4)	-0.93 (3.8)	-2.92 (3.0)
Ambiguous	-0.85 (1.9)	-1.69 (4.4)	-4.31 (4.1)

unit is µV; SD in parentheses

[207] PINV-3: ($F(1,21)=12.11$, $p=.002$; PINV-4: ($F(1,21)=15.48$, $p=.0008$

[208] PINV-4: ($F(1,21)=6.18$, $p=.02$

While the augmentation of the PINV amplitudes due to ambiguous IS is similar at the *frontal* middle ((F7,F8) and inner ((F3,F4)) sagittal leads, along the temporo-central coronal row the ambiguity effect is clearly more pronounced at the inner ((C3,C4)) compared to the middle ((T3,T4))sagittal leads. For PINV-4, the preponderance of the ambiguity at inner compared to outer sagittal leads is more pronounced along the frontal compared to the temporo-central coronal row. These effects can be followed in **table 8.14.1.7**. Results on normalized data are somewhat ambiguous as to the quality of these interaction effects. While a generator effect is proposed for the 4th PINV segment, an amplitude effect is suggested for PINV-3, however, merely since the effect falls just short of significance (p=.06). Referring to the interpretation of the Condition by Coronal Row interaction, a change in the generator structures appears plausible.

To sum: The main results of the statistical analyses of the cortex amplitude data can be summarized as follows. Within the subset of selected frontal and temporo-central leads, the PINV is predominant at frontal (F7, F3, F4, F8) and central (C3, C4) positions. The overall PINV amplitudes do not distinguish schizophrenic patients and healthy controls. However, larger right hemisphere than left hemisphere PINV amplitudes in the control group contrast significantly with a more "balanced", slightly left-hemisphere predominant PINV in the patient group. Maximum group differences as well as the right-frontal PINV maximum in controls evolve about 1.5 s after IS onset (segment 2). The results of the normalized ERP data propose a generator structure interpretation of this effect. According to this interpretation, different bio-electrical source configurations are present during PINV generation in patients and controls. Ambiguous IS augment the PINV in both groups by comparable amounts and most at frontal compared to temporo-central leads. Following the results of normalized data, this effect rises from a change of the underlying source configuration from the clear to the ambiguous condition.

Altogether, these results underline that meaningful statistical analyses using ANOVA procedures can be accomplished for multi-channel recordings, if a selection of topographically "relevant" channels is derived from the map data.

8.14.2 Low-Scoring Versus High-Scoring Subjects[209]

For the 3rd and 4th PINV segments the **Coronal Row** effect is significant, since frontal PINV amplitudes are significantly larger than temporo-central. This difference is descriptively present for PINV-2, but does not reach statistical significance (see **table 8.14.2.2**). Larger temporo-central than frontal amplitudes are present for PINV-1, but only descriptively. Furthermore, the

[209] Since the high-scoring group contains 2 left-handed subjects, all ANOVA were computed with these subjects excluded. The results are almost identical with those presented in **table 8.14.2.1** and are documented in **Appendix A.4**.

significant **Sagittal Row** main effect reveals significant amplitude differences between the inner (F3, F4, C3, C4) and the middle (F7, F8, T3, T4) sagittal row[210] and between the outer and the inner sagittal row[211].
The PINV amplitudes of the inner sagittal rows (F3, C3; F4, C4) are larger than those of the middle (F7, T3; F8, T4) and outer (F9, FT9; F10, FT10) sagittal rows (see **table 8.14.2.2**). A further differentiation of the PINV topography is indicated by the significant **Coronal Row by Sagittal Row** interaction.

For all PINV segments this interaction is due to significant contrasts between the frontal and the temporo-central coronal row concerning the inner versus middle sagittal row differences[212], with increasing F-values along the succession of PINV segments. While for PINV-1 and PINV-2 central amplitudes are larger than frontal at the inner sagittal leads (C3, C4; F3, F4), frontal amplitudes are larger than central at the middle sagittal leads (F7, F8; T3, T4). For PINV-3 and PINV-4 the frontal amplitudes are larger than the temporo-central, but this difference is most pronounced at the mid-sagittal leads (F7, F8; T3, T4; see **table 8.14.2.2**).

Table 8.14.2.1 ANOVA Effects on PINV in Experiment D

		PINV -1		PINV -2		PINV -3		PINV -4	
Effect	**df1, df2**	**F**	**p**	**F**	**p**	**F**	**p**	**F**	**p**
C	1,35	4.45	.04	11.79	.002	14.33	.0006	9.92	.003
C * G	1,35	0.17	.68	0.00	.99	0.25	.60	0.32	.57
Cr	1,35	0.55	.46	2.38	.13	12.55	.001	21.51	.0001
Cr * G	1,35	0.01	.94	0.62	.44	1.39	.25	1.84	.18
H	1,35	0.58	.26	4.82	.03	8.94	.005	13.21	.0009
H * G	1,35	8.98	[G].005[1.1]	9.62	[G].004[1.2]	7.46	[G].01[1.3]	4.47	[G].04[1.4]
Sr	2,70	42.42	.0001	32.15	.0001	16.22	.0001	9.58	.0002
Sr * G	2,70	1.39	.26	1.74	.18	1.51	.22	1.48	.24
C * Cr	1,35	7.37	[G].01[2.1]	6.37	[G].02[2.2]	12.67	[G].001[2.3]	13.43	[G].001[2.4]
C * Cr * G	1,35	0.58	1.13	0.89	.35	1.78	.19	0.12	.73
C * H	1,35	1.02	.32	2.09	.16	2.69	.11	4.35	[A].04[3.4]
C * H * G	1,35	5.77	[A].02[4.1]	6.71	[A].01[4.2]	5.92	[A].02[4.3]	4.36	[A].04[4.4]
Cr * H	1,35	0.43	.51	2.16	.15	3.77	.06	5.20	.03
Cr * H * G	1,35	0.34	.56	0.62	.64	0.97	.33	0.70	.41
C * Sr	2,70	0.87	.39	1.87	.17	0.15	.85	0.68	.50
C * Sr * G	2,70	0.28	.36	1.38	.26	0.89	.41	1.54	.22

[210] PINV-1: F(1,35)=14.72, p=.0005; PINV-2: F(1,35)= 18.23, p=.0001; PINV-3: F(1,35)= 8.65, p=.006; PINV-4: F(1,35)=5.66, p=.02

[211] PINV-1: F(1,35)=66.45, p=.0001; PINV-2: F(1,35)= 48.76, p=.0001; PINV-3: F(1,35)= 26.80, p=.0001, PINV-4: 15.71, p=.0003

[212] PINV-1: F(1,35)= F(1,35)=5.09, p=.03; PINV-2: F(1,35)=14.92, p=.0005; PINV-3: F(1,35)=25.56, p=.0001; PINV-4: F(1,35)= 27.64, p=.0001

Table 8.14.2.1 continued

		PINV -1		PINV -2		PINV -3		PINV -4	
Effect	**df1, df2**	**F**	**p**	**F**	**p**	**F**	**p**	**F**	**p**
Cr * Sr * G	2,70	0.20	.78	0.35	.64	0.01	.39	1.63	.21
H * Sr * G	2,70	0.62	.54	1.10	.33	0.97	.38	0.57	.55
C * Cr * H	1,35	0.66	.42	1.26	.27	3.79	.06	5.79	$^{A}.02^{5.4}$
C * Cr * H * G	1,35	0.05	.83	1.49	.23	2.09	.16	2.35	.13
C * Cr * Sr	2,70	5.32	$^{G}.007^{6.1}$	4.20	$^{A}.02^{6.2}$	5.54	$^{A}007^{63}$	5.88	$^{A}.006^{6.4}$
C * Cr * Sr * G	2,70	3.18	$^{A}.05^{7.1}$	3.54	$^{A}.04^{7.2}$	3.32	$^{A}.05^{7.3}$	4.46	$^{A}.02^{7.4}$
C * H * Sr	2,70	3.38	$^{G}.03^{8.1}$	3.32	$^{G}.04^{8.2}$	2.28	.11	0.36	.70
C * H * Sr * G	2,70	0.52	.39	0.37	.69	0.36	.70	0.23	.79
Cr * H * Sr	2,70	0.29	.74	0.49	.61	1.01	.37	1.58	.21
Cr * H * Sr * G	2,70	1.03	0.32	1.80	.17	1.92	.15	1.54	.22
C * Cr * H * Sr	2,70	0.12	.83	0.24	.75	0.93	.36	1.17	.31
C *Cr*H*Sr* G	2,70	0.72	.46	1.62	.21	1.36	.26	1.06	.34
Group	1,35	0.05	.83	0.08	.78	0.22	.64	0.50	.48

Table legend: ***C****=Condition factor,* ***CR****=Coronal Row factor,* ***H****=Hemisphere factor,* ***SR****=Sagittal Row factor,* ***G****=Group factor, univariate ANOVA, Huynh-Feldt corrected p-values;* $^{G(G)}$=Generator effect, $^{A(A)}$=Amplitude effect as indicated by results of normalization; $^{1.1}$=F(1,35)=12.64,p=.001; $^{1.2}$=F(1,35)=13.48,p=.008; $^{1.3}$=F(1,35)=10.15,p=.003; $^{1.4}$=F(1, 35)=6.14,p=.02; $^{2.1}$=F(1,35)=6.42,p=.02; $^{2.2}$=F(1,35)=5.93,p=.02; $^{2.3}$=F(1,35)=8.28,p=.07; $^{2.4}$=F(1 ,35)=6.16,p=.02; $^{3.4}$=F(1,35)=1.71,p=.20; $^{4.1}$=F(1,35)=1.96,p=.17; $^{4.2}$=F(1,35)=3.52,p=.07; $^{4.3}$=F (1,35)=3.24,p=.08; $^{4.4}$=F(1,35)=2.78,p=.10; $^{5.4}$=F(1,35)=3.60,p=.06; $^{6.1}$=F(2,70)=4.82,p=.01; $^{6.2}$= F(2,70)=2.95,p=.07; $^{6.3}$=F(2,70)=3.03,p=.06; $^{6.4}$=F(2,70)=2.18,p=.13; $^{7.1}$=F(2,70)=1.01,p=.37; $^{7.2}$ =F(2,70)=1.95,p=.15; $^{7.3}$=F(2,70)=1.85,p=.17; $^{7.4}$=F(2,70)=2.48,p=.10; $^{8.1}$=F(2,70)=3.22,p=.05; $^{8.2}$=F(2,70)=3.73,p=.03

Table 8.14.2.2 Coronal Row by Sagittal Row Interaction on PINV

	Outer	Middle	Inner
PINV-1			
Frontal	1.51 (4.6)	-1.00 (5.3)	-2.95 (4.4)
Central	1.01 (3.4)	-0.17 (3.6)	-4.14 (3.8)
PINV-2			
Frontal	0.45 (4.2)	-2.48 (4.7)	-3.26 (3.4)
Central	0.28 (3.1)	-0.61 (3.5)	-3.68 (3.4)
PINV-3			
Frontal	-0.10 (3.6)	-2.66 (4.1)	-2.74 (3.1)
Central	-0.10 (2.8)	-0.10 (3.5)	-2.59 (3.1)
PINV-4			
Frontal	-0.21 (3.3)	-2.61 (3.5)	-2.45 (2.5)
Central	-0.22 (2.4)	-0.22 (3.4)	-1.84 (3.1)

unit is µV; SD in parentheses

Table 8.14.2.3 Condition by Coronal Row Interaction on PINV

	Frontal		Central	
	Clear	Ambiguous	Clear	Ambiguous
PINV-1	-0.18 (3.9)	-1.39 (5.0)	-0.97 (2.9)	-1.19 (2.9)
PINV-2	-0.98 (3.4)	-2.62 (4.4)	-0.97 (2.7)	-1.70 (2.8)
PINV-3	-0.87 (3.1)	-2.87 (3.9)	-0.54 (2.7)	-1.32 (2.8)
PINV-4	-0.87 (2.7)	-2.82 (3.3)	-0.28 (2.4)	-1.00 (2.6)

unit is µV; SD in parentheses

Ambiguous IS, compared to clear IS, augment the PINV amplitudes, as indicated by the **Condition** main effect, which is significant for all 4 PINV segments. This effect is significantly stronger at frontal than at temporo-central locations, as the **Condition by Coronal Row** interaction indicates (see **table 8.14.2.3**).

The significant Condition by Coronal Row interaction of the normalized data, suggests that a change in the generator structure of the ERP takes place from the clear to the ambiguous condition.

Table 8.14.2.4 Condition by Coronal Row by Sagittal Row Interaction on PINV-1 and PINV-2

	Outer	Middle	Inner
PINV-1			
Frontal			
Clear	1.94 (4.2)	-0.46 (4.8)	-2.01 (4.3)
Ambiguous	1.22 (5.1)	-1.53 (6.3)	-3.88 (5.1)
Central			
Clear	1.42 (3.4)	-0.31 (3.7)	-4.03 (4.4)
Ambiguous	0.72 (3.7)	-0.03 (3.9)	-4.25 (4.0)
PINV-2			
Frontal			
Clear	0.83 (3.9)	-1.72 (4.2)	-2.04 (3.5)
Ambiguous	-0.13 (4.8)	-3.24 (5.8)	-4.48 (4.0)
Central			
Clear	0.76 (3.1)	-0.51 (3.6)	-3.17 (3.9)
Ambiguous	-0.21 (3.6)	-0.72 (3.9)	-4.18 (3.6)

units is µV; SD in parentheses; only PINV-1 and PINV-2 are documented, since PINV-3 and PINV-4 enter in the Condition by Coronal Row by Sagittal Row by Group interaction (see below)

A further differentiation of the topography of the Ambiguity effect is revealed by the significant interaction of **Condition by Coronal Row by Sagittal**

Row. This interaction originates from significant contrasts between the middle and the inner sagittal row[213] and the outer and the inner sagittal row[214] concerning the differences in the ambiguity effect between frontal and temporo-central coronal row. While the PINV augmentation subsequent to ambiguous IS (compared to clear IS) is generally larger at frontal than at temporo-central positions, this difference is most pronounced at the inner sagittal row (F3,F4; C3,C4) compared to the middle sagittal row (F7, F8; T3, T4) and larger than at the outer sagittal row (F9, F10; FT9, FT10). Only for the 1st PINV segment an generator structure effect can be assumed following the results of normalized data.

However, the last-mentioned triple interaction must be differentiated according to the groups, as underlined by the significant interaction of **Condition by Coronal Row by Sagittal Row by Group**. While there is no significant contrast for the first two PINV segments, the significance of the 3rd [215] and 4th [216] segments rises from group differences in the ambiguity *effect* for the comparison of the inner and outer sagittal rows along the frontal versus temporo-central coronal rows. In both groups PINV is augmented from the clear to the ambiguous condition at the *inner* sagittal leads of the frontal (F3, F4) and temporo-central (C3,C4) coronal row. In both groups this effect is larger at the frontal as compared to the central leads. The last-mentioned difference is, however, more pronounced in low-scoring compared to high-scoring subjects. In other words, the frontal predominance of the ambiguity effect is larger in low-scoring than in high-scoring subjects. This is the first part of the interaction. At frontal (F9, F10) and temporo-central (FT9, FT10) leads of the *outer* sagittal row there is an augmentation of the PINV from the clear to the ambiguous condition as well. In low-scoring subjects this augmentation is larger at the temporo-central (FT9, FT10) when compared to the frontal (F9, F10) leads, while the reverse holds for the high-scoring group (larger ambiguity effect at frontal compared to temporo-central leads). In other words, along the outer sagittal row the ambiguity effect is most pronounced temporo-centrally in low-scoring subjects, but most pronounced frontally in high-scoring subjects. This is the second part of this complex interaction effect. Altogether, this interaction rises from an “inner-frontal“ and “outer-temporo-central“ predominance of the ambiguity effect in low-scoring subjects, as compared to a less pronounced “inner-frontal“ and “outer-frontal“ predominance of the ambiguity effect in high-scoring subjects. Amplitude rather than generator structure effects are suggested by the results of normalized data.

---

[213] PINV-1: F(1,35)=6.88, p=.01; PINV-2: F(1,35)=6.38, p=.02; PINV-3: F(1,35)=9.63, p=.004; PINV-4: F(1,35)=12.91, p=.001

[214] PINV-1: F(1,35)=11.35, p=.002; PINV-2: F(1,35)=8.73, p=.006; PINV-3: F(1,35)=11.12, p=.002; PINV-4: F(1,35)=8.75, p=.006

[215] F(1,35)=4.32, p=.05

[216] F(1,35)=7.94, p=.008

Table 8.14.2.5 Interaction of Condition by Coronal Row by Sagittal Row by Group on PINV-3 and PINV-4

		Outer	Middle	Inner
PINV-3				
Low-Scoring Group	Frontal / Clear	0.58 (4.1)	-2.63 (3.8)	-1.78 (4.3)
	Frontal / Ambiguous	-1.09 (4.9)	-4.05 (5.5)	-4.56 (3.5)
	Central / Clear	1.17 (3.2)	-0.41 (3.9)	-1.92 (4.7)
	Central / Ambiguous	-0.97 (3.7)	-1.10 (4.5)	-2.68 (3.8)
High-Scoring Group	Frontal / Clear	0.45 (2.6)	-0.58 (3.9)	-1.34 (2.3)
	Frontal / Ambiguous	-0.84 (3.4)	-3.45 (4.8)	-3.32 (3.5)
	Central / Clear	-0.04 (2.0)	0.52 (3.0)	-2.55 (2.8)
	Central / Ambiguous	-0.55 (3.0)	-0.53 (4.2)	-3.18 (2.9)
PINV-4				
Low-Scoring Group	Frontal / Clear	0.50 (4.0)	-2.49 (3.8)	-1.71 (3.7)
	Frontal / Ambiguous	-1.68 (4.7)	-4.07 (4.7)	-4.35 (2.7)
	Central / Clear	1.21 (3.2)	-0.35 (3.7)	-1.52 (4.2)
	Central / Ambiguous	-1.38 (3.2)	-0.64 (4.7)	-1.71 (3.3)
High-Scoring Group	Frontal / Clear	0.27 (2.1)	-0.71 (3.2)	-1.17 (1.8)
	Frontal / Ambiguous	-1.01 (3.1)	-3.24 (4.3)	-2.64 (3.3)
	Central / Clear	-0.04 (1.8)	0.89 (2.9)	-1.89 (2.4)
	Central / Ambiguous	-0.65 (2.7)	0.93 (3.9)	-2.61 (2.8)

unit is µV; SD in parentheses

Table 8.14.2.6 Condition by Hemisphere by Sagittal Row Interaction on PINV-1 and PINV-2

	Outer	Middle	Inner
PINV-1			
Left Hemisphere / Clear	1.78 (3.9)	-0.68 (4.6)	-3.53 (4.4)
Left Hemisphere / Ambiguous	0.80 (4.8)	-0.95 (4.6)	-4.00 (4.7)
Right Hemisphere / Clear	1.58 (4.0)	-0.10 (4.8)	-2.51 (4.1)
Right Hemisphere / Ambiguous	1.16 (4.8)	-0.61 (5.6)	-4.14 (4.2)
PINV-2			
Left Hemisphere / Clear	1.01 (3.5)	-0.67 (4.0)	-2.62 (3.9)
Left Hemisphere / Ambiguous	-0.07 (4.5)	-1.33 (4.1)	-3.73 (3.8)
Right Hemisphere / Clear	0.49 (3.8)	-1.55 (4.3)	-2.60 (3.7)
Right Hemisphere / Ambiguous	-0.27 (4.6)	-2.63 (5.3)	-4.93 (3.7)

unit is µV; SD in parentheses

The significant **Condition by Hemisphere by Sagittal Row** interaction of the PINV-1 and PINV-2 measure is due to significant differences in the lateralization of the ambiguity effect between the outer and the inner sagittal rows[217]. While the PINV augmentation from condition clear to condition

[217] PINV-1: $F(1,35)= 7.62$, $p=.009$; PINV-2: $F(1,35)= 6.78$, $p=.01$

ambiguous is larger at the outer (F9, FT9) than at the inner (F3, C3) sagittal row over the left hemisphere, this effect is larger at the inner (F4, C4) than the outer (F10, FT10) positions over the right hemisphere (see **table 8.14.2.6**). According to the analyses of normalized data, a change in the PINV source configuration takes place from the clear to the ambiguous condition.

The **Hemisphere** main effect, which is significant for the last two PINV measures and falls just short of significance for PINV-2, indicates overall larger right than left hemisphere amplitudes. While there is no overall amplitude difference between the two groups, the **Hemisphere by Group** interaction is significant for all analyzed PINV-segments. In low-scoring subjects PINV is larger over the right than over the left hemisphere. For high-scoring subjects the reverse holds, except for PINV-4. However, the amount of the left-right difference decreases from PINV-1 to PINV-3, reversing its sign with PINV-4 in the high-scoring group. The Hemisphere by Group interactions are illustrated in **table 8.14.2.7**. The analyses of the normalized ERP data suggest that different source configurations are present in the two groups. The group differences in the lateralization of the successive PINV segments are further supported by the within-group ANOVA. In the low-scoring group the Hemisphere effect is *non-significant* for the 1st PINV segment only, but significant for the segments PINV-2 to PINV-4 [218]; in the high-scoring group the Hemisphere effect is significant only for the 1st PINV segment, but *non*-significant for the segments PINV-2 to PINV-4 [219].

Table 8.14.2.7 Hemisphere by Group Interaction on PINV

	Left		**Right**	
	Low-Scoring Group	High-Scoring Group	Low-Scoring Group	High-Scoring Group
PINV-1	-0.37 (3.4)	-1.78 (3.3)	-1.24 (4.0)	-0.32 (3.7)
PINV-2	-0.84 (3.5)	-1.58 (2.5)	-2.59 (3.9)	-1.28 (3.1)
PINV-3	-0.76 (3.4)	-1.16 (2.1)	-2.48 (3.6)	-1.23 (2.3)
PINV-4	-0.68 (2.8)	-0.77 (1.8)	-2.35 (3.0)	-1.21 (1.9)

unit is µV; SD in parentheses

The group differences in PINV lateralization are even larger under the ambiguous than under the clear condition, since the amplitude augmentation due to ambiguous IS is larger over the left hemisphere in the high-scoring, but larger over the right hemisphere in low-scoring subjects. These differential processes

---

[218] PINV-1: F(1,17)=2.07, p=.17; PINV-2: F(1,17)=12.30, p=.003; PINV-3: F(1,17)=17.06; p=.0007; PINV-4: F(1,17)=19.59, p=.0004

[219] PINV-1: F(1,18)=8.71, p=.009; PINV-2: F(1,18)=0.47, p=.50; PINV-3: F(1,18)=0.03, p=.86; PINV-4: F(1,18)=1.02, p=.33

are reflected by the significant **Condition by Hemisphere by Group** interaction and documented in **table 8.14.2.8.** An amplitude effect is suggested by the results of normalized data for this interaction. The within-group ANOVA of the Condition by Hemisphere interaction resemble the within-group ANOVA of the Hemisphere effect. Low-scoring subjects exhibit a consistent asymmetric pattern of the ambiguity effect[220], with larger PINV augmentations over the right than over the left hemisphere. In contrast, high-scoring subjects show no significant hemisphere asymmetry of the ambiguity effect[221]. It should be mentioned, however, that in high-scoring subjects descriptively the ambiguity effect is larger over the left than over the right hemisphere (see **table 8.14.2.8**)

Table 8.14.2.8 Condition by Hemisphere by Group Interaction on PINV

	Low-Scoring Group		High-Scoring Group	
	Left	Right	Left	Right
PINV-1				
Clear	-0.34 (3.3)	-0.70 (3.9)	-1.25 (3.3)	0.00 (3.6)
Ambiguous	-0.40 (4.1)	-1.79 (4.4)	-2.32 (3.6)	-0.64 (4.0)
PINV-2				
Clear	-0.55 (3.5)	-1.70 (3.9)	-0.90 (2.5)	-0.77 (2.9)
Ambiguous	-1.13 (4.0)	-3.48 (4.3)	-2.26 (2.8)	-1.79 (3.5)
PINV-3				
Clear	-0.30 (3.7)	-1.37 (3.7)	-0.49 (2.0)	-0.69 (2.2)
Ambiguous	-1.23 (3.8)	-3.59 (4.0)	-1.83 (2.5)	-1.78 (2.9)
PINV-4				
Clear	-0.29 (3.4)	-1.16 (3.3)	-0.22 (1.6)	-0.66 (1.7)
Ambiguous	-1.08 (3.3)	-3.54 (3.3)	-1.32 (2.5)	-1.76 (2.8)

unit is µV; SD in parentheses

To sum: Among the present selection of frontal, temporal, and central channels, the PINV is predominant at frontal (F7, F3, F4, F8) and central (C3, C4) leads. The PINV augmentation due to ambiguous compared to clear IS is largest at frontal leads, and this effect is compatible with a change in the underlying source configuration of the PINV. Among the frontal leads, the inner sagittal positions (F3, F4) "benefit" most from the ambiguity effect, especially in the low-scoring group. The ambiguity effect differs between the inner and the outer sagittal leads of the hemispheres for the first and the second PINV segment. It is larger at the inner leads of the right hemisphere, but larger at the

[220] F(1,17)=4.06, p=:06; PINV-2: F(1,17)=5.24, p=.04; PINV-3: F(1,17)=4.95, p=.04; PINV-4: F(1,17)=5.20, p=.04

[221] PINV-1: F(1,18)=1.60, p=.22; PINV-2: F(1,18)=1.31, p=.27; PINV-3: F(1,18)=0.80, p=.38; PINV-4: F(1,18)=0.00, p=.99

outer leads of the left hemisphere. While the overall PINV amplitudes do not differ between groups, a different spatio-temporal pattern of the PINV becomes apparent with the succession of Hemisphere by Group interactions. Since the interactions remain significant after data normalization, different source configurations can be assumed to underlie the PINV in low- respectively high-scoring subjects. While in low-scoring subjects a hemisphere asymmetry of the PINV (with larger right hemisphere amplitudes) is confirmed for the PINV segments 2 to 4, in high-scoring subjects a left-hemisphere asymmetry is confirmed only for PINV-1; for the successive PINV segments the descriptive left-hemisphere preponderance of the PINV does not reach statistical significance. Group differences in the lateralization of the PINV are more pronounced under the ambiguous condition than under the clear condition, since ambiguous IS augment the PINV of low-scoring subject more over the right than over the left hemisphere, with no significant lateralization of this effect in high-scoring subjects.

8.15 General Discussion

The discussion of the results of experiments C and D will focus on two aspects. Firstly, a comparison between the different methods used to display the grand average results. Secondly, the differences between experimental and control groups in experiments C and D.

8.15.1 Comparing the Different Types of Maps

With the multi-channel recordings of experiment C and D a comprehensive description of the PINV topography became possible. While grand average ERP curves proved especially sensitive to the detection of different wave forms within the PINV distribution, the spatial aspects of the development of the PINV were clearly better described by the maps compared to the ERP curves. In this sense maps and curves supplemented another.

However, especially due to the "smearing" effects of the highly resistive skull, scalp amplitude data are subject to stimulus degradation (Goff et al., 1978), as mentioned in *paragraph 8.2.2*. In particular, reduced amplitudes, reduced spatial resolution, and increased spatial averaging —typically over an area of at least several cm^2 (Nunez, 1989) —characterize scalp amplitude data. Projecting down to the cortical surface with a bio-physically valid method should result in a better spatial resolution and lower spatial averaging and, hence, a separation of maxima of negative or positive potential deflections which are not separated in the scalp amplitude data. This would also reduce the statistical redundancy of pairs of adjacent electrodes and lead to only moderate correlations between scalp and cortical potentials (Nunez et al., 1994; see below).

The comparison of the scalp and cortex amplitude data of experiments C and D confirm these theoretical predictions. With a few exceptions, the scalp amplitude data gave the impression of a single large area of PINV. This was especially true of the maps of the schizophrenic patient group. An almost concentric, sometimes slightly skewed amplitude reduction surrounding a single, rather "focal" negative maximum over the precentral region suggested one broad area of PINV activity. Although within the maps of the other groups "violations" of this concentric amplitude decrease and extensions of the negative area into different scalp regions, or the transitory emergence of a second local maximum suggested the existence of another negative maximum, this could not be identified, let alone be separated from the first one.

After projecting down to the cortical surface the situation differed. By now it became clear that even in schizophrenic patients, but more so in the other groups, distributed local negative maxima underlie the "broad" PINV scalp topography. These maxima are located precentrally over both hemispheres in both control groups and, with a less lateral location and less separated, in schizophrenic patients. In the high-scoring group a small precentral, bilateral extending band of maximal negativity, separated from a left-hemisphere anterior-frontal second maximum, occurred. Towards the end of the ambiguous condition, these two maxima were supplemented by another right fronto-lateral peak negativity.

Scalp laplacian and cortex amplitude maps revealed very similar results. Current sinks could be found at almost all locations where cortex amplitude maps had indicated areas of negative potential shifts. Negative surface Laplacian L_s and cortical potential V_B are related to each other by the equation $V_B \approx V_S + A_{KS} L_S$, where V_S denotes the scalp potential and A_{KS} is a parameter depending on skull and scalp thickness and resistivity, and must be known if cortical potentials are to be estimated approximately from recorded scalp potentials and the estimated scalp Laplacian (Nunez et al., 1994), which is in most instances not the case.

The convergence of results lends mutual support to each of the two "high-resolution" methods and corresponds with comparisons between cortex amplitude and scalp laplacian maps reported in the literature. The scalp Laplacian has been shown to approximate the cortical potential much more closely than the scalp potential, as indicated by the different sizes of intercorrelations between scalp Laplacian (in $\mu V/cm^2$) and cortical potential data ($.82<r<.92$)[222] on the one hand and scalp and cortical potentials ($.47<r.53$)[218] on the other hand (Nunez et al., 1994). Although scalp Laplacian and cortical imaging methods compared by Nunez et al. (1994) used different model assumptions, they yielded similar maps (see also Nunez & Westdorp, 1994). This convergence is replicated by the scalp laplacian and cortex amplitude maps of experiments C and D.

The results on Laplacian maps are in keeping with the conclusions drawn from computer simulations by Nunez (1989). In the case of multiple sources, the

[222] The first correlation coefficient relates to 4 sphere models (with CSF), the second to 3 sphere model (no CSF).

scalp Laplacian can distinguish cortical sources which are separated by a distance in the order of the average center-to-center spacing of the electrodes, which is about 5 cm with the standard 10-20 system montage. These are not discernible in the scalp amplitude maps (Nunez, 1989). This distance should be kept in mind when interpreting the map data.

8.15.2 The Topography of the PINV in Schizophrenic Patients and Healthy Controls

Schizophrenic patients and healthy controls exhibited different spatio-temporal patterns of the PINV modulation. In *both* groups a left-frontal PINV maximum was present throughout the interval. Similarly, in *both* groups a second right-frontal maximum developed. The crucial difference was that the development of this right-frontal maximum occured not only faster but in particular much more pronounced in controls than in patients, giving the global impression of an asymmetric PINV distribution in controls and a less asymmetric in patients. These qualitative/descriptive results of the map data were statistically confirmed by ANOVA using the cortical data for a subset of 12 frontal and temporo-central leads. Only for the second and the subsequent PINV segments (1.5-3.0 s after IS onset), but not for the first (1.0-1.5 s after IS onset), different sources of the PINV are suggested by the results of normalized data to *distinguish* the groups. In addition, only for controls (segments 2 to 4) the within-group ANOVA revealed a significant right-sided lateralization of the PINV; in patients, a left-sided predominance was present only for the mean amplitudes but did not turn out significant. The problems encountered when trying to interprete these results have been described in *paragraph 5.10.4*. They will be discussed in *paragraph 9.3.2*. Spatio-temporal group differences similar to those obtained for schizophrenic patients and healthy controls were found for the groups scoring high versus low in the German translation of the SPQ. While the spatio-temporal modulation of the PINV in the low-scoring group was very similar to that observed in healthy controls, the high-scoring group exhibited an initial left-sided predominance of the PINV, which turned out significant in the within-group ANOVA, followed by an almost symmetric distribution after about 1.5 s after IS onset (maps 2 to 4). Again, only few interpretational premises can be derived from the literature on the schizotypal personality dimension, in particular, since the DSM-III-R oriented SPQ (Raine, 1991) is a rather new research instrument which is only weakly correlated with "established" scales that claim to measure aspects of the schizotypal personality, such as the "Anhedonia" scale ($r \approx .19$) or the "Psychoticism" scale ($.27 \leq r \leq .37$) (Raine, 1991). None the less, the present results add to the list of reports on differences between groups selected by the SPQ in neuropsychological, anatomical, and psychophysiological variables (Kidd, 1992; Lencz et al., 1993; Raine et al., 1992).

Chapter 9 Result Integration and General Discussion

9.1 Reaction Times

In all experiments and under all experimental conditions, schizophrenic patients exhibited longer reaction times (RT) than controls. This is an almost ubiquitous result of research with schizophrenic patients (c.f. King, 1991).

Ambiguous compared to clear IS resulted in RT augmentations of about 140-181 ms in patients and about 106-143 ms in controls. This indicates that subjects detected the deviant size of the ambiguous IS, even if they, perhaps, did not clearly identify the nature of the deviancy. The increase in RT under the ambiguous condition may be associated with the additional time needed to select a response under the instructed "pressure" to respond as fast as possible.

High compared to low provisional memory load augmented RT also. Under the low memory load condition a choice RT situation was realized, under which responses could be chosen by means of a temporally stable association of the size of the IS diamond and a response side (small - left; large - right). In contrast, under the high memory load condition an additional step (stage) in IS processing was required: the matching of the IS and the identical WS diamond associated with the correct response side. The augmentation of RT of about 150 ms in patients and 53 ms in controls may plausibly be associated with the matching process. Although in schizophrenic patients the average RT prolongation under the ambiguous condition was larger than that of controls, this differential effect did not become statistically significant. Hence, there it may not be concluded that schizophrenic patients needed more time than healthy controls for the retrieval of the position of the appropriate (matching) WS diamond.

9.2 Contingent Negative Variation (CNV)

The CNV was investigated in experiments A and B. In these experiments, tCNV was found to be larger at central than at frontal leads and larger over the left than over the right hemisphere; if mid-sagittal leads were applied (experiment B), the topographical maximum of the tCNV was found there. None of these distributional features occured unexpectedly: a central and left-sided predominance of the tCNV in the present data is consistent with the literature (e.g. van den Bosch, 1984) and the assumption that the RP is a main part of the tCNV. A central to precentral predominance of the RP is known (e.g. Lang et al., 1991). While the RP may be lateralized if a lateralized response must be prepared (e.g. Lang et al., 1991), in the present experiments subjects did never know the side of the future response and had to respond with the left and right hand with equal probability. Furthermore, a left-hemisphere predominance of negativities preceding instruction stimuli and a right-hemisphere predominance of negativities preceding 'knowledge of result' stimuli has been reported (Van Boxtel & Brunia, 1994). The present results are in line with these observations.

In experiment B, tCNV amplitudes were larger over the left than over the right hemisphere, but maximal at mid-sagittal leads. This may be simply an effect of the superposition of distributed potentials (cf. Lutzenberger et al., 1987).

In all 3 experiments there was no "overall" (i.e. averaged over leads or conditions) reduction of iCNV or tCNV amplitudes in schizophrenic patients when compared to healthy controls. Only some descriptive amplitude differences in the expected direction between the groups could be found for the clear condition of experiment B, but even then this difference did not turn out statistically significant. Hence, preparation for stimulus evaluation and response, as presumably reflected by the tCNV (Brunia, 1988, Rösler, 1991), and contingency formation (Rockstroh et al., 1989) and orientation (Rohrbaugh & Gaillard, 1983), as presumably reflected by the iCNV, were *not* grossly impaired or altered in the present samples of schizophrenic patients.

This result is at odds with the frequently reported reduction of CNV amplitudes in schizophrenic patients (e.g., Timsit-Berthier et al., 1984; reviewed by Cohen, 1990). However, for several reasons the "trait" character of CNV amplitude diminutions in schizophrenic patients has been questioned. Firstly, CNV increases have been reported in schizophrenic patients after neuroleptic treatment and in remission (Knott et al., 1976; Rizzo et al. 1983). Also, reduced CNV has been reported for acutely ill patients with florid symptomatology (Abraham & McCallum, 1977; Abraham et al., 1976, 1980; Bachneff & Engelsmann, 1983; McCallum & Abraham, 1973; Roth et al. 1979; Timsit-Berthier et al., 1984; Verhey et al., 1984), possibly due to the distracting effect of such psychotic states (cf. Pritchard, 1986). Amphetamine, capable of inducing psychotic states, has been found to reduce negative slow potentials over the frontal cortex in experimental animals, an effect which can be reduced by haloperidol or barbiturates (cf. Marczynski, 1978, Pirch, 1978). Furthermore, in acutely ill schizophrenic patients fluphenazine may both increase CNV amplitudes and ameliorate psychotic symptoms (cf. Tecce & Cole, 1976).

None of the patients investigated in the present experiments exhibited severe acute symptomatology during testing, and all patients were sufficiently stabilized in order to participate in an experimental session of at least 1.5 hours duration. Furthermore, all patients participating in the present experiments were under neuroleptic medication. Hence, it is conceivable that the lack of group differences in CNV amplitudes is due to the relative clinical stability and the neuroleptic medication of the schizophrenic patients.

The only group difference in the CNV topography was the posterior-to-anterior potential gradient of the tCNV in experiment A, which was steeper in controls than in patients. Only little is known about the distribution of ERP components in schizophrenic patients. Some authors reported on reduced posterior-to-anterior amplitude gradients of the CNV in schizophrenic patients (e.g. Rockstroh et al., 1993, 1994; Wagner et al., 1993, submitted; Van den

Bosch, 1984). In these studies a relative augmentation of the frontal CNV in schizophrenic patients accounted for the diminished amplitude gradients and the "broader" topography. In the present study, group differences were maximal at central and parietal, and minimal at frontal leads; furthermore, amplitude differences between frontal (F3,F4), central (C3,C4), and parietal (P3, P4) leads were clearly smaller in patients than in controls. But the relative augmentation of the frontal CNV was not found in experiment A. While the results of experiment A do not confirm the pattern of a relatively increased frontal CNV in schizophrenic patients, they indicate reduced *spatial information* in the distribution of an ERP component in patients (see also Cohen, 1990).

High as compared to low provisional memory load enlarged iCNV and tCNV amplitudes, in particular at central (C3, C4) and frontal (F3, F4) leads. While the former effect is in line with the observed relatedness of the iCNV to contingency formation (Rockstroh et al., 1989) and the information value of the WS (McCarthy & Donchin, 1978), the latter effect may be explained by the sensitivity of the tCNV to task complexity (McCallum, 1986). Despite the topographical similarity of the two effects, a qualitative difference is suggested by the results of normalized data. While the topography of the provisional memory effect on iCNV is compatible with a change in the generator structure, an amplitude effect on tCNV could not be ruled out. Accordingly, a *qualitative* effect, indicating the "activational" shift towards central and frontal regions, seems to be present only for the iCNV, while a rather *quantitative* effect cannot be rejected for the tCNV. It therefore appears, that the *qualitative* cognitive differences associated with the shift from stable to variable positions of the WS diamonds is better represented by the iCNV than by the tCNV, corresponding to the assumed sensitivity of the iCNV to contingency formation (cf. Rockstroh et al., 1989).

9.3 Post-Imperative Negative Variation (PINV)

9.3.1 PINV in Healthy Subjects

One of the main results of the experiments presented here is the observation of a PINV in healthy controls with an experimental task including ambiguous matching processes. Since this was found in a total of 4 experiments, it is a replicable and apparently stable result.

To my knowledge, there is only one other *replicated* variation of cognitive task elements that is apt to induce a PINV in healthy controls. This is the "loss of control" paradigm (Delaunoy et al., 1978; Elbert et al., 1982; Kathmann et al., 1990; Rockstroh et al., 1979), which had been discussed briefly in *chapter 3* and will be reconsidered in the following paragraphs.

Delaunoy et al. (1978) reported an increase of PINV duration from less than 2 s when the IS (light flashes) could be terminated by a button press to a duration of 4.5-6.0 s when the IS continued for 1 s independent of the subjects' response in one-third of the trials. A PINV in healthy subjects was also induced by Rockstroh et al. (1979) and Elbert et al. (1982) using aversive IS that could not be terminated during one block of trials. Finally, Kathmann et al. (1990) observed a PINV in healthy subjects under different conditions of control over the termination of an aversive IS.

Several explanations can be given in order to explain these results, which shall be reviewed briefly in the following.

Firstly, in the studies of Rockstroh et al. (1979), Elbert et al. (1982), and Kathmann et al. (1990) the uncontrollable IS was motivationally significant due to its aversive nature. This might indicate that the perception of aversive (highly intensive) stimulation per se, but especially in combination with the experience of loss or lack of control accounts for the occurence of the PINV. The results of Delaunoy et al. (1978) clearly argue against this hypothesis, since they observed effects comparable to those observed by the other authors with non-aversive stimuli (light flashes). In addition, a PINV was found in healthy subjects also following the "loss of control" over a *neutral* IS, although with a lower amplitude than when compared to the aversive IS condition (Elbert et al., 1982; Rockstroh et al., 1982, cited after Birbaumer et al., 1986).

Secondly, a common feature of the designs used by Delaunoy et al. (1978), Rockstroh et al. (1979), and Elbert et al. (1982) might indicate that PINV is related to the experience of a *loss* of control. In all of these experiments control over the IS was present during a first block of trials and was lost during a second one. Consequently, it is conceivable that the experience of control (during the first block) is a necessary condition for the occurence of a PINV under conditions of uncontrollability (during the second block). The results of Kathmann et al. (1990) argue against this hypothesis. In this study, PINV amplitudes were the same irrespective of the order of the control / no control conditions. Hence, the prior experience of control is not a necessary condition for the development of a PINV following loss of control.

Thirdly, an explanation for the irrelevance of the *order* of the control / no control conditions is given by Kathmann et al. (1990) by emphasizing the possible role of the *instruction*, which was the same for both orders of conditions. Due to the instruction "control may have always been expected" (Kathmann et al., 1990). Instructions might provide expectations and the modalities for a kind of stimulus processing similar to the "top-down processing" in Cognitive Psychology (Anderson, 1985[223]) into which the occurence of uncontrollable IS does not fit.

In this context it is interesting to note that *not all* of the subjects investigated by Delaunoy et al. (1978) developed a PINV. When the previously controllable IS (light flashes) could no longer be terminated by a button press, a

---

[223] "When context or general world knowledge guides perception, we refer to the processing as top-down processing, because high-level general knowledge determines the interpretation of the low-level perceptual units." (Anderson, 1985, p. 60)

PINV occured only in those subjects who "did not understand what had happened or could not provide an explanation", while the subjects rationalizing "that the experimenter had interfered with the apparatus or that an equipment malfunction had occured" did not develop a PINV (Delaunoy et al., 1978, p. 356). The authors therefore associated the PINV in healthy subjects with the unability "to resolve the ambiguity of the experimental intervention" (Delaunoy et al., 1978, p. 356).

What processes gave rise to the ambiguity that could not be resolved by those subjects who developed a PINV ? In the absence of experimental investigations, this question can be answered only speculatively.

The term "working memory" as used a neurophysiologists refers to the ability to guide behavior by representations (instead of being driven by stimuli). Working memory is generally held to be subserved by the prefrontal cortex (Fuster, 1989; Goldman-Rakic, 1991, 1994). Fuster (1989) assumes working (or "provisional") memory to be supramodal, and defined primarily by a contextual basis, i.e. the "need to bridge a temporal gap in the course of behavioral action". Hence, prefrontal working memory is *action based*, including retrospective (provisional memory) and prospective (anticipatory/motor set) functions, as well as interference control. According to Fleming et al. (1994) "the prefrontal cortex guides behavior through the access, maintainance, and updating of representations (memories) over time, when the stimuli are no longer present."

Working memory functions have been investigated extensively with delay tasks (see *chapter 5*). Unit recordings reveal the onset of prefrontal activity with the perception of the cue (comparable to the WS), during the delay (comparable to the ISI), and towards the perception of the sample (comparable to the ISI). Decelerating activity during the delay has been associated with "short-term memory" functions, accelerating activity with preparatory set.

It is important to note that the critical condition for the occurrence of the delay-bridging activity is the existence of a *cross-temporal contingency between cue (WS) and response*. If the cue is experimentally deprived of its "significance as the indicator of a particular choice and motor response" (Fuster, 1989, p. 109) firing changes during the delay disappear.

Hence, during delay tasks any event of the behavioral sequence is referred to both preceding events and the scheme of action (the task structure). Both, scheme of action and preceding events must therefore be *represented and retained temporally* in order to make this future reference possible. This temporary retention of action-relevant or action-oriented information amounts to the short-term or provisional memory function, the crucial feature of which is its operational character.

Accordingly, a (mnemonic) *representation of the scheme of action* or the task structure, including experimental stimuli, response, and contingencies between them is provided by the prefrontal cortex (in part cooperating with thalamic and inferotemporal cortex). This representation of the scheme of action or the task structure has retrospective and prospective aspects.

It has been argued in *paragraph 6.10.4* that a common feature of the present design and the random lack of control condition in the study of Kathmann et al. (1990) is the experimental untying of contingencies between elements of the task structure (WS, IS, response, response outcome) by ambiguous or uncontrollable IS.

In the study of Kathmann et al. (1990) two factors affected the PINV: Firstly, the *presence or absence* of a contingency between response and response outcome (called "control" by Kathmann and colleagues), and secondly, the *reliability* of the contingency / discontingency, which was high under constant presence or absence of control, but low under changing presence or absence of control (reliability of contingency is called "contingency" by Kathmann and colleagues). Absence of contingency and low reliability of contingency contributed to an augmentation of the PINV.

In the present experiment trials with clear and ambiguous contingencies were randomly interspersed, comparable to the *random* control / random lack of control condition of Kathmann et al. (1990). In the case of ambiguous IS there was neither a contingency between WS and IS, nor between IS and response, since the IS could not be matched with the WS and no response alternative was provided for that case. This condition is comparable with the (random) lack of control condition of Kathmann et al. (1990). In the case of clear IS there was a contingency between WS and IS, and IS and response, similar to the (random) control condition of Kathmann et al. (1990). Altogether, contingencies were unreliable.

It has been argued that in the case of delay tasks, working memory enables the guidance of behavior by representations of an action scheme. It is at least plausible to assume that the representation of an action scheme is adaptive only if it contains exclusively reliable contingencies (in particular of stimulus and response, or of stimulus (WS), stimulus (IS), and response in the case of delayed matching-to-sample tasks) since otherwise the representation would not accomplish its role in guiding behavior.

The present experiments and the study of Kathmann et al. (1990) suggest that a PINV develops whenever the task provides discontingencies or unreliable contingencies between elements of the task structure. These *dis*contingencies or unreliable contingencies are perceived by the subject as contrasting with a representation of the action scheme that is comprised of reliable contingencies between the same elements of the task structure.

Unreliable contingencies can be expected to have a more enduring effect (over a succession of trials) than discontingencies, since a *consistent* *dis*contingency will change into a *new* contingency (e.g. a *stable* association between response and *no* IS termination). As Kathmann et al. (1990) have argued: "(...), it does not seem plausible that subjects would continue to process non-control trials (...) as a loss of control after 10, 20, or more repetitions, ..." (p.257). This hypothesis implies that there is a decline in PINV amplitudes over the trials, concomittant with the assumed formation of the new reliable contingency "response-*no* IS termination", which might explain the lower PINV under the condition of constant lack of control compared to random lack of control in the study of Kathmann et al. (1990).

It can be argued that under the constant lack of control condition *no* PINV should have been present since with the assumed formation of the new contingency "response-*no* IS termination" the factual task structure was correctly represented. While this argument seems correct, in principle, it should be reminded that a PINV of similar amplitude can be observed even when subjects listen *passively* and are not engaged in any task with aversive tones (compatible with a sustained potential; Kathmann et al., 1990). Furthermore, the changing of the representation of the action scheme can be expected to take place only gradually with the continual perception of the "response-no IS termination" contingency. If subjects continue to believe in the instruction and continue to expect the instructed contingency, this assumed change will take longer. This assumption corresponds with the observation of Delaunoy et al. (1978), that only those subjects exhibited a PINV (averaged over the entire block of trials with uncontrollable IS) who could not resolve the ambiguity of the situation (by assuming an interference of the experimenter and, hence, developing a rationale for the assumption of a reliable "response-no IS termination" contingency).

In the present experiments, a situation similar to the random control/random lack of control condition of Kathmann et al. (1990) is assumed. Interspersed ambiguous IS reduce the "overall" (for the entire trial block) reliability of the WS-IS and IS-response contingencies, since they cannot be matched with one of the WS diamonds and since no response alternative is provided for ambiguous IS.

At first sight the observation that there were hardly any "missing responses" under the ambiguous condition seems to argue against this interpretation. However, it should be recalled that the instruction *forced* subjects to respond as quickly and as correctly as possible in each trial. Hence, subjects *had* to respond. If, on the other hand, subjects had developed a coping strategy such as "respond always with the right hand when ambiguous IS occur", thereby establishing a response category for ambiguous IS, reaction times should not have been augmented under the ambiguous condition, as was the case in all experiments.

An important subsequent question is how the representation of the action scheme develops and what determines its contents.

The experiments of Delaunoy et al. (1978), Rockstroh et al. (1979), and Elbert et al. (1982) suggest that prior experience of reliable contingencies between response and response outcome (controllability of the aversive or neutral IS) might be the basis on which discontingencies are perceived as contrasting with the representation of the action scheme. Yet, as Kathmann et al. (1990) argue, the observation of the same PINV amplitudes under the lack of control condition (perceived discontingencies) irrespective of the presence or absence of the prior experience of control renders the assumption of a prior experience of control doubtful. If "prior experience of control" means "expectation of control", this function may be substituted by the instruction, as Kathmann et al. (1990) have argued. It can be assumed that by the instruction a representation of the action scheme can be established, since the instruction enables the subjects to accomplish the task requirements. As a consequence, subjects develop expectations about contingencies (for instance the expectation that a response is followed by a termination of the IS). Since in the "loss of control" studies the instruction delineated reliable contingencies between the elements of the task structure, in particular between response and reponse outcome, lack (and loss) of control was presumably perceived as a deviation from the representation of the action scheme. The possible role of the instruction can easily be tested by its variation for subjects naive to the aims of the experiment.

In any event, the present experiments have demonstrated that a PINV can be reliably observed in healthy subjects under experimental conditions including ambiguity of the matching of experimental stimuli. The modified delayed matching-to-sample task used here is non-aversive and appears, for that reason, more suitable than the "uncontrollability" task for the investigation of clinical samples. Hence, comparative investigations between schizophrenic patients and healthy controls of cognitive, topographical, and other aspects of the generation of the PINV became feasible with this task. The main results of these investigations will be summarized and discussed in the following paragraphs.

9.3.2 Topography of the PINV—Averaged Over the Experimental Conditions

Before discussing the experimental task effects and their topography, the topography of the PINV as revealed by the topographical main effects in the left-right or anterior-posterior direction as well as the statistical interaction between these main effects and the group main effect, averaged over the levels of the experimental task factors, shall be considered. Since these topographical main effects result from averaging over the levels of the experimental task factors, it should be kept in mind that they can not be regarded as independent of the topographical effects associated with an experimental task factors discussed below.

9.3.2.1 Lateralization of the PINV

The consistent lateralization of the PINV in healthy subjects is the second replicated result of the present experiments. The healthy control groups of the experiments A to C as well as the low-scoring subjects of experiment D exhibited a right-sided predominance of the PINV. This right-sided predominance was reflected by the "Hemisphere" or "Laterality" simple effects in the within-group ANOVA, which became significant in experiments B to D and fell short of significance (p=.06) in experiment A[224]. Altogether a stable right-sided predominance of the PINV in healthy subjects under a task condition including ambiguous matching of geometrical shapes (diamonds) is suggested by these results.

The map data of experiments C and D, and in particular the maps generated by the "high-resolution" (Nunez & Westdorp, 1994) techniques (cortical and current source density mapping) revealed a clear picture of the right-hemisphere predominance of the PINV in showing that there are two *separate* PINV maxima in the control groups, one over the left and one over the right frontal cortex, and that the right-sided maximum is larger than the left-sided.

As consistent as the right-sided predominance of the PINV in healthy controls, was the observation of a *divergent* lateralization of the PINV in schizophrenic patients. All "Hemisphere by Group" (experiments A and C) and the "Laterality by Group" (experiment B) interactions comparing the overall lateralization in patients and controls, as well as the "Hemisphere by Group" interaction for the topographical comparison between low- and high-scoring subjects in experiment D revealed significant results. And except for one measure (the first PINV segment in experiment C) these interaction effects remained significant after normalization of the data according to McCarthy & Wood (1985) in order to control for amplitude effects, suggesting that different "generator structures" are underlying the PINV in the experimental (patients, high-scoring subjects) and control (healthy and low-scoring subjects) groups.

While these results indicate that "real" *topographical* differences (instead of amplitude effects generating topography by group ANOVA interactions) between schizophrenic patients and healthy controls can be observed and replicated, the precise quality of the distribution of the PINV in schizophrenic patients could not unambiguously be identified within the present series of experiments.

On the one hand, in experiments A and C a left-hemisphere predominance of the PINV could be found in schizophrenic patients, which, however, became statistically significant in the within-group ANOVA only for the experiment A but not experiment C. Similarly, the high-scoring group of experiment D exhibited a left-hemisphere predominance of the PINV which became statistically significant only for the first PINV segment (1.0-1.5 s after IS onset). On the other hand, while the PINV of the patient group in experiment B was descriptively larger at mid-sagittal and/or right-hemisphere compared to left-hemisphere leads, this distributional pattern did not turn out statistically significant in the within-group ANOVA (simple effect "Laterality" of the "Group by Laterality" interaction).

One possible explanation for these divergencies, especially between the results of experiments A and C on the one side and experiment B on the other

[224] The course of the PINV lateralization in the control groups will be discussed in detail below.

side, is the different electrode montage used in the experiments. As can be seen in the cortical data of the patient group of experiment C (**table 8.10.3**), providing the most comprehensive picture of the PINV topography, PINV extends far into lateral regions. Especially under the ambiguous condition this lateral extension appears to be slightly more pronounced over the left than over the right hemisphere at frontal and temporal positions. While in experiments A and C fronto-lateral (F7, F8) and temporal (T3, T4) leads were applied, this was not the case in experiment B (with F3, F4, C3, C4 as the most lateral fronto-central leads applied). It is therefore possible that the electrode montage of experiment B missed some lateralized topographical information. But even if this were the case, the contribution of fronto-lateral and temporal leads to the PINV lateralization in patients cannot be too important for several reasons. Firstly, if the fronto-lateral or temporal leads contributed particularly to the PINV lateralization, the within-group ANOVA, for instance of experiment A, should have revealed a significant "Hemisphere by Electrode" interaction in the patient group. This was not the case. Secondly, experiments A and C should have yielded similar "Hemisphere" effects in the patient group, since in both experiments the lateral positions were covered by the electrode montages entering the statistical analyses. But the "Hemisphere" effect in the patient group of experiment C did not even approach statistical significance due to the small F-values (≤ 1.38) and not due to the relatively small sample size.

Table 9.1 Hypothesis and Error Mean Squares of the "Hemisphere"[1] and "Laterality"[2] Simple Effects in Patients and Controls

Experiment		A		B			C/D			
Segments		#1	#2	#1	#2	#3	#1	#2	#3	#4
Patients	Hypothesis	215	109	7.5	8.4	15.7	69.6	27.8	10.9	5.8
	Error	5.9	4.3	6.6	6.1	4.8	50.6	27.9	17.7	12.9
Controls	Hypothesis	118.	167.	61.2	82.8	102.	113.	246.	224.	203.
	Error	29.9	23.5	7.6	5.8	6.1	33.8	21.6	14.3	11.3
Ratio Pat/Con	Hypothesis	1.82	0.65	0.12	0.10	0.15	0.61	0.11	0.05	0.03
	Error	0.20	0.18	0.87	1.05	0.79	1.50	1.29	1.13	1.14
Low-Scoring Group	Hypothesis						82.2	329.4	317.2	300.8
	Error						39.8	26.8	18.6	15.3
High-Scoring Group	Hypothesis						238.5	10.8	1.33	27.6
	Error						30.8	24.0	23.5	23.9

[1]=experiments A, C, and D; [2]=experiment B; hypo=hypothesis mean square; error=error mean square

Another possible explanation for the divergency of lateralizational results refers to the possible heterogeneity of the PINV topography within the patient groups. Heterogeneity within samples of schizophrenic patients is a result that pertains to almost all variables that have been investigated in this group. The homogeneity/heterogeneity of the samples investigated across the experiments A to C might have differed because of random fluctuations of the subjects. As a consequence, within-group "Hemisphere" effects may remain non-significant due to such heterogeneity. This hypothesis can easily be checked by comparing the hypothesis and error mean squares of the within-group "Hemisphere" or "Laterality" effects.

In **table 9.1** the hypothesis and error mean squares for the "Hemisphere" and "Laterality" simple effects of experiments A to D are documented. Due to the different electrode montages in experiments A to C (D) hypothesis and error terms will be compared between the groups of each experiment but not between experiments. These comparisons are highlighted by the patient : control - ratio for the hypothesis and error mean squares, expressing the mean squares of the patient group as a proportion of that of the control group.

The comparisons documented in **table 9.1** indicate that patients and controls in experiment A differ consistently concerning the error mean squares (with those of controls being larger than those of patients) and the hypothesis mean squares for the first PINV segment. In experiment B and C (except for PINV-1), groups differ primarily in the hypothesis mean squares with less differences in the error mean squares, except for PINV-1 in experiment C.

In sum, these results indicate that the lack of lateralization of the PINV in schizophrenic patients in experiments B and C is primarily (not exclusively) due to relatively small hypothesis mean squares, reflecting the variance of PINV group means due to the left vs. right hemisphere differentiation. By contrast, the patient group of experiment A exhibited smaller error mean squares than controls for both PINV segments, but larger hypothesis mean squares only for PINV-1, but not for PINV-2.

Hence, disproportionately large heterogeneity within the patient as compared to the control group is apparently *not* the main source of a lack of a statistically significant PINV lateralization in the patient groups of experiments B and C. Instead, it appears as if there was indeed a lower degree of lateralization of the PINV in patients compared to controls, suggesting a "broader" distribution of the PINV in schizophrenic patients. That schizophrenic patients "are less completely lateralized than normal individuals" (Weinberger, 1994, p. 23) is a common finding of studies on lateralized cerebral functions and cognitive tasks. The results from experiments B and C confirm this trend. The clear lateralization of the PINV to the left frontal region in the patient group of experiment A, however, is not in accordance with this notion. In discussing the general *inconsistency* of the literature on cerebral anatomical asymmetries in schizophrenia, Weinberger (1994) offers uncontrolled overrepresentation of mixed dominance in samples of schizophrenic patients (versus samples of healthy controls) as possible explanation. Since in all experiments groups were matched for handedness, the hypothesis put forward by Weinberger (1994) does not plausibly explain the present results[225].

The question remains why the patient group of experiment A exhibited a stronger lateralization of the PINV than controls, while the reverse held for experiments B and C.

Assuming that differences in procedural experimental features such as the presentation or absence of acoustical probe stimuli do not plausibly account for these differences (since in this respect experiments A and B were more similar than experiments B and C), and excluding an effect of the electrode montange (see above), the random composition of the patient samples remains the most salient difference between the experiments. It should be noted that this question focuses on individual differences *between* the subjects of experimental patient groups that were not disproportionately heterogeneous. The sociodemographic and clinical variables assessed for the patients of experiments A to C do not grossly differ between groups. Since no further clinically relevant variables were assessed for the patient groups, the question of what *individual* differences might account for the *group* differences in PINV lateralization cannot be answered here.

None the less, the investigation of individual differences in PINV lateralization is suggested by the heterogeneity of PINV lateralization accross the *groups* of experiments A to C. This approach should be guided by specific hypotheses on the causes of PINV lateralization. In order to generate such hypotheses it is useful to question why a right-sided PINV lateralization was obervable and could be replicated in healthy subjects.

The specification of conditions *preceding* a PINV (which has been attempted above and will be continued below) does not provide answers to the questions of which processes take place *during* the development of the PINV and what accounts for its topography. In the present context, this distinction between "antecedent conditions" and "internal information processing" (see Donchin & Coles, 1988, p. 359) means that specifying some kind of a mismatch detection as one of the antecedent conditions of the PINV does not specify what kind of information processing takes place during the PINV and where in the cortex this occurs.

For that reason the question of why healthy controls exhibit a right-sided PINV predominance can be dealt with only speculatively.

The most plausible interpretation of the right-sided predominance of the PINV in healthy controls is the rather spatial nature of the matching task. The right hemisphere more than the left hemisphere seems to be involved in the processing of spatial information (cf. Bradshaw & Nettleton, 1981). In particular the frontal lobes have been associated with spatial discrimination (Kolb & Wishaw, 1990, p. 653). "Response uncertainty" (Cohen, 1990) or "contingency reappraisal" (Elbert et al., 1982) as hypothesized processes concomitant to the

---

[225] Excluding all left-handed subjects of experiment D from the statistical analyses (ANOVA) did not alter the results critically (see **Appendix A.4**).

PINV, might be localized according to the modality of the stimulus material they refer to, i.e. the spatial modality.

While this hypothesis is speculative at present, it is testable if processes of mismatch detection are inducable with material that would be processed in the left hemisphere in comparison with material that might be expected to be processed in the right-hemisphere, too. While acoustical stimuli would pertain to the latter category and should result in a right-hemisphere predominance of the PINV, *verbal* material should alter, and perhaps reverse, the PINV asymmetry in healthy subjects.

The research on cognitive prototypes (e.g. Rosch, 1977) indicates that there are different degrees of certainty, as reflected by response times, with which instances of certain classes of objects are attached to their category. For instance, verifying that a robin is a bird takes significantly less time than verifying that a penguin is a bird. Hence, a robin is a more prototypical instance of the bird category than a penguin (cf. Hussy, 1984). This schematic structure of categories warrants the prediction that categories do not have fixed boundaries (cf. Anderson, 1985, p. 127). Hence, there should be objects that are ambiguous with respect to superordinate categories. For instance, if two different categories such as "furniture and music instrument"or "vehicle and means of transportation" are presented as WS, the attaching of terms like "piano" or "elevator" should be ambiguous and apt to induce a PINV. Perhaps PINV induction would be even more powerful with stimuli that cannot be attached to any of the WS categories, such as "bell" or "tire" in the examples presented above. If PINV lateralization corresponds to the hemisphere primarily involved in the processing of the stimulus material, it should be predominant over the left hemisphere during this kind of task.

Assuming that the right-sided predominance of the PINV in healthy controls and low-scoring subjects is related to the rather spatial character of the stimulus material, the question arises why schizophrenic patients do not exhibit the same topographical pattern of PINV modulation. Again, at present this question can be treated only speculatively and must focus on what is known about "laterality" in schizophrenia.

While there is evidence of right-frontal dysfunctions in schizophrenia as assessed by EEG (e.g. Guich et al., 1989; Karson et al., 1987) and PET (Buchsbaum et al. 1982, 1984) techniques, the majority of evidence points to the left hemisphere, including structural (e.g. DeMyer et al., 1988; Golden et al. 1981) and physiological (e.g. Gur, 1978, 1985, 1994) evidence, as reviewed in *chapter 2*, as well as results from e.g. dichotic listening and tachistoscopic studies (reviewed by Taylor, 1987). Most of these studies suggest an overactivation of the left hemisphere in schizophrenic patients. This conclusion, derived from the (selected) literature survey presented in *chapter 2*, is in keeping with recent reviews of literature which suggest that "when asymmetric findings have been reported, they usually involve the left hemisphere" (Weinberger, 1994, p. 20).

Another aspect of such functional or physiological interpretations of the PINV lateralization in patients refers to the concepts of "cognitive style" and "hemisphericity" (Bradshaw & Nettleton, 1981, pp. 62; Ehrlichman & Weinberger, 1978). Some evidence indicates that the lateralization of information processing is not only determined by nature of the stimulus material but also by individual preferences in processing. It has been postulated that the direction of Conjugate Lateral Eye Movements (CLEM) in response to stimuli triggering information processing, such as questions, indicates which hemisphere is preferentially activitated (cf Ehrlichman & Weinberger, 1978). CLEM have been suggested to be generated in the frontal area 8 (cf. Schweitzer et al., 1978). Some results indicate that schizophrenic patients are predominantly "right movers", suggesting that they prefer the activitation of the left hemisphere (Schweitzer et al., 1978). These and related studies suggest that "schizophrenics overutilze their left hemisphere" (Schweitzer, 1982).

It is interesting to note that Rohrbaugh & Gaillard (1983) cite a study (Rohrbaugh et al., in preparation) in which a reversal of the "normal" right-sided predominance of the O-wave (which the authors identified with the PINV; but see **figures 8.12.1a,b**) was moderately correlated (r=.40) with the strength and direction of CLEM. This observation, together with the reported directions of CLEM in schizophrenic patients (Schweitzer et al., 1978), or at least subgroups of schizophrenic patients (Tomer et al., 1981), and the different lateralization of the PINV reported here, would suggest a differentiation of patients and controls according to the prefered CLEM direction. It must be questioned, however, whether CLEM are a valid indicator of lateral hemisphere activation (Ehrlichman & Weinberger, 1978).

None the less, there are other hints as to an overactiviation of the left hemisphere in schizophrenic patients under a variety of task conditions (reviewed by Walker & McGuire, 1982). These observations are in line with the structural and physiological left-hemisphere abnormalities in schizophrenic patients reported in *chapter 2*. Furthermore, loss of automatism frequently reported of schizophrenic patients (Peralta & Cuesta, 1991) and excessive self-awareness of information processing healthy subjects are not aware of (Frith, 1979) may propagate conscious verbalization —and hence left-hemisphere processing— in task situations in which healthy subjects cope with the right hemisphere. Recent rCBF observations of a failure of schizophrenic patients in exhibiting the pattern of a focal right-hemisphere activation seen in healthy controls during a task of facial memory (Gur et al., 1994) are compatible with this interpretation and correspond to the observations reported here.

9.3.2.2 Anterior-Posterior Distribution of the PINV

Another replicated result of the present experiments is the frontal predominance of the PINV. In experiment A, PINV was largest at the frontal (F3,

F4) sites in patients and at (F7, F8) in controls, in experiment B, C, and D the frontal PINV was larger than the central (experiment B) and the temporo-central PINV (experiments C and D). The Current Source Density (CSD) analyses of the data of experiments C and D indicated current sinks over frontal to fronto-lateral regions of both hemispheres. These results, particularly the CSD maps, make a frontal origin of the PINV likely.

Except for experiment A, where the "Electrode by Group" interaction remained significant after normalization of the data, there was no evidence of topographical group differences in the anterior-posterior direction (i.e. "Anteriority by Group" (experiment B) or "Coronal Row by Group" (experiment C) interactions). These results suggest that both groups exhibit a frontal predominance of the PINV and that there is no replicated topographical difference between groups except for the lateralization of the PINV, as discussed above.

A frontal predominance of the PINV has been reported by some authors (e.g. Elbert et al., 1982; Kathmann et al., 1990; Lutzenberger et al., 1981) but has not always been found (e.g. Van den Bosch, 1983). Some studies reveal different fronto-central gradients of the PINV in schizophrenic patients (larger at frontal sites) and healthy controls (larger at central sites) (Eikmeier et al., 1992; Wagner et al., submitted).

It has been argued above that a mismatch process precedes the PINV, and terms such as "response uncertainty" and "contingency reappraisal" have been suggested as concomitants of the PINV. Luria (1966) has suggested that the frontal lobes contain a system for the "verification of activity", a view which is in keeping with recent neuropsychological theories of the frontal lobes (e.g. Shallice, 1990). It is tempting to speculate that the perception of the ambiguous and to-be-responded IS triggered the verification of activity in the prefrontal cortex.

The topography of the PINV differed from that of the tCNV (see **figures 8.12.1a,b**), and this topographical difference became very salient if the tCNV was taken as the baseline for PINV parametrization, as was shown with the cortex amplitude maps of experiments C and D (**figures 8.12.2a-d**). With the tCNV as baseline it became apparent that the topographical changes from tCNV to PINV are predominant over the lateral frontal cortex. Hence, these frontal and fronto-lateral topographical proportions of the PINV cannot be considered as mere "CNV prolongations". In contrast, a topographical overlap between tCNV and PINV exists over centro-parietal regions (see **figures 8.12.1a,b**) and does not reject the possibility that these proportions of the PINV are related to the preceding tCNV. The differentiation of frontal and centro-parietal "facets" of the PINV is suggested by these topographical considerations and will be further discussed within the context of the "nogo" effect.

9.3.2.3 Temporal Course of the Lateralization of the PINV

One of the advantages of the EEG is its high temporal resolution of relatively fast neuronal processes (depolarization of apical dendrites, see *chapter 8*). In contrast, popular imaging techniques such as PET or fMRI provide only limited temporal resolution of neurophysiological processes (e.g. blood flow) that may be delayed with respect to the neuronal activity of interest (cf. Rösler et al., submitted).

The application of multi-channel EEG recordings therefore allows the combination of the advantages of high-resolution EEG techniques such as CSD with the fine-grained temporal resolution of certain aspects of cortical activity. Altogether, the delineation of the spatio-temporal pattern of the modulation of an ERP component and of group differences in these patterns becomes feasible.

While the grand average curves presented in **figures 8.10.1a,b** and **8.11.1a,b** give a fine-grained impression of the temporal course of the potential shifts at each recording site, yet provide poor power for the topographical integration of these shifts, the subsequently presented scalp and cortical amplitude, and CSD maps focus on the temporal succession of topographical patterns of the PINV activity.

The cortex amplitude maps were very similar to the CSD maps and provided better spatial resolution than the scalp amplitude maps. The cortical amplitudes of a selection of 12 frontal and temporo-central leads were analyzed statistically.

These maps revealed two separate PINV maxima in the control group: A left-frontal PINV maximum, supplemented by a second, even larger, and rapidly increasing right-frontal maximum. In schizophrenic patients, two separate PINV maxima were present, as well. However, the initial left-frontal PINV maximum was only weakly supplemented by the second right-frontal maximum. The clear and fast-rising predominance of the right-frontal maximum that was present in the maps of the control group did not develop in patients. Therefore, the two maxima of the PINV appeared more balanced, respectively less asymmetric in schizophrenic patients.

These qualitative differences between the spatio-temporal patterns of PINV modulation in patients and controls was confirmed statistically. A significant Hemisphere by Group interaction was present throughout the PINV epoch, except for the first PINV segment. Corresponding to this, the Hemisphere effect in controls was present for all PINV segments except for the first one. In contrast, there was no significant hemisphere effect in patients. Furthermore, the Hemisphere by Group interaction remained significant after normalization of the data, except for the first PINV segment. Finally, while the F-values of the Hemisphere effects remained rather similar over the entire PINV segment in patients (below 1.38), a drastic increase from 3.36 to 11.39 could be observed in controls from the first to the second segment, with only moderate subsequent decreases towards the end of the epoch.

Altogether these qualitative and statistically-quantitative results indicate different spatio-temporal patterns *of the modulation of the PINV in patients and controls. While in controls there is a rapid shift from a relative balance between the two frontal PINV maxima to a predominance of the right-frontal maximum, as expressed by the increase in F-values of the within-group hemisphere effect, this shift towards a right-frontal predominance of the PINV does not take place to that extent in schizophrenic patients. Instead, the right-frontal PINV maximum develops only slowly and, in particular, much less pronounced than in controls. Accordingly, topographical differences between patients and controls, as indicated by the significant results of normalized data, evolve only about 1.5 s after the IS and due to the topographical "dynamics" in the control group.*

While high spatial resolution and a comprehensive assessment of the PINV topography was feasible only with the multi-channel recordings of experiment C (and D), it is meaningful to ask whether the statistical results of experiments A and B confirmed those of experiment C.

At least for the control groups, the "delayed" development of the pattern of maximum (right-frontal) lateralization can be followed in the F-values and hypothesis mean squares presented in **paragraph 8.14.1** and **table 9.1**. The within-group hemisphere effects of experiments A and B indicate a monotonous increase in hypothesis mean squares from the first to the last PINV segments analyzed; a similar sequence is obtained for the low-scoring group of experiment D (see **table 9.1**). Although perfect congruence of results between the experiments cannot be expected, for instance because of the difference of the intervals chosen for PINV parametrization, these comparisons suggest that even the temporal sequencing of the spatial PINV modulation is replicable in healthy subjects.

In schizophrenic patients the spatio-temporal patterns of PINV modulation differed between the experiments. In experiments A and C maximum hypothesis mean squares were found for the first PINV segments, but in experiment B the reverse was found. However, none of the within-group Hemisphere or Laterality effects for the patients of experiments B and C became statistically significant. As a consequence, no clear lateralization of the PINV developed in these experiments. Only in experiment A a left-frontal predominance developed which decreased in strength towards the end of the epoch, and this pattern was at least descriptively found in experiment C.

Similar results were obtained for the groups of experiment D. The low-scoring group also showed a drastic increase in hypothesis mean squares (and F-values for the Hemisphere simple effect) from PINV-1 (non-significant as in the control group of experiment C) to PINV-2, with somewhat decreasing values towards the end of the epoch. Furthermore, the high-scoring group exhibited a strong and statistically significant intial left-frontal PINV maximum, which rapidly changed into a balanced pattern with PINV-2 and the subsequent segments. Therefore, the spatio-temporal patterns of PINV modulation in low-scoring and high-scoring subjects resembled those of healthy controls and schizophrenic patients, respectively.

Altogether these results suggest that schizophrenic patients and high-scoring subjects, fail to develop the right-frontal PINV maximum seen in healthy controls and low-scoring subjects. The hypothesis and error mean squares documented in **table 9.1** indicate that this is not an effect of grossly augmented within-group heterogeneities.

9.3.3 Functional Significance of the PINV

The results of the probe-evoked potentials during tCNV and PINV of experiments A and B indicate at most small and overall non-significant group differences between schizophrenic patients and healthy controls. While this result seems to be at odds with results reported in the literature (e.g. Rockstroh et al., 1994), several procedural differences, for instance the task irrelevance of the probes in the present experiments, might account for the differences. In any event, the present results do not lend support to the notion of an atypical functional significance of slow negative potential shifts in schizophrenic patients, despite their atypical spatio-temporal modulation. The present experiments replicate reports of smaller N100 amplitudes of the probe-evoked potentials during PINV, compared to the probe-evoked potentials during tCNV (Rockstroh et al., 1994; Timsit-Berthier et al., 1971) and suggest an inhibitory or dysfacilitatory interpretation of the PINV, at least as to those aspects of inhibitory neurophysiological states that are reflected by the N100.

9.3.4 The Ambiguity Effect

One of the replicated results of this dissertation is the effect of ambiguous IS. In all experiments presented here ambiguous IS were associated with a significant augmentation of PINV amplitudes. It is interesting to note that this effect was "overall" (i.e. irrespective of topographical differentiations) of comparable strength in the experimental and control groups of all 4 experiments. This suggests that this effect taps a rather general condition of the development of the PINV. The putative nature of this condition, the detection of a mismatch between the representation of the task, and the perception of one of its elements in a concrete instance of that task, has been discussed in *paragraph 9.3.1*.

In the present paragraph, the topography of the ambiguity effect and possible group specific variants of this effect shall be discussed.

In experiment A the ambiguity effect was most pronounced at frontal (F3, F4), minimal at parietal (P3, P4), and of similar strength at the other leads ("Ambiguity by Electrode" interaction). This clear frontal dominance was replicated in experiments C and D ("Condition by Coronal Row" interaction), but not in experiment B ("Condition by Anteriority" interaction and planned contrast clear vs ambiguous). Whether the failure to replicate the frontal predominance of the ambiguity effect in experiment B was a consequence of the electrode montage was scrutinized by recomputing the ANOVA for the F3, F4, C3, C4, P3, P4 configuration of the data of experiment B and by checking the Condition by Anteriority interaction and the clear versus ambiguous planned contrast which both included the comparison between the frontal (F3, F4) and the central (C3, C4) electrode pairs, as was the case in experiment A. Neither of the Condition by Anteriority interactions (Fs $\leq$ 1.48) or of the planned contrasts (Fs $\leq$ 2.33) was statistically significant. This indicates that the failure to replicate the frontal predominance of the ambiguity effect —as far as the (F3, F4) versus (C3, C4) is meant— is not a consequence of a different electrode montage.

In any event these topographical changes are compatible with a change in the generator structure underlying the PINV, since they remain significant after normalization of the data. As a consequence, different cortical regions appear to be involved subsequent to the perception of clear versus ambiguous IS. The frontal predominance of the ambiguity effect as a presumed effect of the underlying generator structure and not a mere amplitude effect, emphasizes that an anterior proportion of the PINV may be experimentally and functionally distinct from a posterior proportion.

A frontal predominance has also been reported for the PINV augmentation from "constant lack of control" to "random lack of control" conditions by Kathmann et al. (1990), suggesting that the effect of contingency changes is most pronounced frontally. The topo-graphical similarity between the ambiguity and the "contingency change" effects suggest that similar processes may account for the two seemingly different effects, as has been postulated above.

The association between a *frontal* PINV and "the loss of expected response-dependent control of relevant outcome" (Birbaumer et al., 1986) is considered as another example of the postulated mismatch process: the "response-dependent control of relevant outcome" is part of the representation of the (expected) task structure —as is the occurence of clear IS and the contingency between IS and response categories— into which the perception of a trial (or a block of trials) with uncontrollable IS does not fit. It is the frontal proportion of the PINV that seems to be most sensitive to these postulated processes.

There is only weak evidence that schizophrenic patients and controls differ in the non-lateralizational (see below) topography of the ambiguity effect. In experiment A the "Ambiguity by Electrode by Group" interaction became significant because of the stronger frontal (F3, F4) accentuation of the ambiguity effect in controls and the stronger parietal accentuation in patients. The normalization of the data indicated that an amplitude effect cannot be rejected as the source of this interaction. Beyond this, no evidence of topographical differences in the ambiguity effect between schizophrenic patients and controls was found in any of the experiments.

For that reason, schizophrenic patients are apparently as sensitive as controls to the effects of ambiguous IS. This result does not occur unexpectedly, since there had been no sound arguments to assume the contrary. Impairments in visual perception or in elementary recognition memory for geometrical shapes, two deficiencies that might have leveled the effects of ambiguous IS, were certainly not present in the patient samples who responded as correctly as healthy controls. Similarly, augmented sensitivity to ambiguous IS was not expected and would have resulted in an augmentation of the ERP response in schizophrenic patients: a rare ERP phenomenon in this group.

A differentiation of the topography of the ambiguity effect between groups similar to that reported for experiment A was found in experiment D. The "Condition by Coronal Row by Sagittal Row by Group" interaction, including significant planned contrasts only for the third and fourth PINV segment, revealed a larger ambiguity effect in the low-scoring compared to the high-scoring group at the (F3, F4) leads. Hence, a diminished response at the "inner" frontal leads was present in schizophrenic patients and high-scoring subjects. While this observation is certainly worth mentioning, its importance should not be overemphasized since the diminished frontal responsivity in schizophrenic patients was not replicated in the experiments B and C.

While the non-lateralizational topography of the ambiguity effect is similar in schizophrenic patients and healthy controls, some differences in the lateralization of the ambiguity effect were found.

In experiment A the "Ambiguity by Hemisphere by Group" interaction was significant for all PINV segments analyzed, indicating a stronger right- than left-sided response in controls and, with lower strength, the reverse lateralization in schizophrenic patients. Only for controls was the within-group "Hemisphere by Ambiguity" interaction significant. A comparable result was obtained in experiment B only for the first PINV segment, indicating that although the midline-to-right predominance of the ambiguity effect was present in both groups, it was more accentuated in controls than in schizophrenic patients. Finally, in experiment C a stronger right- than left-sided accentuation of the ambiguity effect in controls, and the reverse lateralization in patients was found. This group-specific lateralization of the ambiguity effect began as a weak trend ($p=.15$) with PINV-1 but increased monotonously in strength towards the end of the interval ($p=.05$ for PINV-4).

Altogether these results are not entirely conclusive and suggest that a different lateralization of the ambiguity effect in patients and controls is at best a weak and inconsistent result. The results of normalized data, in any event, suggest that a simple amplitude effect cannot be ruled out for this effect.

The lateralization of the ambiguity effect in the low-scoring group of experiment D resembled that of the control groups in that it was maximal over the right-hemisphere. In contrast, the high-scoring group did not exhibit an asymmetry of the ambiguity effect. Again, a simple amplitude effect could not be rejected for this aspect of a group-specific topography of the ambiguity effect.

To summarize: The results of 3 experiments with schizophrenic patients and healthy controls and 1 experiment with subjects scoring high or low in the German translation of the SPQ suggest that the effect of ambiguous stimulus-stimulus contingencies during a visual matching task is a general one. In all experiments the experimental and control groups exhibited a similar "overall" PINV augmentation after ambiguous compared to clear IS. The topographical maximum of this effect was at frontal leads in 3 out of 4 experiments, but included central (C3, C4) leads in one experiment. Evidence for reduced frontal (F3, F4) ambiguity effects was found for schizophrenic patients and high-scoring subjects, but this effect was not replicated in schizophrenic patients. While the lateralization of the ambiguity effect differed inconsistently within the succession of PINV segments between the experiments, in no instance could a simple amplitude effect be rejected.

Hence, a frontal predominance of the ambiguity effect in all groups investigated can be concluded from these considerations. This topography resembles the topography reported by other authors for effects such as "contingency change" (Kathmann et al., 1990) and "loss of expected response-dependent control of relevant outcome" (cf. Birbaumer et al., 1986). It has been suggested that a commonality between these seemingly different experimental task variations is the induction of a mismatch between the representation of the task structure, including relaible stimulus-stimulus, stimulus-response, and response-outcome contingencies, and the perception of unreliable contingencies or discontingencies between elements of that task structure not fitting into this representation. Schizophrenic patients (and high-scoring subjects) do not differ from controls in responding to this presumed mismatch detection.

9.3.5 The Provisional Memory Effect

Increasing the demands on provisional memory functions increased the PINV amplitudes in schizophrenic patients, while there was no effect upon the PINV of healthy controls. The selectivity of the provisional memory effect for schizophrenic patients contrasts with the lack of specifity of the ambiguity and nogo effects. While both ambiguity and nogo effects were considered *general* factors of the PINV, the provisional memory factor appears to evoke psychophysiological processes that are specific to schizophrenic patients (when compared to healthy controls).

According to the theory of the prefrontal cortex by Fuster (1989) and investigations by other authors (e.g. Goldman-Rakic, 1987, 1991, 1994), the execution of delay tasks requires the involvement of the prefrontal cortex, interacting with other brain structures, most noteworthy infero-temporal cortex and hippocampus (see *paragraph 5.1*). While structural abnormalities are frequently reported for the temporo-limbic system in schizophrenia, pathophysiological or dysfunctional processes have been observed repeatedly in the prefrontal cortex of schizophrenic patients using different measurement techniques (see *paragraph 2.2*). The latter group of findings pertains in particular to task situations which increase prefrontal cortex metabolic parameters in healthy subjects. One of these task situations is the Wisconsin Card Sorting Test (WCST). While different psychological functions are obviously required in order to execute this —like any— neuropsychological task, e.g. shifting of response strategies (Kolb & Wishaw, 1990), the involvement of working or representational memory functions in the WCST performance has been proposed by several authors (e.g. Goldman-Rakic, 1991, 1994; Weinberger, 1991). As has been argued, the selection of a response during the WCST cannot be guided by the stimulus material but must be guided by a representation of a concept (of color, shape, etc.) and the strategy of problem solving.

Since different authors have used different terms for the description of the last mentioned function, with partially but not perfectly compatible semantics, it is noteworthy that some memory functions required by the WCST appear to warrant a provisional memory qualification in the sense of Fuster (1989). Each choice during the WCST must be related both to the preceding choices and to the task solving strategy, which must be represented and retained over time in order to guide the current choice. Almost similar words were used in *paragraph 5.1* in order to characterize the provisional memory function required by delay tasks as elaborated by Fuster (1989).

Goldman-Rakic (1991, 1994) assumes schizophrenia to be associated with a loss of *working memory functions*, resulting in a "deficit in the regulation of behavior by interalized schemata, symbolic representations, and ideas". The term "working memory" refers to the "ability to bring to mind events without direct stimulation". Working memory is therefore necessary in order to guide behavior by representations rather than stimulations.

In line with this reasoning Park & Holzman (1992, 1993) have demonstrated deficits of schizophrenic patients in an oculomotor delay task consistent with a working memory dysfunction. These working memory functions are hypothetically subserved by the joined activity of a network including the PFC and the hippocampus (Friedman et al., 1990).

The stressing of provisional memory functions should therefore stress one of the functions of the prefrontal cortex interacting with the temporo-limbic system. The results of experiment A indicate that the PINV in schizophrenic patients is related to this stressing of provisional memory functions. Although provisional or working memory are considered to be prefrontal cortical functions, this result does not necessarily support the "hypofrontality" hypothesis of schizophrenia. Envisaging the abundancy of functional connections between the prefrontal cortex and other brain structures (cf. Fuster, 1989), some of which, like the temporo-limbic system, are assumed to be afflicted in schizophrenia (see *chapter 2*), "hypofrontality" can not be taken as an isolated phenomenon. Instead, the "information networking" between certain regions of the brain should be assumed to be deficient in schizophrenic patients.

It has been suggested in *paragraph 9.3.1* that a PINV develops whenever the task provides discontingencies or unreliable contingencies between elements of the task structure, contrasting with a representation of the action scheme (task structure) in the subject which is comprised of reliable contingencies between these elements. It is tempting to speculate whether the PINV augmentation due to high provisional memory load might be related to these processes. Again, it is assumed that the subject normally sustains a representation of the scheme of action (task structure), which is comprised of reliable contingencies, and perceives the (trial-wise) empirical task structure in a top-down fashion against this background. While a *mismatch* between the (perceived) task structure and its representation is assumed to occur with interspersed ambiguous IS because of the experimentally induced discontingency or unreliability of contingencies between the experimental stimuli, the provisional memory effect should be explained by a weakening of the contingency formation between WS and IS, i.e. by a deficit in provisional or working memory functions in schizophrenic patients. In other words: the crucial requirement of the task, the sustaining of a representation of the task structure for the purpose of matching WS and IS, is assumed to be deficient in schizophrenic patients, weakening the ability to match the perceived task struture (experimental stimuli and their contingencies) with the representation of the task structure. Only in schizophrenic patients was there an augmentation of the IS-LPC amplitude under the high compared to the low provisional memory condition. Does this differential effect which approaches conventional significance levels ($p=.08$ for the Provisional Memory by Group interaction) reflect greater "context updating" due to a faded representation of the WS in schizophrenic patients ? The appropriateness of this post hoc-explanation cannot be evaluated at present. In any event, this putative incomplete matching shall also be called a *mismatch.*

In this sense the Ambiguity and the Provisional Memory effects represent two instances of a mismatch between perception and representation of the scheme of action or the task structure, which is more than a representation of the summation of the task elements but critically includes their contingencies. While the Ambiguity effect results from an experimental manipulation of the reliability of contingencies on the stimulus side, the Provisional Memory effect results from a weakening of trial-wise contingency formation on the (schizophrenic) subject side. This trial-wise contingency formation is sustained by the provisional or working memory functions that have been suggested to be deficient in schizophrenic patients (e.g. Goldman-Rakic, 1994; Weinberger, 1991, 1994), and have been shown to impair oculomotor delay task performance in this group (Park & Holzman, 1992, 1993).

9.3.6 The Nogo Effect

Nogo, compared to clear IS resulted in diminished PINV amplitudes, in particular at central and parietal leads. This contrasts with the anterior PINV augmentation following ambiguous IS. A common feature of the nogo and the ambiguity effect, distinguishing both effects from the provisional memory effect, is the similarity of their strenghts in schizophrenic patients and healthy controls. Hence, those functions influenced by the nogo effect are a common factor of the PINV.

Several interpretations of the nogo effect are conceivable and have been presented in *paragraph 6.10.4*. Some of these different interpretations are not mutually exclusive but possibly stress different aspects of the nogo effect.

It has been argued in *paragraph 6.10.4* that a PINV diminution following nogo as compared to clear IS is probably not related to the abolishing of the matching process by nogo IS.

The signal function of the nogo IS, to withhold the response, may have triggered the inhibition of motor structures preparing for the response. Response preparation and preparation for stimulus evaluation are considered to be reflected by the centro-parietally predominant tCNV (see *chapter 3*). While the tCNV was maximum at parietal leads, the nogo effect on the PINV was most pronounced centrally. This might indicate that primarily those processes reflected by the anterior proportion of the tCNV were inhibited by the nogo IS. These anterior proportions can plausibly be identified with the readiness potential (see *chapter 3*). Hence, the centro-parietal predominance of the nogo effect on PINV might indicate the inhibition or dampening of excited motor systems. This interpretation is supported by the IS-LPC augmentation and topographical shift towards anterior regions after nogo compared to clear IS (Roberts et al., 1994). This interpretation suggests that the posterior proportions of the PINV reflect a non-inhibited tCNV, lending support to the interpretation of the PINV as "delayed CNV resolution" (Mantanus et al., 1981; Roth et al., 1977) or "´prolongated duration´ CNV" (Timsit-Berthier et al., 1987) as a sign of ongoing motor preparation.

The nogo IS differed from the other IS not only with respect to their signal function but also concerning their perceptual discernibility which was, by color and shape, certainly superior to that of the clear and ambiguous IS. Hence, uncertainty about stimulus characteristics should be least for the the nogo IS. This aspect of uncertainty has been related to the posterior proportions of the PINV by Birbaumer et al. (1986). Again, the IS-LPC augmentation following nogo IS may be seen to support this suggestion. The amount of equivocation is inversely related to the amplitude of the P300 within the Triarchic model of Johnson, Jr. (1986). Reduced uncertainty about stimulus characteristics of the nogo IS amounts to reduced equivocation and should augment the IS-LPC following nogo as compared to clear IS.

What kind of process might be reflected by the posterior proportions of the PINV ? It has been argued in *paragraph 7.8* that the anterior proportions of the

PINV might reflect ruminations about the appropriateness of the response, affected by task features such as the quality of the matching process (Ambiguity effect). Complementary to the anterior PINV, the posterior PINV might reflect ruminations about responding *per se* which are suspended by the instruction not to respond (nogo effect).

These ruminations related to the responding per se may be related to Frith's neuropsychological theory of schizophrenia (Frith, 1992, 1994; Frith & Done, 1988), for which experimental evidence has been reported by several authors (e.g. Malenka et al. 1982; Mlakar et al., 1994). A lack of corollary discharges may be present in schizophrenic patients during termination of the forewarned reaction task and should result in a failure of appreciating "having completed their task" (Cohen, 1990). Removing the motor response from the task structure, as in the present nogo condition, should circumvent impaired corollary discharge generation, avoiding the triggering of response uncertainty and PINV. In this sense, the smaller PINV under the nogo condition is compatible with the hypothesis that impaired corollary discharges during the motor response in CNV standard tasks are triggering response uncertainty and PINV in schizophrenic patients.

Summary

This dissertation presents a series of experiments on the Post-Imperative Negative Variation (PINV) in schizophrenic patients and groups of healthy subjects.

Chapter 1 introduces the schizophrenic patients participating in the experiments by explaining the diagnostic concept (DSM-III-R) used for their selection.

Chapter 2 gives a survey over some of the anatomical, physiological, and cognitive abnormalities reported for schizophrenic patients in the literature, focusing on the neuropsychological theory of schizophrenia of Frith (1992, 1994; Frith & Done, 1988) and on the neurodevelopmental theory of schizophrenia of Weinberger (1991, 1994). Both theories "converge" in stressing the implications of a deficient communication between prefrontal and temporo-limbic (hippocampal) sites for the development and for cognitive abnormalities of schizophrenia, in particular deficits in "working memory" processes.

In *chapter 3* a survey of the literature on the Event-Related Potential (ERP) components investigated in the present experiments is given. In particular, results on the Contingent Negative Variation (CNV) and the Post-Imperative Negative Variation (PINV) in healthy subjects and schizophrenic patients are reviewed. Some arguments for the working hypotheses that the development of the PINV is related to the "stressing" of prefrontal cortex functions are presented.

Chapter 4 describes the methodological "requisites" used in the experiments, including the criteria for subject selection and clinical description of patients; the basic experimental design, which is a modified delayed matching-to-sample task in the visual modality, and its variations in particular experiments; the physiological recordings and recording equipment; and the procedures used for data reduction and data analysis. The basic experimental design was the presentation of two horizontally arranged diamonds of different sizes for 100 ms on a PC monitor as the warning stimulus (WS) within the task structure of the two-stimulus paradigm. With a stimulus onset asynchrony of 4.0 s a single diamond, being identical to one of the WS diamonds, appeared in the middle of the screen for 100 ms as the imperative stimulus (IS). The subjects were instructed to match the IS diamond with the identical WS diamond by pressing the response button of the left or right hand as fast and as accurately as possible.

Chapter 5 deals with the first out of 4 experiments of this dissertation. Experiment A had 4 goals. The induction of a PINV in healthy subjects, the selective experimental modulation of the PINV in schizophrenic patients, the explorative assessment of the topography of the PINV in schizophrenic patients and healthy controls, and the investigation of the functional significance of the CNV and PINV by means of the "probe" technique. Data of 19 patients with a DSM-III-R diagnosis of a chronic schizophrenic disorder and of 19 healthy controls matched for age, sex, and education with the patient group are reported. The results indicate that a PINV can be observed in healthy subjects with an experimental task including ambiguity in the matching of the imperative and warning stimuli (by interspersing IS with a size intermediate between those of the WS diamonds ("ambiguity effect")). Furthermore, the differential "stressing" of provisional (or working) memory functions, considered to be subserved primarily by the prefrontal cortex (Fuster, 1989) (by holding the arrangement of the WS diamonds constant in one session, but varying this arrangement in the other session from trial to trial; "provisional memory effect"), modulates the amplitude of the PINV only in schizophrenic patients but not in healthy controls. Schizophrenic patients exhibit an overall larger PINV than healthy subjects. While the PINV is predominant over the left hemisphere in patients, the reverse asymmetry can be seen in controls.

Chapter 6 deals with experiment B, which investigated whether the PINV develops if the motor response is abolished from the task structure by the inclusion of "nogo" IS (an ellipse signalling not to respond). Another goal of experiment B was the investigation of the functional significance of CNV and PINV by the "probe" technique. Data of 11 patients with a DSM-III-R diagnosis of a chronic schizophrenic disorder and of 13 healthy controls, matched for age, sex, and eduction with the patient group are reported. The results show a significant diminution of PINV amplitudes especially at central and parietal sites following response withholding due to "nogo" IS in both groups. Again a significantly larger PINV was found in schizophrenic patients than in healthy controls. While the former group exhibited no significant lateralization of the PINV, for the latter group a right-sided predominance could be replicated.

In c*hapter 7* the results of analysing the probe-evoked potentials from experiments A and B are reported. Acoustical probe stimuli of 20 ms duration and 50 (experiment A) and 80 (experiment B) dB intensity were delivered by headphones (a) one per second continually throughout the trial (experiment A) or (b) two per trial during the pretrial baseline, initial or late CNV, or PINV (experiment B) in one out of six possible pairwise combinations. Results of both experiments indicate no group differences in the overall amplitudes of the N100, P200, or N1/P2 components of the probe-evoked potentials, and smaller amplitudes of these components during PINV when compared to the late CNV.

Chapter 8 deals with two separate 32-channel recordings of the PINV. Data of 12 patients with DSM-III-R diagnosis of a chronic schizophrenic disorder and of 11 healthy controls matched for age, sex, and eduction are reported for experiment C. Data of 18 subjects with "low" scores and of 19 subjects with "high" scores (defined by a median split of the total sample) in a German translation of the Schizotypal Personality Questionnaire (SPQ, Raine, 1991) are reported for experiment D. In addition to the grand average curves, scalp and cortical amplitude as well as scalp laplacian maps and statistical ANOVA analyses using cortical data of a subset of 12 frontal and temporo-central leads are reported. Cortical amplitude and scalp laplacian maps indicate that the PINV consists of two separate maxima, located over the left and right frontal cortex in both groups, which are less discernible in the scalp amplitude maps. In healthy controls and in low-scoring subjects a predominance of the right-frontal maximum develops about 1.5 s after IS onset, but this spatio-temporal pattern is not seen in schizophrenic patients and high-scoring subjects. These groups, instead, exhibit a rather symmetric pattern of the two PINV maxima, slightly favoring the left side. The qualitative description of the maps is confirmed by the ANOVA results.

Chapter 9 provides a general summary and integration of results, focusing on (a) the replicated induction of a PINV in healthy subjects with an experimental task that includes ambiguity in the matching of IS and WS, (b) the replicated PINV augmenting effect of ambiguously as compared to clearly matchable IS and its topography, (c) the replicated right-frontal predominance of the PINV in healthy (and low-scoring) subjects and the divergency of the PINV topography in schizophrenic patients (and high-scoring subjects), (d) the effect of stressing provisional memory functions on PINV in schizophrenic patients, (e) the PINV diminishing effect of nogo IS and its topography, (f) the functional significance of the PINV as revealed by the "probe" technique. Furthermore, speculations about the unity of the antecedent conditions of the PINV are undertaken.

References

1. (1980). Methodology: Repeated Measures F Tests and Psychophysiological Research: Controlling the Number of False Positives. *Psychophysiology, 17 (5),* 499-503.

2. Abraham, P., Docherty, T.B., Spencer, S.C., et al (1980). *An International Pilot Study of CNV in Mental Illness.* In H.H. Kornhuber& L. Deecke (Eds.).*Motivation, Motor and Sensory Processes of the Brain. Electrical potentials and Clinical Use. Progress in Brain Research.* Amsterdam: Elsevier, pp.535-542.

3. Abraham, P.& McCallum, W.C. (1976). *The CNV and its Relation to Specific Psychiatric Syndroms.* In W.C. McCallum& J.R. Knott (Eds.).*The Responsive Brain.* Bristol: Wright, pp.144-149.

4. Abraham, P.& McCallum, W.C. (1977). A Permanent Change in the EEG (CNV) of the Schizophrenics. *Electroencephalography and Clinical Neurophysiology, 43,* 553.

5. Abraham, P., McCallum, W.C., Docherty, T., Fox, A.& Newton, P. (1974). The CNV in Schizophrenia. *Electroencephalography and Clinical Neurophysiology, 36,* 217-219.

6. Aizawa, H., Inase, M., Mushiake, H., Shima, K.& Tanji, J. (1991). Reorganization of Activity in the Supplementary Motor Area Associated with Motor Learning and Functional Recovery. *Experimental Brain Research, 84,* 668-671.

7. Alexander, G.E., DeLong, M.R.& Strick, P.L. (1986). Parallel Organization of Functionally Segregated Circuits Linking Basal Ganglia and Cortex. *Annual Review of Neuroscience, 9,* 357-381.

8. Alexander, G.E.& Goldman, P.S. (1978). Functional Development of the Dorsolateral Prefrontal Cortex: An Analysis Utilizing Reversible Cryogenic Depression. *Brain Research, 143,* 233-249.

9. Allen, W.J. (1994). Effects of Ablating the Frontal Lobes, Hippocampi, and Occipito- Paretio-Temporal (Excepting Pyriform Areas) Lobes on Positive and Negative Olfactory Conditioned Reflexes. *American Journal of Physiology, 128,* 754-771.

10. Altshuler, L.L., Casanova, M.F., Goldberg, T.E.& Kleinman, J.E. (1990). The Hippocampus and Parahippocampus in Schizophrenic, Suicide, and Control Brains. *Archives of General Psychiatry, 47*, 1029-1034.

11. Amabile, G., Fattapposta, F., Pozzessere, G., et al (1986). Parkinson Disease: Electrophysiological (CNV) Analysis Related to Pharmacological Treatment. *Electroencephalography and Clinical Neurophysiology, 64*, 521-524.

12. Amassian, V.E., Cracco, R.Q., Maccabee, P.J.& Cracco, J.B. (1992). Cerebello-Frontal Cortical Projections in Humans Studies with the Magnetic Coil. *Electroencephalography and Clinical Neurophysiology, 85*, 265-272.

13. American Encephalographic Society, (1991). American Electroencephalographic Society Guidelines for Standard Electrode Position Nomenclature. *Journal of Clinical Neurophysiology, 8 (2)*, 200-203.

14. Andreasen, N.C., Erhardt, J.C., Swayze, V.W., et al (1990). Magnetic Resonance Imaging of the Brain in Schizophrenia. *Archives of General Psychiatry, 47*, 35-44.

15. Andreasen, N.C., Nasrallah, H.A., Dunn, V., et al (1986). Structural Abnormalities in the Frontal System in Schizophrenia. *Archives of General Psychiatry, 43*, 136-144.

16. Andreasen, N.C.& Olson, S. (1982). Negative versus Positive Schizophrenia. *Archives of General Psychiatry, 39*, 789-794.

17. Andreasen, N.C., Rezai, K., Alliger, R., et al (1992). Hypofrontality in Neuroleptic-Naive Patients and in Patients With Chronic Schizophrenia. *Archives of General Psychiatry, 49*, 943-958.

18. Andreasen, N.C., Swayze, V., Flaum, M., Alliger, R.& Cohen, G. (1990). Ventricular Abnormalities in Affective Disorder: Clinical and Demographic Corelates. *American Journal of Psychiatry, 147 (7)*, 893-900.

19. Andresen, B. (1980). *Persönlichkeitsstruktur und Psychosetendenz: Entwicklung, Psychopathologische Konstruktvalidierung und Diagnostische Operationalisierung eines Integrativen 6-Dimensionalen Modell der Persönlichkeit.* In *6 Basisdimensionen des Temperamenst und Charakters auf der Grundlage des Hamburger-Persönlichkeits-Fragebogen (HPF-G) (Habilitationsschrift, in Vorbereitung).*

20. Arndt, S., Alliger, R.J.& Andreasen, N.C. (1991). The Distinction of Positive and Negative Symptoms. *British Journal of Psychiatry, 158*, 317-322.

21. Arnstein, A.F.T& Goldman-Rakic, P.S. (1984). Selective Prefrontal Cortical Projections to the Region of the Locus Coeruleus and Raphe Nuclei in the Rhesus Monkey. *Brain Research, 306*, 9-18.

22. Bachneff, S.A.& Engelsmann, F. (1980). Contingent Negative Variation, Postimperative Negative Variation, and Psychopathology. *Biological Psychiatry, 15 (2)*, 323-328.

23. Bachneff, S.A.& Engelsmann, F. (1983). Correlates of Cerebral Event-Related Slow Potentials and Psychopathology. *Psychological Medicine, 13*, 763-770.

24. Bakay Pragay, E., Mirsky, A.F.& Nakamura, R.K. (1987). Attention-Related Unit Activity in the Frontal Association Cortex During a Go/No-Go Visual Discrimination Task. *Experimental Neurology, 96*, 481-500.

25. Barrett, K., McCallum, W.C.& Pocock, P.V. (1986). Brain Indicators of Altered Attention and Information Processing in Schizophrenic Patients. *British Journal of Psychiatry, 148*, 414-420.

26. Barta, P.E., Pearlson, G.D., Powers, R.E., Richards, S.S.& Tune, L.E. (1990). Auditory Hallucinations and Smaller Superior Temporal Gyrus Volume in Schizophrenia. *American Journal of Psychiatry, 147*, 1457-1462.

27. Barta, P.E., Pearlson, G.D., Powers, R.E., Richards, S.S.& Tune, L.E. (1990). Reduced Volumes of Superior Temporal Gyrus in Schizophrenia; Relationship to Auditory Hallucinations. *American Journal of Psychiatry, 147*, 1457-1462.

28. Bartlett, E.J., Brown, J.W., Wolf, A.P.& Brodie, J.D. (1987). Correlations between Glucose Metabolic Rates in Brain Regions of Healthy Male Adults at Rest and During Language Stimulation. *Brain and Language, 32*, 1-18.

29. Bauer, H. (1984). *Regulation of Slow Potentials Affect Task Performance*. In T. Elbert, B. Rockstroh, W. Lutzenberger& N. Birbaumer (Eds.).*Self Regulation of the Brain and Behavior*. Heidelberg,Berlin: Springer, pp.216-226.

30. Bauer, H., Korunka, C.& Leodolter, M. (1993). *Possible Glial Contribution in the Electrogenesis of SPs*. In W.C. McCallum& S.H. Curry (Eds.).*Slow Potential Changes in the Human Brain*. New York: Plenum Press, pp.2334.

31. Becker, W.& Kristeva, R. (1980). *Cerebral Potentials Prior to Various Force Deployments*. In H.H. Kornhuber& L. Deecke (Eds.).*Motivation, Motor*

and Sensory Processes of the Brain. Electrical Potentials, Beahavior and Clinical Use. Amsterdam: Elsevier, pp.189-194.

32. Benes, F.M., Davidson, J.& Bird, E.D. (1986). Quantitative Cytoarchitectural Studies of the Cerebral Cortex of Schizophrenics. *Archives of General Psychiatry, 43*, 31-35.

33. Benton, A.L. (1968). Differential Behavioral Effects in Frontal Lobe Disease. *Neuropsychologia, 6*, 53-60.

34. Berg, P. (1986). The Residual After Correcting Event-Related Potentials for Blink- Artifacts. *Psychophysiology, 23*, 354-364.

35. Berg, P.& Scherg, M. (1990). *Modelling the Ocular Dipoles*. In C.H.M. Brunia, A.W.K. Gaillard& A. Kok (Eds.).*Psychophysiological Brain Research*. Tilburg Univerity Press, pp.23-25.

36. Berg, P.& Scherg, M. (1991). Dipole Models of Eye Movements and Blinks. *Electroencephalography and Clinical Neurophysiology, 79*, 36-44.

37. Berg, P.& Scherg, M. (1991). Dipole Modelling of Eye Activity and its Application tp the Removal of Eye Artefacts from the EEG and MEG. *Clinical Physiology and Physiological Meassurement, 12*, 49-54.

38. Berg, P.& Scherg, M. (1994). A Multiple Source Approach to the Correction of Eye Artifacts. *Electroencephalography and Clinical Neurophysiology*, 1-13.

39. Berman, K.F. (1987). Cortical "Stress Tests" in Schizophrenia: Regional Cerebral Blood Flow Studies. *Biological Psychiatry, 22*, 1304-1326.

40. Berman, K.F., Doran, A.R., Pickar, D.& Weinberger, D.R. (1993). Is the Mechanism of Prefrontal Hypofunction in Depression the Same as in Schizophrenia ? *British Journal of Psychiatry, 162*, 183-192.

41. Berman, K.F., Gold, C.R.J., Abi-Dargham, A., et al (1991). PET Studies of Frontal Lobe Function During Cognition. *Schizophrenia Research, 4*, 399-415.

42. Berman, K.F., Illowsky, B.P.& Weinberger, D.R. (1988). Physiological Dysfunctions of Dorsolateral Prefrontal Cortex in Schizophrenia / IV. Further Evidence for Regional and Behavioral Specifity. *Archives of General Psychiatry, 45*, 616-622.

43. Berman, K.F.& Weinberger, D.R. (1990). Lateralisation of Cortical Function During Cognitive Tasks: Regional Cerebral Blood Flow Studies of Normal

Individuals and Patients with Schizophrenia. *Journal of Neurology,Neurosurgery,and Psychiatry, 53*, 150-160.

44. Berman, K.F., Weinberger, D.R., Shelton, R.C.& Zec, R.F. (1987). A Relationship Between Anatomical and Physiological Brain Pathology in Schizophrenia: Lateral Cerebral Ventricular Size Predicts Cortical Blood Flow. *American Journal of Psychiatry, 144 (10)*, 1277-1282.

45. Berman, K.F., Zec, R.F.& Weinberger, D.F. (1986). Physiological Dysfunction of Dorsolateral Prefrontal Cortex in Schizophrenia / II. Role of Neuroleptic Treatment, Attention, and Mental Effort. *Archives of General Psychiatry, 43*, 126-135.

46. Bilder, R.M., Wu, H., Bogerts, B., et al (1994). Abnormal Cortical Asymmetry and Sulcal Prominence Are Independent Predictors of Schizophrenia. *Schizophrenia Research, 11 (2)*, 131.

47. Birbaumer, N.& Elbert, T. (1988). Byproduct of a Byproduct. *Behavioural and Brain Science, 11 (3)*, 375.

48. Birbaumer, N., Elbert, T., Rockstroh, B.& Lutzenberger, W. (1986). *On the Dynamics of the Post-Imperative Negative Variation.* In W.C. McCallum, R. Zappoli& F. Denoth (Eds.).*ERPs Associated with Preparatory and Movement-Related Processes. A Review*. Elsevier, pp.212-219.

49. Björklund, A., Divac, I.& Lindvall, O. (1978). Regional Distribution of Catecholamines in Monkey Cerebral Cortex, Evidence for a Dopaminergic Innervation of the Primate Prefrontal Cortex. *Neuroscience Letters, 7*, 115-119.

50. Blackwood, D.H.R, Clair, D.M.S.& Kutcher, S.P. (1986). P300 Event-Related Potential Abnormalities in Borderline Personality Disorder. *Biological Psychiatry, 21*, 557-560.

51. Blackwood, D.H.R& Muir, W.J. (1990). Cognitive Brain Potentials and their Application. *British Journal of Psychiatry, 157 (9)*, 96-101.

52. Blackwood, D.H.R., Whalley, L.J., Christie, J.E., Blackburn, I.M., St Clair, D.M.& McInnes, A. (1987). Changes in Auditory P3 Event-Related Potential in Schizophrenia and Depression. *British Journal of Psychiatry, 150*, 154-160.

53. Bliss, E.L.& Ailion, J. (1971). Relationship of Stress and Activity to Brain Dopamine and Homovanillic Acid. *Life Sciences, 10 (1)*, 1161-1169.

54. Bogerts, B., Ashtari, M., Degreef, G., Alvir, J.M.J., Bilder, R.M.& Liebermann, J.A. (1990). Reduced Temporal Limbic Structure Volumes on Magnetic Resconance in First Episode Schizophrenia. *Psychiatry Research, 35*, 1-13.

55. Bogerts, B., Falkai, P.& Haupts, M. (1990). Post-Mortem Volume Measurements of Limbic System and Basal Ganglia Structures in Chronic Schizophrenia. *Schizophrenia Research, 3*, 295-301.

56. Bogerts, B., Häntsch, J.& Herzer, M. (1983). A Morphometric Study of the Dopamine-Containing Cell Groups in Mesencephalon of Normals, Parkinson Patients, and Schizophrenics. *Biological Psychiatry, 18 (9)*, 951-969.

57. Bogerts, B., Lieberman, J.A., Ashtari, M., et al (1993). Hippocampus-Amygdala Volumes and Psychopathology in Chronic Schizophrenia. *Biological Psychiatry, 33*, 236-246.

58. Bogerts, B., Meertz, E.& Schönfeldt-Bausch, R. (1985). Basal Ganglia and Limbic System Pathology in Schizophrenia. *Archives of General Psychiatry, 42*, 784-791.

59. Bolz, J.& Giedke, H. (1981). Controllability of an Aversive Stimulus in Depressed Patients and Healthy Controls: A Study Using Slow Brain Potentials. *Biological Psychiatry, 16 (5)*, 441-452.

60. Borst, U.& Cohen, R. (1987). Impact of Time Estimation on the Crossover Effect in Schizophrenia. *Psychiatry Research, 22*, 331-340.

61. Bortz, J. (1993). *Statistik für Sozialwissenschaftler*. Berlin, Heidelberg: 3rd Ed., Springer Verlag.

62. Box, G.E.P. (1954). Some Theorems on Quadratic Forms Applied in the Study of Analysis of Variance Prblems: II. Effects of Inequality of Variance and Correlation between Errors in Two-Way Classification. *Annals of Mathematical Statistics, 25*, 484-498.

63. Bötzel, K., Plendl, H., Paulus, W.& Scherg, M. (1993). Bereitschaftspotential: Is There a Contribution of the Supplementary Area. *Electroencephalography and Clinical Neurophysiology, 89*, 187-196.

64. Bracale, M., Kemali, D., Perris, C., et al (1984). Evoked Potential in Psychiatry. *Advances in Biological Psychiatry, 15*, 30-45.

65. Bracha, H.S., Lyden, P.D.& Khansarinia, S. (1989). Delayed Emergence of Striatal Dopaminergic Hyperactivity After Anterolateral Ischemic Cortical Lesions in Humans; Evidence From Turing Behavior. *Biological Psychiatry, 25*, 265-274.

66. Bradshaw, J.L.& Nettelton, N.C. (1981). The Nature of Hemispheric Specialization in Man. *Behavioural and Brain Science, 4*, 51-91.

67. Braitenberg, V. (1978). *Cell Assemblies in the Cerebral Cortex.* In R. Heim& G. Palm (Eds.).*Theoretical Approach to Complex Systems.* Heidelberg: Springer, pp.171-188.

68. Braun, C. (1991). *Analyseverfahren ereigniskorrelierter Potentiale bei Vielkanalableitungen.* (UnPub)

69. Braun, C., Lutzenberger, W., Miltener, W.& Elbert, W. (1990). *Can Subcortical Structures Generate Potentials Large in Amplitude.* In C.M.H. Brunia, A.W.K. Gaillard& A. Kok (Eds.).*Psychophysiological Brain Research.* Tilburg: Tilburg University Press, pp.31-35.

70. Brecher, M.& Begleiter, H. (1983). Event-Related Brain Potentials to High-Incentive Stimuli in Unmedicated Schizophrenic Patients. *Biological Psychiatry, 18 (6)*, 661-674.

71. Brecher, M., Porjesz, B.& Begleiter, H. (1987). Late Positive Component Amplitude in Schizophrenics and Alcoholics in Two Different Paradigms. *Biological Psychiatry, 22*, 848-856.

72. Brown, R., Colter, N., Corsellis, N., et al (1986). Postmortem Evidence of Structural Brain Changes in Schizophrenia. *Archives of General Psychiatry, 43*, 36-42.

73. Brown, T.J. (1990). Temporal Lobe Asymmetry as the Key to the Etiology of Schizophrenia. *Schizophrenia Bulletin, 16*, 433-443.

74. Brozoski, T.J., Brown, R.M., Rosvold, H.E.& Goldman, P.S. (1979). Cognitive Deficit Caused by Regional Depletion of Dopamine in Prefrontal Cortex of Rhesus Monkey. *Science, 205 (31)*, 929-932.

75. Brunia, C.H.M. (1988). Movement and Stimulus Preceding Negativity. *Biological Psychology, 26*, 165-178.

76. Brunia, C.H.M. (1993). Waiting in Readiness: Gating in Attention and Motor Preparation. *Psychophysiology, 30*, 327-339.

77. Brunia, C.H.M. (1993). *Stimulus Preceding Negativity: Arguments in Favour of Non Motoric Slow Waves.* In W.C. McCallum& S.H. Curry (Eds.).*Slow Potential Changes in the Human Brain.* New York: Plenum Press, pp.147-161.

78. Bruyant, P., García-Larrea, L.& Mauguière, F. (1993). Target Side and Scalp Topography of the Somatosensory P300. *Electroencephalography and Clinical Neurophysiology, 88*, 468-477.

79. Buchanan, R.W., Breier, A., Kirkpatrick, B., et al (1993). Structural Abnormalities in Deficit and Nondeficit Schizophrenia. *American Journal of Psychiatry, 150 (1)*, 59-65.

80. Buchsbaum, M.S. (1990). The Frontal Lobes, Basal Ganglia, and Temporal Lobes as Sites for Schizophrenia. *Schizophrenia Bulletin, 16 (3)*, 379-389.

81. Buchsbaum, M.S., Nuechterlein, K.H., Haier, R.J., et al (1990). Glucose Metabolism Rate in Normals and Schizophrenics During the Continuous Performance Test Assessed by Positron Emission Tomography. *British Journal of Psychiatry, 156*, 216-227.

82. Buchsbaum, M.S., Wu, J.C., DeLisi, L.E., et al (1987). Positron Emission Tomography Studies on Basal Ganglia and Somatosensory Cortex Neuroleptic Drug Effects: Differences Between Normal Controls and Schizophrenic Patients. *Biological Psychiatry, 22*, 479-494.

83. Bugbee, N.M.& Goldman-Rakic, P.S. (1981). Functional 2-Deoxyglucose Mapping in Association Cortex: Prefrontal Activation in Monkeys Performing a Cognitive Task. *Society of Neurosciences Abstracts, 7*, 239.

84. Butler, S. (1993). *Methodological Issues in the Measurement of DC Potentials.* In W.C. Callum& S.H. Curry (Eds.).*Slow Potential Changes in the Human Brain.* New York: Plenum Press, pp.63-77.

85. Callaway, E.& Nagdhi, S. (1982). An Information Processing Model for Schizophrenia. *Archives of General Psychiatry, 39*, 339-347.

86. Callaway, E.I. (1970). Schizophrenia and Interference. *Archives of General Psychiatry, 22*, 193-208.

87. Carnahan, H., Chua, R., Eliot, D., Velamoor, V.R.& Carnahan, C.J (1994). Effects of Schizophrenia and Prefrontal Leukotomy on Movement Preparation and Generation. *Journal of Clinical and Experimental Neuropsychology, 16 (2)*, 253-260.

88. Casaglia, D.C.& Pantaleo, G.G. (1993). Brain Mapping: A Contribution to Linear Interpolation. *Brain Topography, 5 (3)*, 283-288.

89. Caspers, H., Speckmann, E.-J.& Lehmenkühler, A. (1980). *Electrogenesis of Cortical DC Potentials.* In K.K. Kornhuber& L. Deecke (Eds.).*Motivation, Motor and Sensory Processes of the Brain. Electrical Potentials, Behaviour and Clinical Use. Progress in Brain research.* Amsterdam: Elsevier, pp.3-15.

90. Caspers, H., Speckmann, E.-J.& Lehmenkühler, A. (1984). *Neurophysiological Mechanisms That Regulate Brain Potential Changes: Electrogenesis of Slow Potentials of the Brain.* In T. Elbert, B. Rockstroh, W. Lutzenberger& N. Birbaumer (Eds.).*Self-regulation of the Brain and Behaviour.* Berlin: Springer-Verlag, pp.26-42.

91. Catts, S.V., Ward, P.B., Armstrong, M.S., Afox, A.M.& McConaghy, N. (1986). *CNV/PINV in Psychiatric Patients and Their Parents.* In J.W. Rohrbough, R. Johnson& R. Parasuraman (Eds.).*Eight International Conference on Event-Related Potentials of the Brain.* Stanford: Research Reports, pp.57-59.

92. Chang, W.-H., Yeh, E.-K., Hu, W.-H., Tseng, Y.-T., Chung, M.-C.& Chang, H.-F. (1986). Prefrontal Cortex: Possible Site of Antipsychotic Action of Haloperidol. *Biological Psychiatry, 21*, 422-424.

93. Charbol, H., Guell, A., Bes, A.& Morom, P. (1986). Cerebral Blood Flow in Schizophrenic Adolescents. *Archives of General Psychiatry, 143*, 130.

94. Chiarenza, G.A. (1986). *Development of Sensory Motor and Cognitive Processes in Children. Movement Related Brain Macropotentials in Children.* In V. Gallai (Ed.). *Maturation of the CNS and Evoked Potentials.* Elsevier Science Publishers B.V., pp.236-246.

95. Chiarenza, G.A., Papakostopoulos, D., Dini, M.& Cazzullo, C.L. (1985). Neurophysiological Correlates of Psychomotor Activity in Chronic Schizophrenics. *Electroencephalography and Clinical Neurophysiology, 61*, 218-228.

96. Chouinard, G., Annable, L.& Dongier, M. (1977). Differences in Psychopathology of Schizophrenic Patients with Normal and Abnormal Postimperative Negative Variation (PINV). *Comprehensive Psychiatry, 18 (1)*, 83-87.

97. Chouinard, G., Annable, L., Dubrovsky, B.& Dongier, M. (1975). Postimperative Negative Variation (PINV) in Ambulatory Schizophrenic Patients. *Comprehensive Psychiatry, 16 (5)*, 457-460.

98. Christison, G.W., Casanova, M.F., Weinberger, D.R., Rawlings, R.& Kleinman, J.E. (1989). A Quantitative Investigation of Hippocampal Pyramidal Cell Size, Shape, and Variability of Orientation in Schizophrenia. *Archives of General Psychiatry*, *46*, 1027-1032.(Abstract)

99. Chwilla, D.J.& Brunia, C.H.M. (1991). Event-Related Potential Correlates of Non-Motor Anticipation. *Biological Psychology*, *32*, 125-141.

100. Classen, W.& Fritze, J. (1988). Ventricular Size, Cognitive and Psychomotor Performance, and Laterality in Schizophrenia. *Neuropsychobiology*, *20*, 87-90.(Abstract)

101. Cleghorn, J.M., Franco, S., Szechtman, B., et al (1992). Toward a Brain Map of Auditory Hallucinations. *American Journal of Psychiatry*, *149*, 1062-1069.(Abstract)

102. Cleghorn, J.M., Garnett, E.S., Nahmias, C., et al (1990). Regional Brain Metabolism During Auditory Hallucinations in Chronic Schizophrenia. *British Journal of Psychiatry*, *157*, 562-570.

103. Cleghorn, J.M., Garnett, E.S., Nahmias, C., et al (1989). Increased Frontal and Reduced Parietal Glucose Metabolism in Acute Untreated Schizophrenia. *Psychiatry Research*, *28*, 119-133.(Abstract)

104. Cleghorn, J.M., Kaplan, R.D., Nahmias, C., Garnett, E.S., Szechtman, H.& Szechtman, B. (1989). Inferior Parietal Region Implicated in Neurocognitive Impairment in Schizophrenia. *Archives of General Psychiatry*, *46*, 758-759.(Abstract)

105. Cohen, R. (1990). *Event-Related Potentials and Cognitive Dysfunction in Schizophrenia*. In H. Häfner& W.F. Gattaz (Eds.).*Search for the Causes of Schizophrenia*. Berlin: Springer, pp.342-360.

106. Cohen, R., Berg, P.& Hopman, G. (1991). *Slow Response-Related Negatives in Chronic Schizophrenics*. In E. Racagni (Ed.). *Biological Psychiatry*. Elsevier Science Publishers B.V., pp.550-552.

107. Cohen, R., Berg, P.& Hopmann, G. (1991). Slow Response-Related Negativities in Chronic Schizophrenics. *Biological Psychiatry*, *1*, 550-552.

108. Cohen, R., Hermanutz, M.& Rist, F. (1984). *Zur Spezifität Sequentieller Effekte in den Reaktionszeiten und Ereignisbezogenen Potentialen Chronisch Schizophrener*. In A. Hopf& H. Beckmann (Eds.).*Forschungen zur Biologischen Psychiatrie*. Berlin: Springer, pp.24-34.

109. Cohen, R., Rist, F., Finger, L.& Mussagy, L. (1990). *Modality-Dependent Post-Imperative Negative Variations to Auditory Stimuli in Schizophrenic Patients.* In C.H.M. Brunia, A.W.K. Guillard& A. Kok (Eds.).*Psychophysiology and Brain Research.* Tilburg University Press, pp.203-208.

110. Cohen, R.M., Semple, W.E., Gross, M., et al (1987). Dysfunction in a Prefrontal Substrate of Sustained Attention in Schizophrenia. *Life Sciences, 40,* 2031-2039.

111. Colebatch, J.G., Debier, M.-P., Passingham, R.E., Friston, K.J.& Frackowiak, R.S.J. (1991). Regional Cerebral Blood Flow during Voluntary Arm and Hand Movements in Human Subjects. *Journal of Neurophysiology, 65 (6),* 1392-1401.

112. Connolly, J.F., Gruzellier, J.H., Manchanda, R.& Hirsch, S.R. (1983). Visual Evoked Potentials in Schizophrenia. *British Journal of Psychiatry, 142,* 152-155.

113. Connolly, J.F., Manchanda, R.& Gruzelier, J.H. (1985). Pathway and Hemisphere Differences in the Event-Related Potential (ERP) to Mono Aural Stimulation, A Comparison of Schizophrenic Patients with Normal Controlls. *Biological Psychiatry, 20,* 293-303.

114. Connolly, J.F., Manchanda, R., Gruzellier, J.H.& Hirsch, S.R. (1983). Auditory Event-Related Potentials in Scizophrenic Patients. *Advances in Biological Psychiatry, 13,* 93-97.

115. Cooper, J.E., Andrews, H.& Barber, C. (1985). Stable Abnormalities in the Lateralisation of Early Cortical Somatosensory Evoked Potentials in Schizophrenic Patients. *British Journal of Psychiatry, 146,* 585-593.

116. Cooper, R., Winter, A.L., Crow, H.J.& Walter, W.G. (1965). Comparison of Subcortical, Cortical and Scalp Activity. *Electroencephalography and Clinical Neurophysiology, 18,* 217-228.

117. Cooper, R., Winther, A.L., Crow, H.J.& Grey Walter, W. (1965). Comparison of Cortical, Subcortical and Scalp Activity Using Chronically Indwelling Electrodes in Man. *Electroencephalography and Clinical Neurophysiology, 18,* 217-228.

118. Corwin, J.V., Kanter, S., Watson, R., Heilman, K.M., Valenstein, E.& Hashimoto, A. (1986). Apomorphie has a Therapeutic Effect on Neglect Produced by Unilateral Dorsomedial Prefrontalcortex Lesions in Rats. *Experimental Neurology, 94,* 683-698.

119. Coryell, W.& Zimmermann, M. (1987). Progress in the Classification of Functional Psychoses. *American Journal of Psychiatry, 144 (11)*, 1471-1474.

120. Courchesne, E., Elmasian, R.& Courchesne, R. (1987). *Electrophysiological Correlates of Cognitive Processing: P3b and Nc, Basic, Clinical, and Developmental Research.* In S.R. Holiday& R. Butler (Eds.).*A Textbook to Clinical Neurophysiology*. Chichester: Wiley, pp.645-676.

121. Crow, T.J. (1990). Temporal Lobe Asymmetries as the Key to the Etiology of Schizophrenia. *Schizophrenia Bulletin, 16 (3)*, 433-443.

122. Crow, T.J., Bal, J.& Bloom, S.R. (1989). Scizophrenia as an Anomally of Developmentof Cerebral Asymmetry. A Psotmortem Study and a Proposal Concerning the Genetic Basis of the Disease. *Archives of General Psychiatry, 46*, 1145-1150.

123. Crow, T.J., Ball, J.& Bloom, S.R. (1989). Schizophrenia as an Anomaly of Development of Cerebral Asymmetry. *Archives of General Psychiatry, 46*, 1145-1150.

124. Curry, S.H. (1993). *An Examination of the Relationships Between CNV Amplitude and Preceding Steady Potential Level.* In W.C. McCallum (Ed.). *Slow Potential Changes in the Human Brain*. New York: Plenum Press, pp.121-145.

125. Cutting, J.& Murphy, D. (1990). Preferences for Denotative as Opposed to Conotative Meanings in Schizophrenics. *Brain and Language, 39*, 459-468.

126. Daniel, D.G., Weinberger, D.R., Jones, D.W., et al (1991). The Effect of Amphetamine on Regional Cerebral Blood Flow During Cognitive Activation in Schizophrenia. *The Journal of Neuroscience, 11 (7)*, 1907-1917.

127. Daniel, D.G., Weinberger, D.R., Jones, D.W., et al (1991). The Effect of Amphetamine on Regional Blood Flow during Cognitive Activation Schizophrenia. *The Journal of Neuroscience, 11(7)*, 1907-1917.

128. Dauphinais, D., DeLisi, L.E.& Crow, T.J. (1990). Reduction in Temporal Lobe Size in Siblings with Schizophrenia. *Psychiatry Research, 35*, 137-147.

129. Davidson, M.L. (1972). Univariate Versus Multivariate Tests in Repeated Measures Experiments. *Psychological Bulletin, 77*, 446-452.

130. Deakin, J.F.W., Slater, P., Simpson, M.D.C., et al (1989). Frontal Cortical and Left Temporal Glutamatergic Dysfunction in Schizophrenia. *Journal of Neurochemistry, 52*, 1781-1786.

131. Deecke, L., Kornhuber, H.H., Lang, W., Lang, M.& Schreiber, H. (1985). Timing Function of the Frontal Cortex in Sequential Motor Learning Tasks. *Human Neurobiology, 4*, 143-154.

132. Deecke, L.& Lang, W. (1988). P300 as the Resolution of Negative Cortical DC Shifts. *Behavioural and Brain Science, 11 (3)*, 379-381.

133. Degreef, G., Ashtari, M., Bogerts, B., et al (1992). Volumes of Ventricular System Subdivisions Measured From Magnetic Resonance Images in First-Episode Schizophrenic Patients. *Archives of General Psychiatry, 49*, 531-537.

134. Delaunoy, J., Gerono, A.& Rosseau, J.C. (1978). *Experimental Production of Post-Imperative Negative Variation in Normal Subjects.* In D.A. Otto (Ed.). *Multidisciplinary Perspectives in Event-Related Brain Potential Research.* Washington,D.C.: Environmental Protection Agency, pp.355-357.

135. DeLisi, L.E., Buchsbaum, M.S., Holcomb, H.H., et al (1989). Increased Temporal Lobe Glucose Use in Chronic Schizophrenic Patients. *Biological Psychiatry, 25*, 835-851.

136. DeLisi, L.E., Hoff, A.L., Neale, C.& Kushner, M. (1994). Asymmetries in the Superior Temporal Lobe in Male and Female First-Episode Schizophrenic Patients: Measures of the Planum Temporale and Superior Temporal Gyrus by MRI. *Schizophrenia Research, 12*, 19-28.

137. DeLisi, L.E.& Holcomb, H.H. (1984). Anteroposterior Gradients in Cerebral Glucose Use in Schizophrenia and Affective Disorders. *Archives of General Psychiatry, 41*, 1159-1166.

138. DeLisi, L.E., Holcomb, H.H., Cohen, R.M., et al (1985). Positron Emission Tomography in Schizophrenic Patients With and Without Neuroleptic Medication. *Journal of Cerebral Blood Flow and Metabolism, 5*, 201-206.

139. DeLisi, L.E., Stritzke, P.H., Holan, V., et al (1991). Brain Morphological Changes in 1st Episode Cases of Schizophrenia: Are They Progressive ? *Schizophrenia Research, 5*, 206-208.

140. DeMyer, M.K., Gilmor, R.L., Hendrie, H.C., DeMyer, W.E., Augustyn, G.T.& Jackson, R.K. (1988). Magnetic Resonance Brain Images in Schizophrenic and Normal Subjects: Influence of Diagnosis and Education. *Schizophrenia Bulletin, 14 (1)*, 21-37.

141. Desmedt, J.E.& Tomberg, C. (1990). Topographic Analysis in Brain Mapping Can Be Compromised by the Average Reference. *Brain Topography, 3 (1)*, 35-42.

142. Deutch, A.Y., Tam, S.-Y.& Roth, R.H. (1985). Footshock and Conditioned Stress Increase 3,4- Dihydroxyphenylacetic Acid (DOPAC) in the Ventral Tegmental Area But Not Substantia Nigra. *Brain Research, 333*, 143-146.

143. Diamond, A.& Goldman-Rakic, P.S. (1989). Comparison of Human Infants and Rhesus Monkeys on Piaget's AB Task: Evidence for Dependence on Dorsolateral Prefrontal Cortex. *Experimental Brain Research, 74*, 24-40.

144. DiMattia, B.D., Posley, K.& Fuster, J.M. (1990). Crossmodal Short Term Mempry of Haptic and Visual Information. *Neuropsychologia, 28 (1)*,

145. Divac, I. (1972). Neostratium and Functions of Prefrontal Cortex. *Acta Neurobiologica Exp., 32*, 461-477.

146. Donchin, E.& Coles, M.G.H. (1988). Is the P300 Component a Manifestation of Context Updating. *Behavioural and Brain Science, 11*, 357-374.

147. Donchin, E.& Heffley, E. (1978). *Multivariate Analysis of ERP Data: A Tutorial View*. In D.A. Otto (Ed.). *Multidisciplinary Perspectives in Event-Related Brain Potential Research*. Washington D.C.: U.S. Environmental Protection Agency, pp.555-572.

148. Dongier, M. (1969). Separation of the Various Independent Phenomena Among the Slow Potential Changes (Contingent Negative Variations). *Electroencephalography and Clinical Neurophysiology, 27*, 104-109.

149. Dongier, M., Dubrovsky, B.& Engelsmann, F. (1977). *Event-Related Slow Potentials in Psychiatry*. In C. Shagass, S. Gershon& A.J. Friedhoff (Eds.).*Psychopathology and Brain Dysfunction*. New York: Rawen Press, pp.339-352.

150. Dongier, M., Dubrovsky, B.& Garcia-Rill, E. (1974). Les Potentieles Cerebraux Lents en Psychiatrie. *Canadian Psychiatric Association Journal, 19*, 177-183.

151. Doyere, V., Burette, F., Redini-Del Negro, C.& Laroche, S. (1993). Long-Term Potentation of Hippocampal Afferents and Efferents to Prefrontal Cortex: Implications for Associative Learning. *Neuropsychologia, 31 (19)*, 1031-1053.

152. Drew, G.C. (1951). Variations in Reflex Blink Rate During Visual Motor Tasks. *Quarterly Journal of Experimental Psychology, 3*, 73-88.

153. Dubrovsky, B.& Dongier, M. (1976). *Evaluation of Event-Related Slow Potentials in Selected Groups of Psychiatric Patients.* In W.C. McCallum& J.R. Knott (Eds.).*The Responsive Brain.* Bristol: John Wright & Sons Ltd., pp.150-153.

154. Dubrovsky, B.& Dongier, M. (1978). *Some Methodological and Theoretical Issues of ERP in Psychiatric Populations.* In D.A. Otto (Ed.). *Multidisciplinary Perspectives in Event-Related Brain Potential Research.* Washngton D.C.: Environmental Protection Agency, pp.387-388.

155. Dubrovsky, B., Garcia-Rill, E., Simkus, R.& Dongier, M. (1976). Effects of Changes in Carbon Dioxide Tension in Abnormally Prolonged Contingent Variation (CNV). *Biological Psychiatry, 11 (5)*, 535-541.

156. Dubrovsky, B., Solyom, L.& Barbas, H. (1983). Characteristics of the Contingent Negative Variation in Patients Suffering from Specific Phobias. *Biological Psychiatry, 13 (5)*, 531-540.

157. Duncan, C (1988). Event-Related Brain Potentials: A Window on Information Processing in Schizophrenia. *Schizophrenia Bulletin, 14 (2)*, 199-203.

158. Duncan, C.C., Morihisa, J.M., Fawcett, R.W.& Kirch, D.G. (1987). P300 in Schizophrenia: State or Trait Marker. *Psychopharamacology Bulletin, 23 (3)*, 497-501.

159. Duncan, C.C., Perlstein, W.M.& Morihisa, J.M. (1987). The P300 Metric in Schizophrenia: Effects of Probability and Modality. *Electroencephalography and Clinical Neurophysiology, 40*, 670-674.

160. Early, T.S., Haller, J.W., Posner, M.I.& Raichle, M. (1994). *The Left Striato-Pallidal Hyperactivity Model of Schizophrenia.* In A.S. David& J.C. Cutting (Eds.).*The Neuropsychology of Schizophrenia.* Hilsdale: Lawrence Erlbaum Associates, pp.15-39.

161. Early, T.S., Reiman, E.M., Raichle, M.E.& Spitznagel, E.L. (1987). Left Globus Pallidus Abnormality in Never-Medicated Patients With Schizophrenia. *Proceedings of the National Academy of Science,U.S.A., 84*, 561-563.

162. Ehrlichman, H.& Weinberger, A. (1978). Lateral Eye Movements and Hemispheric Asymmetry: A critical Review. *Psychological Bulletin, 85 (5)*, 1080-1101.

163. Eikmeier, G., Lodemann, E., Olbrich, H.M., Pach, J., Zerbin, D.& Gastpar, M. (1992). Altered Fronto-Central PINV Topography and the Primary Negative Syndrome in Schizophrenia. *Schizophrenia Research, 8*, 251-256.

164. Elbert, T. (1993). *Slow Cortical Potentials Reflect the Regulation of Cortical Excitability*. In W.C. McCallum& S.H. Curry (Eds.).*Slow Potential Changes in the Human Brain*. New York: Plenum Press, pp.235-251.

165. Elbert, T. (1993). *Slow Cortical Potentials Reflect the Regulation of Cortical Excitability*. In W.C. McCallum& S.H. Curry (Eds.).*Slow Potential Changes in the Human Brain*. New York: Plenum Press, pp.235-251.

166. Elbert, T., Lutzenberger, W., Rockstroh, B.& Birbaumer, N. (1983). When Regulation of Slow Brain Potentials Fails - A Contribution to the Psychophysiology of Perceptual Aberration and Anhedonia. *Advances in Biological Psychiatry, 13*, 98-106.

167. Elbert, T., Lutzenberger, W., Rockstroh, B.& Birbaumer, N. (1985). Removal of Ocular Artifacts from the EEG - A Biophysical Approach to the EOG. *Electroencephalography and Clinical Neurophysiology, 60*, 455-463.

168. Elbert, T., Lutzenberger, W., Rockstroh, B.& Birbaumer, N. (1986). *Responce Outcome Influences the Bereitschaftspotential*. In W.C. McCallum, R. Zappoli& F. Denoth (Eds.).*Cerbral Psychophysiology: Studies in Event-Related Potentials*. Amsterdam: Elsevier Science Publishers B.V., pp.248-250.

169. Elbert, T., Rockstroh, B., Lutzenberger, W.& Birbaumer, N. (1982). Slow Brain Potentials After Withdrawal of Control. *Archiv für Psychiatrie und Nervenkrankheiten, 232*, 201-204.

170. Elkashef, A.H., Buchanan, R.W., Gellad, F., Munson, R.C.& Breier, A. (1994). Basal Ganglia Pathology in Schizophrenia and Tardive Dyskinesia: An MRI Quantitative Study. *American Journal of Psychiatry, 151*, 752-755.

171. Emson, P.C.& Koob, G.F. (1978). The Origin and Distribution of Dopamine-Containing Afferents to the Rat Frontal Cortex. *Brain Research, 142*, 249-267.

172. Endicott, J., Nee, J., Fleiss, J., Cohen, J., Williams, J.B.W.& Simon, R. (1982). Diagnostic Criteria for Schizophrenia. *Archives of General Psychiatry, 39*, 884-889.

173. Ettlinger, G.& Wegener, J. (1958). Somaesthetic Alternation, Discrimination and Orientation after Frontal and Parietal Lesions in Monkeys. *Journal of Experimental Psychology, 10 (4)*, 176-186.

174. Falkai, P., Bogerts, B.& Greve, B. (1992). Loss of Sylvian Fissure Asymmetry in Schizophrenia. A Quantitative Post Mortem Study. *Schizophrenia Research, 7*, 23-32.

175. Falkai, P., Bogerts, B., Kleinschmidt, A., et al (1994). The Planum Temporale in Schizophrenia. A Comparison of Post- Mortem and MRI Data. *Schizophrenia Research, 11 (2)*, 131.

176. Farkas, T., Wolf, A.P., Jaeger, J., Brodie, J.D., Christman, D.R.& Fowler, J.S. (1984). Regional Brain Glucose Metabolism in Chronic Schizophrenia. *Archives of General Psychiatry, 41*, 293-300.

177. Faux, S., Shenton, M., McCarley, R.W., Rorello, M.W.& Duffy, F.H. (1987). *P200 Topographic Alterations in Schizophrenia: Evidence for Left Temporal-Centroparietal Region Deficits.* In R. Johnson, J.W. Rohrbaugh& R. Parasuraman (Eds.).*Current Trends in Event Related Potential Research.* Elsevier, pp.681-687.

178. Faux, S.F., McCarley, R.W., Nestor, P.G., et al (1993). P300 Topographic Asymetries are Present in Unmedicated Schizophrenics. *Electroencephalography and Clinical Neurophysiology, 88*, 32-41.

179. Faux, S.F., Torello, M.W., McCarley, R.W., Shenton, M.E.& Duffy, F.H. (1988). P300 in Schizophrenia: Confirmation and Statistical Validation of Temporal Region Deficit in P300 Topography. *Biological Psychiatry, 23*, 776-790.

180. Feighner, J.P., Robins, E., Guze, S.B., Woodruff, R.A., Winokur, G.& Munoz, R. (1994). Diagnostic Criteria for Use in Psychiatric Research. *Archives of General Psychiatry, 26*, 57-63.

181. Feinberg, I. (1978). Efference Copy and Corollary Discharge: Implications for Thinking and its Disorders. *Schizophrenia Bulletin, 4*, 636-640.

182. Fleming, K., Goldberg, T.E.& Gold, J.M. (1994). *Applying Working Memory Constructs to Schizophrenic Cognitive Impairment.* In A.S. David&

J.C. Cutiing (Eds.).*The Neuropsychology of Schizophrenia*. Hillsdale: Lawrence Erlbaum Associates, pp.197-220.

183. Ford, J.M., Pfefferbaum, A.& Roth, W. (1992). P3 and Schizophrenia. *Psychophysiology and Experimental Psychopathology*, *685*, 146-162.

184. Ford, R.A., Lewis, S.W., Syed, G.M.S., Reveley, A.& Toone, B.K. (1992). Ventricular Size and Regional Cerebral Blood Flow in Schizophrenia: An Attempted Replication. *Psychiatry Research*, *45*, 209-213.

185. Franzen, G.& Ingvar, D.H. (1975). Abscence of Activation in Frontal Structures During Psychological Testing of Chronic Schizophrenics. *Journal of Neurology,Neurosurgery and Psychiatry*, *38*, 1027-1032.

186. Franzén, G.& Ingvar, D.H. (1975). Absence of Activation in Frontal Structures During Psychological Testing of Chronic Schizophrenics. *Journal of Neurology,Neurosurgery,and Psychiatry*, *38*, 1027-1032.

187. Freedman, M.& Oscar-Berman, M. (1986). Bilateral Frontal Lobe Disease and Selective Delayed Response Deficits in Humans. *Behavioral Neuroscience*, *100 (3)*, 337-342.

188. Freund, H.-J. (1985). Clinical Aspects of Premotor Function. *Behavioral Brain Research*, *18*, 187-191.

189. Freund, H.-J.& Hummelsheim, H. (1985). Lesions of Premotor Cortex in Man. *Brain*, *108*, 697-733.

190. Friedman, D., Erlenmeyer-Kimling, L.& Vaughan, H.,Jr. (1985). Auditory Event-Related Potentials in Children at Risk for Schizophrenia Revisited: Rediagnosis of the Patient Parents and Inclusion of the Psychiatric Control Group. *Psychophysiology*, *22 (5)*, 590.

191. Friedman, H.R.& Goldamn-Rakic, P.S. (1988). Activation of the Hippcampus and Dentate Gyrus by Working-Memory Activation: A 2-Dexoyglucose Study of Behaving Rhesus Monkeys. *The Journal of Neuroscience*, *8 (12)*, 4693-4706.

192. Friedman, H.R.& Goldman-Rakic, P.S. (1988). Activation of the Hippocampus and Dentate Gyrus by Working- Memory: A 2-Deoxyglucose Study of Behaving Rhesus Monkeys. *The Journal of Neuroscience*, *8 (12)*, 4693-4706.

193. Friedman, H.R., Janas, J.D.& Goldman-Rakic, P.S. (1990). Enhancement of Metabolic Activity in the Diencephalon of Monkeys Performing Working Memory Tasks: A 2-Deoxyglucose Study in Behaving Rhesus Monkeys. *Journal of Cognitive Neuroscience, 2 (1)*, 18-31.

194. Frith, C. (1994). *Positron Emission Tomography Studies of Frontal Lobe Function: Relevance to Psychiatric Disease.* In Ciba Fondation Symposium 163 (Ed.). *Exploring Brain Functional Anatomy with Positron Tomography.* Chichester: Wiley, pp.181-197.

195. Frith, C. (1994). *Theory of Mind in Schizophrenia.* In A.S. David& J.C. Cutting (Eds.).*The Neuropsychology of Schizophrenia.* Hillsdale: Lawrence Erlbaum Associates, pp.147-159.

196. Frith, C.D. (1979). Consciousness, Information Processing and Schizophrenia. *British Journal of Psychiatry, 134*, 225-235.

197. Frith, C.D. (1984). Schizophrenia, Memory and Anticholinergic Drugs. *Journal of Abnormal Psychology, 93 (3)*, 339-341.

198. Frith, C.D. (1987). The Positive and Negative Symptoms of Schizophrenia Reflect Impairments in the Perception and Initiation of Action. *Psychological Medicine, 17*, 631-648.

199. Frith, C.D. (1992). *The Cognitive Neuropsychologie of Schizophrenia.* Hillsdale: Lawrence Erlbaum Associates.

200. Frith, C.D.& Done, D.J. (1988). Experiences of Alien Control in Schizophrenia Reflect a Disorder in the Central Monitoring of Action. *Psychological Medicine, 19*, 359-363.

201. Frith, C.D.& Done, D.J. (1988). Towards a Neuropsychology of Schizophrenia. *British Journal of Psychiatry, 153*, 437-443.

202. Frith, C.D.& Done, D.J. (1989). Experiences of Alien Control in Schizophrenia Reflect a Disorder in the Central Monitoring of Action. *Psychological Medicine, 19*, 359-363.

203. Frith, C.D., Friston, K., Liddle, P.F.& Frackowiak, R.S.J. (1991). Willed Action and the Prefrontal Cortex in Man: A Study with PET. *Proceedings of the Royal Society in London B, 244*, 241-246.

204. Frith, C.D., Friston, K.J., Liddle, P.F.& Frackowiak, R.S.J. (1991). A PET Study of Word Finding. *Neuropsychologia, 29 (12)*, 1137-1148.

205. Funashi, S., Bruce, C.J.& Goldman-Rakic, P.S. (1989). Mnemonic Coding of Visual Space in the Monkey's Dorsolateral Prefrontal Cortex. *Journal of Neurophysiology, 61 (2)*, 331-349.

206. Fuster, J. (1984). Behavioural Electrophysiology of the Prefrontal Cortex. *Trends in Neurosciences*, 408-414.

207. Fuster, J.M. (1985). The Prefrontal Cortex, Mediator of Cross-Temporal Contingencies. *Human Neurobiology, 4*, 169-179.

208. Fuster, J.M. (1989). *The Prefrontal Cortex: Anatomy, Physiology and Neuropsychology of the Frontal Lobe*. New York: 2nd Ed., Raven Press.

209. Fuster, J.M. (1991). *Role of Prefrontal Cortex in Delay Tasks: Evidence from Reversible Lesion and Unit Recording in the Monkey*. In H.S. Levin, A.L. Eisenberg& A.L. Benton (Eds.).*Frontal Lobe Function and Dysfunction*. New York: Oxford University Press, pp.59-71.

210. Fuster, J.M.& Alexander, G.E. (1971). Neuron Activity Related to Short-Term Memory. *Science, 173*, 652-654.

211. Fuster, J.M., Bauer, R.H.& Jervey, J.P. (1981). Effects of Cooling Inferotemporal Cortex on Performance of Visual Memory Tasks. *Experimental Neurology, 71*, 398-409.

212. Fuster, J.M.& Jervey, J.P. (1982). Neuronal Firing in the Inferotemporal Cortex of the Monkey in a Visual Memory Task. *The Journal of Neuroscience*,

213. Gage, F.H., Kely, P.A.T.& Björklund, A. (1984). Regional Changes in Brain Glucose Metabolism Reflect Cognitive Impairments in Aged Rats. *The Journal of Neuroscience, 4 (11)*, 2856-2865.

214. Gaillard, A.W.K. (1986). *The CNV as an Index of Responce Preparation*. In W.C. McCallum, R. Zappoli& R. Denoth (Eds.).*ERPs Associated with Preparatory and Movement-Related Processes. A Review*. Elsevier, pp.196-206.

215. Gaillard, A.W.K. (1988). Problems and Paradigms in ERP Research. *Biological Psychology, 20*, 91-109.

216. Gattaz, W.F., Mayer, S., Ziegler, P., Platz, M.& Gasser, T. (1994). Hypofrontality on Topographic EEG in Schizophrenia. *European Archives of Psychiatry and Clinical Neuroscience, 241*, 328-332.

217. Gauthier, P.& Gottesmann, C. (1976). Etude de la Variation Contingente Negative et de L'Onde Post- Imperative en Presence D'Interferences. *Electroencephalography and Clinical Neurophysiology, 40*, 143-152.

218. Gauthier, P.& Gottesmann, C. (1983). Influence of Total Sleep Deprivation on Event-Related Potentials in Man. *Psychophysiology, 20 (3)*, 351-355.

219. Gauthier, P.& Gottesmann, C. (9999). Study of Post-Imperative Negativity (PIN) Induction. *Electroencephalography and Clinical Neurophysiology, 43*, 534-535.

220. Geisser, S.& Greenhouse, S.W. (1958). An Extension of Box's Results on the Use of the F Distribution in Multivariate Analysis. *Annals of Mathematical Statistics, 29*, 885-891.

221. Gemba, H.& Sasaki, K. (1984). Studies on Cortical Field Potentials Recorded During Learning Processes of Visually Initiated Hand Movements in Monkeys. *Experimental Brain Research, 55*, 26-32.

222. Gerez, M.& Tello, A. (1992). Clinical Significance of Focal Topographic Changes in the Electroencephalogram (EEG) and Evoked Potentials (EP) of Psychiatric Patients. *Brain Topography, 5 (1)*, 3-10.

223. Gevins, A.& Cutillo, B. (1993). Spatiotemporal Dynamics of Component Processes in Human Working Memory. *Electroencephalography and Clinical Neurophysiology, 87*, 128-143.

224. Gevins, A.S.& Cutillo, B.A. (1986). *Clinical Applications of Computer Analysis EEG and other Neuropsychological Signals*. In F.H. Lopes da Silva, W. Storm van Leeuwen& A. Rémond (Eds.).*Handbook of Electroencephalography and Clinical Neuropsychology*. Asterdam: Elsevier, pp.335-382.

225. Géraud, G., Arné-Bès, M.C., Güell, A.& Bes, A. (1987). Reversibility of Hemodynamic Hypofrontality in Schizophrenia. *Journal of Cerebral Blood Flow and Metabolism, 7*, 9-12.

226. Giard, M.-H., Perin, F., Pernier, J.& Bouchet, P. (1990). Brain Generators Implicated in the Processing of Auditory Stimulus Deviance: A Topographic Event-Related Study. *Psychophysiology, 27 (6)*, 627-640.

227. Giard, M.H., Perrin, F., Echallier, J.F., Thévenet, M., Froment, J.C.& Pernier, J. (1994). Dissociation of Temporal and Frontal Components in the Human Auditory N1 Wave: A Scalp Current Density and Dipole Model Analysis. *Electroencephalography and Clinical Neurophysiology, 92*, 238-252.

228. Giedke, H.& Bolz, J. (1980). *Pre- and Postimperative Negative Variation (CNV and PINV) under Different Conditions of Controllability in Depressed Patients and Healthy Controlls.* In K.K. Kornhuber& L. Deecke (Eds.).*Motivation, Motor and Sensory Processes of the Brain. Electrical potentials, Behavior and Clinical Use. Progress in Brain Research.* Amsterdam: Elsevier, pp.579-582.

229. Godschalk, M., Lemon, R.N., Kuypers, H.G.J.M.& Van Der Steen, J. (1985). The Involvement of Monkey Premotor Cortex Neurons in Preparation of Visually Cued Arm Movements. *Behavioral Brain Research, 18*, 143-157.

230. Goff, W.R., Allison, T.& Vaughan Jr., H.G. (1978). *The Functional Neuroanatomy of Event Related Potentials.* In E. Callaway, P. Tueting& S. Koslow (Eds.).*Event-Related Potentials in Man.* New York: Academic Press, pp.1-80.

231. Golberg, T.E., Weinberger, D.R., Pliskin, N.H., Berman, K.F.& Podd, M.H. (1989). Recall Memory Deficit in Schizophrenia. *Schizophrenia Research, 2*, 251-257.

232. Gold, J.M., Herman, B.P., Randolph, C., Wyler, A.R., Goldberg, T.E.& Weinberger, D.R. (1994). Schizophrenia and Temporal Lobe Epilepsy. *Archives of General Psychiatry, 51*, 265-272.

233. Goldberg, E. (1985). Akinesia, Tardive Dysmentia, and Frontal Lobe Disorder in Schizophrenia. *Schizophrenia Bulletin, 11 (2)*, 255-263.

234. Goldberg, G. (1985). Supplementary Motor Area Structure and Functions: Review and Hypotheses. *Behavioural and Brain Science, 8 (4)*, 567-616.

235. Goldberg, G., Mayer, N.H.& Toglia, J.U. (1981). Medial Frontal Cortex and the Alien Hand Sign. *Archives of Neurology, 38*, 683-686.

236. Goldberg, T., Bigelow, L., Weinberger, D., Daniel, D.& Kleinman, J. (1991). Cognitive and Behavioral Effects of the coadministration of Dextroamphetamine and Halperidol in Schizophrenia. *American Journal of Psychiatry, 148(1)*, 78-84.

237. Goldberg, T.E., Kelsoe, J.R., Weinberger, D.R., Pliskin, N.H., Kirwin, P.D.& Berman, K.F. (1988). Performance of Schizophrenic Patients on Putative Neuropsychological Tests of frontal Lobe Funktion. *International Journal of Neuroscience, 42*, 51-58.

238. Goldberg, T.E., Ragland, D., Torrey, E.F., Gold, J.M., Bigelow, L.B.& Weinberger, D.R. (1990). Neuropsychological Assesment of Monozygotic Twins Discordant for Schizophrenia. *Archives of General Psychiatry, 47*, 1066-1072.

239. Goldberg, T.E.& Weinberger, D.R. (1988). Probing Prefrontal Function in Schizophrenia with Neuropsychological Paradigms. *Schizophrenia Bulletin, 14 (2)*, 179-183.

240. Goldberg, T.E., Weinberger, D.R., Berman, K.F., Pliskin, N.H.& Podd, M.H. (1987). Further Evidence for Dementia of the Prefrontal Type in Schizophrenia. *Archives of General Psychiatry, 44*, 1008-1014.

241. Goldberg-Rakic, P.S.& Friedman, H.R. (1991). *The Circuity of Working Memory Revealed by Anatomy and Metabolic Imaging*. In H.S. Levin, H.M. Eisenberg& A.L. Benton (Eds.).*Frontal Lobe Function and Dysfunction*. New York: Oxford University Press, pp.72-91.

242. Golden, C., Graber, B., Coffman, J., Berg, R.A., Newlin, D.B.& Bloch, S. (1981). Structural Brain Deficits in Schizophrenia. *Archives of General Psychiatry, 38*, 1014-1017.

243. Goldman, P.S. (1971). Functional Development of the Prefrontal Cortex in Early Life and the Problem of Neuronal Plasticity. *Experimental Neurology, 32*, 366-387.

244. Goldman-Rakic, P.S. (1984). Modular Organisation of Prefrontal Cortex. *Trends in Neurosciences*, 419-424.

245. Goldman-Rakic, P.S. (1987). *Circuitry of Primate Prefrontal Cortex and Regulation of Behaviour by Represaentational Memory*. In F. Plum (Ed.). *Handbook of Physiology: The Nervous System*. Bethesda: American Physiological Society, pp.373-417.

246. Goldman-Rakic, P.S. (1987). *Circuitry of Primate Prefrontal Cortex and Regulation of Behavior by Representational Memory*. In F. Plum& V. Mountcastle (Eds.).*Handbook of Physiology: The nervous System*. Baltimore: Wiliams & Wilkins, pp.373-417.

247. Goldman-Rakic, P.S. (1991). The Frontal lobes: Uncharted Provinces in the Brain. *Trends in Neurosciences*, 425-429.

248. Goldman-Rakic, P.S. (1991). *Prefrontal Dysfunctions in Schizophrenia: The Relevance of Working Memory*. In B.J. Carroll& J.E. Barrett (Eds.).*Psychopathology and the Brain*. New York: Raven Press, pp.1-23.

249. Goldman-Rakic, P. (1994). Cerebral Cortical Mechanisms in Schizophrenia. *Neuropsychophrmacology, 10 (35)*, 22-27.

250. Goldman-Rakic, P.S.& Friedman, H.R. (1991). *The Circuitry of Working Memory Revealed by Anatomy and Metabolic Imaging.* In H.S. Levin, H.M. Eisenberg& A.L. Benton (Eds.).*Frontal Lobe Function and Dysfunction.* New York: Oxford University Press, pp.72-91.

251. Goldman-Rakic, P.S., Selemon, L.D.& Schwartz, M.L. (1984). Dual Pathways Connecting the Dorsolateral Prefrontal Cortex With the Hippocampal Formation and Parahippocampal Cortex in the Rhesus Monkey. *Neuroscience, 12 (3)*, 719-743.

252. Gonzales-Andino, S.L., Pasqual-Marqui, R.D., Grave de Peralta, R., Lütkenhöhner, B., Menninghaus, E.& Hohke, M. (1992). *A Strategy for the Solution of the Inverse Problem Using Simultaneous EEG and MEG Measurements.* In M. Hoke (Ed.). *Biomagnetism: Clinical Aspects.* Elsevier, pp.711-715.

253. Gopher, D.& Donchin, E. (1985). *Workload-An Examination of the Concept.* In K.R. Boff, L. Kaufmann& S.R. Thomas (Eds.).*Handbook of Perception and Human Performance.* Chichester: Wiley, pp.41-48.

254. Grafman, J., Vance, S.C., Weingartner, H., Salazar, A.M.& Amin, D. (1986). The Effects of Lateralized Frontal Lesions on Mood Regulation. *Brain, 109*, 1127-1148.

255. Grasby, P., Friston, K., Bench, C., et al (1992). The Effect of Apomorphine on Regional Cerebral Blood Flow in Normal Volunteers. *Schizophrenia Research, 6 (2)*, 149.

256. Grasby, P.M., Friston, K.J., Bench, C.J., et al (1993). The Effect of the Dopamine Agonist, Apomorphine, On Regional Cerebral Blood Flow in Normal Volunteers. *Psychological Medicine, 23*, 605-612.

257. Greenhouse, S.W.& Geisser, S. (1959). On Methods in the Analysis of Profile Data. *Psychometrica, 24*, 95-112.

258. Grillon, C., Ameli, R., Courchesne, E.& Braff, D.L. (1991). Effects of Task Relevance and Attention on P3 in Schizophrenic Patients. *Schizophrenia Research, 4*, 11-21.

259. Grillon, C., Courchesne, E., Rezvan, A., Geyer, M.A.& Braff, D.L. (1990). Increased Distractibility in Schizophrenic Patients. *Archives of General Psychiatry, 47*, 171-179.

260. Gruzelier, J., Liddiard, D., Davis, L.& Wilson, L. (1990). Topographical EEG Differences Between Schizophrenic Patients and Controls during Neuropsychological Functional Activation. *International Journal of Psychophysiology, 8*, 275-282.

261. Guich, S.M., Buchsbaum, M.S., Burgwald, L., et al (1989). Effect of Attention on Frontal Distribution of Delta Activity and Cerebral Metabolic Rate in Schizophrenia. *Schizophrenia Research, 2*, 439-448.

262. Gunne, L.M., Growdon, J.& Glaeser, B. (1982). Oral Dyskinesa in Rats Following Brain Lessions and Neuroleptic Drug Administration. *Psychopharamacology Bulletin, 77*, 134-139.

263. Gur, R. (1977). Motoric Laterality Imbalance in Schizophrenia: A Possible Concomitant of Left Hemisphere Dysfunction. *Archives of General Psychiatry, 34*, 33-37.

264. Gur, R.E. (1977). Motoric Laterality Imbalance in Schizophrenia: A Possible Concomitant of Left Hemispheric Dysfunction. *Archives of General Psychiatry, 34*, 33-37.

265. Gur, R.E. (1978). Left Hemispheric Dysfunction and Left Hemisphere Overactivation in Schizophrenia. *Journal of Abnormal Psychology, 87 (2)*, 226-238.

266. Gur, R.E. (1979). Cognitive Concomitants of Hemispheric Dysfunction in Schizophrenia. *Archives of General Psychiatry, 36*, 269-274.

267. Gur, R.E., Gur, R.C.& Saykin, A.J. (1990). Neurobehavioral Studies in Schizophrenia: Implications for Regional Brain Dysfunction. *Schizophrenia Bulletin, 16 (3)*, 445-451.

268. Gur, R.E., Gur, R.C., Skolnick, B.E., et al (1985). Brain Function in Psychiatric Disorders / III. Regional Cerebral Blood Flow in Unmedicated Schizophrenics. *Archives of General Psychiatry, 42*, 329-334.

269. Gur, R.E., Jaggi, J.L., Shatsel, D.L., Ragland, J.D.& Gur, R.C. (1994). Cerebral Blood Flow in Schizophrenia: Effects of Memory Processing on Regional Activation. *Biological Psychiatry, 35*, 3-15.

270. Gur, R.E., Jaggi, J.L., Shtasel, D.L., Ragland, J.D.& Gur, R.C. (1994). Cerebral Blood Flow in Schizophrenia: Effects of Memory Processing on Regional Activation. *Biological Psychiatry, 35*, 3-15.

271. Gur, R.E., Mozley, P.D., Shtasel, D.L., et al (1994). Clinical Subtypes of Schizophrenia: Differences in Brain and CSF Volumes. *American Journal of Psychiatry, 151 (3)*, 343-350.

272. Gur, R.E.& Pearlson, G.D. (1993). Neuroimaging in Schizophrenia Research. *Schizophrenia Bulletin, 19 (2)*, 337-353.

273. Gur, R.E., Resnick, S.& Gur, R.C. (1989). Laterality and Frontality of Cerebral Blood Flow and Metabolism in Schizophrenia: Relationship to Symptom Specificty. *Psychiatry Research, 27*, 325-334.

274. Gur, R.E., Resnick, S.M., Gur, R.C., et al (1987). Regional Brain Functions in Schizophrenia: II. Repeated Evaluation with Positron Emission Tomography. *Archives of General Psychiatry, 44*, 126-129.

275. Gur, R.E., Skolnic, B.E., Gur, R.C., et al (1983). Brain Functions in Psychiatric Disorders: I. Reginonal Cerbral Blood Flow in Medicated Schizophrenics. *Archives of General Psychiatry, 40*, 1250-1254.

276. Gustafson, B., Wigström, H., Abraham, W.C.& Huang, Y.Y. (1987). Long Term Potentation in the Hippocampus Using Depolarising Current Pulses as the Conditioning Stimulus to Single Volley Synaptic Potentials. *The Journal of Neuroscience, 7(3)*, 774-780.

277. Güell, A., Géraud, G., Jauzac, P., Victor, G.& Arné-Bès, M.C. (1982). Effects of a Dopaminergic Agonist (Piribedil) on Cerebral Blood Flow in Man. *Journal of Cerebral Blood Flow and Metabolism, 2*, 255-257.

278. Halsband, U.& Passingham, R. (1982). The Role of Premotor and Parietal Cortex in the Direction of Action. *Brain Research, 240*, 368-372.

279. Haroutunian, V., Knott, P.& Davis, K.L. (1988). Effects of Mesocortical Dopaminergic Lesions Upon Subcortical Dopaminergic Function. *Psychopharamacology Bulletin, 24 (3)*, 341-344.

280. Harvey, P.D. (1985). Reality Monitoring in Mania and Schizophrenia. *The Journal of Nervous and Mental Disease, 173*, 67-73.

281. Hashimoto, S., Gemba, H.& Saski, K. (1980). Premovement Slow Cortical Potentials and Required Muscle Force in Self-Paced Hand Movements in the Monkey. *Brain Research, 197*, 415-423.

282. Häfner, H., Riecher-Rössler, A., Maurer, K., Fätkenheuer, B.& Löffler, W. (1992). First Onset and Early Symptomatology of Scizophrenia: A Chapter of Epidemological and Neurobiological Research into Age and Sex Differences. *European Archives of Psychiatry and Clinical Neuroscience, 242*, 109-118.

283. Hemmingsen, R., Vorstrup, S., Clemmensen, L., et al (1988). Cerebral Blood Flow During Delirium Tremens and Related Clinical States Studied with Xenon-133 Inhalation Tomography. *American Journal of Psychiatry, 145 (11)*, 1384-1390.

284. Hennighausen, E., Heil, M.& Rösler, F. (1993). A Correction Method for DC Drift Artifacts. *Electroencephalography and Clinical Neurophysiology, 86*, 199-204.

285. Herman, J.P., Guilloneau, D., Dantzer, R., Scatton, B., Semerdjian-Rouquier, L.& Le Moal, M. (1982). Differential Effects of Inescapable Footshocks and of Stimuli Previously Paired With Inescapable Footshocks On Dopamine Turnover in Cortical and Limbic Areas of the Rat. *Life Sciences, 30*, 2207-2214.

286. Herrmann, W.M., Kubicki, S., Künkel, H., et al (1989). Empfehlungen der Deutschen EEG Geselschaft für das Mapping von EEG-Parametern (EEG- und EP-Mapping). *Zeitschrift für EEG-EMG, 20*, 125-132.

287. Hershberger, W.& Misceo, G. (1983). A Conditioned Weight Illusion: Reafference Learning without a Correlation Store. *Perception and Psychophysics, 33 (4)*, 391-398.

288. Hikosaka, O., Sakamoto, M.& Usui, S. (1989). Functional Properties of Monkey Caudate Neurons. I. Activities Related to Saccadic Eye Movements. *Journal of Neurophysiology, 61*, 780-798.

289. Hjorth, B. (1975). An On-Line Transformation pf EEG Scalp Potentials into Orthogonal Source Derivations. *Electroencephalography and Clinical Neurophysiology, 39*, 526-530.

290. Hjorth, B. (1986). *Physical Aspects of EEG Data as a Basis for Topographic Mapping*. In F.H. Duffy (Ed.). *Topographic Mapping of Brain Electric Activity*. Butterworths, pp.175-193.

291. Hoff, A.L., Riordan, H., O'Donnell, D., et al (1992). Anomalous Lateral Sulcus Asymmetry and Cognitive Function in First-Episode Schizophrenia. *Schizophrenia Bulletin, 18,* 257-269.

292. Hoffman, R.E (1986). Verbal Hallucinations and Production Processes in Schizophrenia. *Behavioural and Brain Science, 9,* 503-548.

293. Holinger, D.P., Faux, S, Shenton, M.E., et al (1992). Reversed Temporal Region Asymmetries of P300 in Left and Right- Handed Schizophrenic Subjects. *Electroencephalography and Clinical Neurophysiology, 84,* 532-537.

294. Holm, S. (1979). A Simple Sequentially Rejective Multiple Test Procedure. *Scandinavian Journal of Statistics, 6,* 65-70.

295. Hussy, W. (1984). *Denkpsychologie.* Stuttgart: Kohlhammer.

296. Huynh, H.& Feldt, L.S. (1970). Conditions Under Which Mean Square Ratios in Repeated Messurements Designs Have Exact F-Distributions. *Journal of the American Statistical Association, 65,* 1582-1585.

297. Idiman, F.& Okyay, Y. (1987). Postimperative Negative Variation in Psychiatric Disorders. *Advances in Biological Psychiatry, 16,* 191-198.

298. Ikeda, A., Shibasaki, H., Nagamine, T., et al (1994). Dissociation Between Contingent Negative Variation and Bereitschaftspotential in A Patient with Cerebellar Efferent Lesion. *Electroencephalography and Clinical Neurophysiology, 90,* 359-364.

299. Isseroff, A., Rosvold, H.E., Galkin, T.W.& Goldman-Rakic, P.S. (1982). Spatial Memory Impairments Following Damage to the Mediodorsal Nucleus of the Thalamus in Rhesus Monkeys. *Brain Research, 232,* 97-113.

300. Jackob, H.& Beckmann, H. (1989). Gross and Histological Criteria for Developmental Disorders in Brains of Schizophrenics. *Journal of Royal Social Medicine, 82,* 466-469.

301. Jahn, T.& Mussgay, L. (1989). Die statistische Kontrolle möglicher Medikamenteneinflüsse in experimentalpsychologischen Schizophreniestudien: Ein Vorschlag zur Berechnung von Chlorpromazinäquivalenten. *Zeitschrift für Klinische Psychologie, 18,* 257-267.

302. Jakob, H.& Beckman, H. (1986). Prenatal Developmental Disturbances in the Limbic Allocortex in Schizophrenics. *Journal of Neural Transmission, 65,* 303-326.

303. Javitt, D.C., Doneschka, P., Zylberman, I., Ritter, W.& Vaughan, H.G. (1993). Impairement of Early Cortical Processing in Schizophrenia: An Event-Related Potential Confirmation Study. *Biological Psychiatry, 33*, 513-519.

304. Jennings, J.R. (1987). Methodology: Editorial Policy on Analyses of Variance with Repeated Measures. *Psychophysiology, 24 (4)*, 474-478.

305. Jernigan, T.L., Sargent, T.I., Pfefferbaum, A., Kusubov, N.& Stahl, S.M. (1985). 18-Fluorodeoxyglucose PET in Schizophrenia. *Psychiatry Research, 16*, 317-329.

306. Jernigan, T.L., Zisook, S., Heaton., R.K., Moranville, J.T., Hesselink, J.R.& Braff, D.L. (1991). Magnetic Resonance Imaging Abnormalities in Lenticular Nuclei and Cerebral Cortex in Schizophrenia. *Archives of General Psychiatry, 48*, 881-890.

307. Johnson, R. (1993). On the Neural Generators of the P300 Component of the Event- Related Potential. *Psychophysiology, 30*, 90-97.

308. Johnson, R.A. (1986). A Triarchic Model of P300 Ampltude. *Psychophysiology, 23*, 367-384.

309. Kakigi, S., Matsuda, T.& Shiva, S. (1982). Contingent Negative Variation (CNV) Measurement and the CNV Wave Forms: An Attempt to Obtain Normalized Data. *Hiroshima Shudo University Research Jl, 23*, 173-182.

310. Kakigi, S., Matsuda, T.& Ueda, M. (1985). Our 10 Year Research on the CNV (Contingent Negative Variation). *Neuroscience, 11*, 69-78.

311. Kameyama, T., Niwa, S., Hiramatsu, K., et al (1989). *Event-Related Potential Correlates of Psychotropic Drug Effects on Attentional and Hemispheric Dysfunction in Schizophrenia.* In R. Takahsahi, R. Flor-Henry, J. Gruzelier& J. Niwa (Eds.).*Cerebral Dynamics, Laterality and Psychopathology.* Amsterdam: Elsevier, pp.221-230.

312. Karlin, L., Martz, M.J.& Mordkoff, A.M. (1970). Motor Performance and Sensory-Evoked Potentials. *Electroencephalography and Clinical Neurophysiology, 28*, 307-313.

313. Karniski, W., Blair, C.R.& Snider, A.D. (1994). An Exact Statistical Method for Comparing Topographic Maps, with Any Number of Subjects and Electrodes. *Brain Topography, 6 (3)*, 203-210.

314. Karson, C.N., Coppola, R., Morihisa, J.M.& Weinberger, D.R. (1987). Computed Electroencephalographic Activity Mapping in Schizophrenia. *Archives of General Psychiatry, 44*, 514-517.

315. Kathmann, N., Jonitz, L.& Engel, R.R. (1990). Cognitive Determinatns of the Postimperative Negative Variation. *Psychophysiology, 27 (3)*, 256-263.

316. Kawasaki, Y., Maeda, Y., Suzuki, M., et al (1993). SPECT Analysis of Regional Cerebral Blood Flow Changes in Patients With Schizophrenia During the Wisconsin Card Sorting Test. *Schizophrenia Research, 10*, 109-116.

317. Kawasaki, Y., Suzuki, M., Maeda, Y., et al (1992). Regional Cerebral Blood Flow in Patients with Schizophrenia. *European Archives of Psychiatry and Clinical Neuroscience, 241*, 195-200.

318. Keilp, J.G., Sweeney, J.A., Jacobsen, P., et al (1988). Cognitive Impairment in Schizophrenia: Specific Relations to Ventricular Size and Negative Symptomatology. *Biological Psychiatry, 24*, 47-55.

319. Kenemans, J.L., Molenaar, P.C.M.& Verbaten, M.N. (1991). *Models for Estimation and Removal of Artifacts in Biological Signals*. In R. Weitkunat (Ed.). *Digital Biosignal Processing*. Amsterdam: Elsevier Science Publishers, pp.213-250.

320. Keselman, H.J.& Keselman, J.C. (1988). Methodology: Compared Repeated Measures Means in Factorial Designs. *Psychophysiology, 25 (5)*, 612-618.

321. Keselman, H.J.& Rogan, J.C. (1980). Methodology: Repeated Measures F-Tests and Psychophysiological Research: Controlling the Number of False Positives. *Psychophysiology, 17(5)*, 499-503.

322. Kidd, R.T. (1992). Schizotypal Personality and Hemispheric Processing. *(under review)*, (In Press)

323. King, H.E. (1991). *Psychomotor Dysfunction in Schizophrenia*. In S.R. Steinhauer& J.M. Gruzelier (Eds.).*Neuropsychology, Psychophysiology and Information Processing*. Elsevier, pp.273-303.

324. Kishimoto, H., Kuwahara, H., Ohno, S., et al (1987). Three Subtypes of Chronic Schizophrenia Identified Using 11C- Glucose Positron Emission Tomography. *Psychiatry Research, 21*, 285-292.

325. Klein, C.& Andresen, B. (1991). On the Influence of Smoking Upon Smooth Pursuit Eye Movements of Schizophrenics and Normal Controls. *Journal of Psychophysiology*, *5*, 361-369.

326. Klein, C., Müller, M., Heinz, A.& Rockstroh, B. (1994). Slow Cortical Potentials in Schizophrenic Patients During Association Formation. *Journal of Psychophysiology*, *8*, 314-327

327. Klein, C., Rockstroh, B., Cohen, R., Berg, P.& Heinz, A. (1994). Topography of CNV, PINV, and LPC in Schizophrenic Patients. *Journal of Psychophysiology*, *8 (3)*, 275 (Abstract)

328. Klein, C., Rockstroh, B., Cohen, R., Berg, P., Heinz, A.& Dressel, M. (1994). Functional Significance of Post-Imperative Negative Variation (PINV). *Journal of Psychophysiology*, *8 (3)*, 275 (Abstract)

329. Klein, D. (1982). Relation Between Current Diagnostic Criteria for Schizophrenia and the Dimensions of Premorbid Adjustment, Paranoid Symptomatology, and Chronicity. *Journal of Abnormal Psychology*, *91 (5)*, 319-325.

330. Klimidis, S., Stuart, G., Klimidis, G.W., Jackson, H.J., Singh, B.J.& Copolov, D.L. (1992). Positive and Negative Symptoms in Psychosis: Multidimensional Scaling of SAPS and SANS Items. *Schizophrenia Research*, *8*, 143-156.

331. Kling, A.S., Metter, E.J., Riege, W.H.& Kuhl, D.E. (1986). Comparison of PET Measurement of Local Brain Glucose Metabolism and CAT Measurement of Brain Atrophy in Chronic Schizophrenia and Depression. *American Journal of Psychiatry*, *143 (2)*, 175-180.

332. Knott, J.R., Peters, J.F., Robinson, M.D., Smith, A.& Andreasen, N.J.C. (1976). The Contingent Negative Variation (CNV) in Schizophrenic and Depressed Patients. *Electroencephalography and Clinical Neurophysiology*, *40*, 329-330.

333. Knott, J.R.& Tecce, J.J. (1978). *Event-Related Potentials and Psychopathology*. In D.A. Otto (Ed.). *Multidisciplinary Perspectives in Event-Related Brain Potential Research*. Environmental Protection Agency, pp.355-357.

334. Kochi, K., Koga, Y., Murasaki, M.& Takemasa, K. (1993). Event-Related Potentials in Schizophrenia During Pattern Discrimination Procedures. *Brain Topography*, *5 (3)*, 303.

335. Koga, Y.& Narita, T. (1987). Information Processing in Schizophrenia. *Advances in Biological Psychiatry, 16*, 146-155.

336. Kolb, B.& Whishaw, I.Q. (1983). Performance of Schizophrenic Patients on Tests Sensitive To Left or Right Frontal, Temporal, or Parietal Function in Neurological Patients. *The Journal of Nervous and Mental Disease, 171 (7)*, 435-443.

337. Kolb, B.& Whishaw, I.Q. (1990). *Fundamentals of Human Neuropsychology*. New York: W.H. Freeman and Company.

338. Kopp, B.& Rist, F. (1994). Error-Correcting Behaviour in Schizophrenic Patients. *Schizophrenia Research, 13*, 11-22.

339. Kowalska, D.M., Bachevallier, J.& Mishkin, M. (1991). The Role of the Inferior Prefrontal Convexity in Performance of Delayed Nonmatching-To-Sample. *Neuropsychologia, 29 (6)*, 583-600.

340. Kraut, M.A., Steinschneider, M., Arezzo, J.C.& Vaughan, H.G. (1986). *The Use of Current Source Density Analysis in the Interpretation of Cortical Evoked Potentials*. In W.C. McCallum, R. Zappoli& F. Denoth (Eds.).*Cerebral Psychophysiology: Studies in Event-Related Potentials*. Elsevier, pp.25-27.

341. Krieger, S., Tegler, J., Schweizer, A., Bertling, R.& Kalveram, K.T. (1980). *Ereignisbezogene Potentiale und elementare Informationsverarbeitungsprozesse bei schizophrenen Patienten*. In K. Heinrich& B. Bogerts (Eds.).*Pathophysiologische und pathomorphologische Aspekte bei schizophrenen Psychosen*. Stutgart(in press): Schattauer,

342. Kubota, K., Tonoike, M.& Mikami, A. (1980). Neuronal Activity in the Monkey Dorsolateral Prefrontal Cortex During a Discriminatin Task with Delay. *Brain Research, 183*, 29-42.

343. Kurachi, M., Kobayashi, K., Matsubara, R., et al (1985). Regional Cerebral Blood Flow in Schizophrenics. *European Neurology, 24*, 176-181.

344. Kutscher, S.P., Blackwood, D.H.R., Gaskell, D.F., Muir, W.J.& St.Clair, D.M. (1989). Auditory P300 Does Not Differentiate Borderline Personality Disorder from Schizotypal Personality Disorder. *Biological Psychiatry, 26*, 766-774.

345. Lagerlund, T.D., Sharbrough, F.W., Jack, C.R., et al (1993). Determination of 10-20 System Electrode Locations Using Magnetic resconance Image scanning with Markers. *Electroencephalography and Clinical Neurophysiology, 86*, 7-14.

346. Laming, P.R. (1993). *Slow Potential Shifts as Indicants of Glial Activation and Possible Neuromodulation.* In W.C. McCallum& S.H. Curry (Eds.).*Slow Potential Changes in the Human Brain.* New York: Plenum Press, pp.35-47.

347. Lang, M., Lang, W., Uhl, F., Kornhuber, A., Deecke, L.& Kornhuber, H.H. (1987). Slow Negativ Potential Shifts Indicating Verbal Cognitive Learning in a Concept Formation Task. *Human Neurobiology, 6*, 183-190.

348. Lang, W., Cheyne, D., Kristeva, R., Lindinger, G.& Deecke, L. (1991). *Functional Localisation of Motor Processes in the Human Cortex.* In C.M.H. Brunia, G. Mulder& M.N. Verbaten (Eds.).*Event-Related Brain Research.* Amstedram: Elsevier Science Publishers, pp.97-115.

349. Lang, W., Lang, M., Podreka, I., et al (1988). DC-Potential Shift and Regional Cerebral Blood Flow Reveal Frontal Cortex Involvement In Human Visumotor Learning. *Experimental Brain Research, 71*, 353-364.

350. Lang, W., Lang, M., Uhl, F., Koska, C., Kornhuber, A.& Deecke, L. (1988). Negative Cortical DC Shifts Proceding and Accompanying Simultaneous and Sequential Finger Movements. *Experimental Brain Research, 71*, 579-587.

351. Lang, W., Obrig, H., Lindinger, G., Cheyene, D.& Deecke, L. (1990). Supplementary Motor Area Activation While Tapping Bimanually Different Rhythms in Musicians. *Experimental Brain Research, 79*, 504-514.

352. Larbig, W., Elbert, T., Lutzenberger, W., Rockstroh, B., Schnerr, G.& Birbaumer, N. (1982). EEG and Slow Brain Potentials During Anticipation and Control of Painful Stimulation. *Electroencephalography and Clinical Neurophysiology, 53*, 298-309.

353. Lassen, N.A. (1959). Cerebral Blood Flow and Oxygen Consumption in Man. *Psychological Review, 39 (2)*, 183-238.

354. Laurian, S., Bryois, C., Gaillard, J.-M., Le, P.K.& Schöpf, J. (1984). Some Aspects of Brain Electrical Activity in Schizophrenia. *Advances in Biological Psychiatry, 15*, 60-68.

355. Law, S.K., Nunez, P.L.& Wijesinghe, R.S. (1993). High Resolution EEG using Spline Generated Surface Laplacians on Spherical and Ellipsoidal Surfaces. *IEEE Trans.Biomed.Engl., 40*, 517-528.

356. Lehmann, D. (1986). *Spatial Analysis of EEG and Evoked Potential Data.* In F.H. Duffy (Ed.). *Topographic Mapping of the Brain and Evoked Potential Data.* Butterworth, pp.29-61.

357. Lehmann, D. (1986). *Mapping, Spatial Analysis, and Adaptive Time Segmentation of EEG/ERP Data.* In C. Shagas, R.C. Josiassen& R.A. Roemer (Eds.).*Brain Electrical Potentials and Psychopathology*. Elsevier, pp.27-46.

358. Lencz, T., Raine, A.& Scerbo, A. (1993). Eye Tracking Abnormalities in Schizotypal Personality Disorders. *American Journal of Psychiatry, 150*, 152-154.

359. Lencz, T., Raine, A., Scerbo, A., et al (1993). Impaired Eye Tracking in Undergraduates with Schizotypal Disorders. *American Journal of Psychiatry, 150(1)*, 152-154.

360. Lenzenweger, M.F., Dworkin, R.H.& Wethington, E. (1989). Models of Positive and Negative Symptoms in Schizophrenia: An Empirical Evaluation of Latent Structures. *Journal of Abnormal Psychology, 98*, 62-70.

361. Leudar, I., Thomas, P.& Johnston, M. (1994). Self-Monitoring in Speech Production: Effects of Verbal Hallucinations and Negative Symptoms. *Psychological Medicine, 24*, 749-761.

362. Levander, S.E., Batrfai, A.& Schalling, D. (1985). Regional Cortical Dysfunctioning in Schizophrenic Patients Studied by Computerized Neuropsychological Methods. *Perc.Mot.Skills, 61*, 479-495.

363. Levin, S. (1984). Frontal Lobe Functions in Schizophrenia. I. Eye Movement Impairments. *Journal of Psychiatric Research, 18 (1)*, 27-55.

364. Levit, A.R., Sutton, S.& Zubin, J. (1973). Evoked Potential Correlates of Information Processing in Psychiatric Patients. *Psychological Medicine, 3*, 487-494.

365. Levy, D.L., Holzmann, P.S., Matthysse, S.& Mendell, N.R. (1993). Eye Tracing Dysfunction in Schizophrenia: A Critical Perspektive. *Schizophrenia Bulletin, 19(4)*, 461-476.

366. Lewis, S.W., Ford, R.A., Syed, G.M., Reveley, A.M.& Toone, B.K. (1992). A Controlled Study of 99m-Tc-HMPAO Single-Photon Emission Imaging in Chronic Schizophrenia. *Psychological Medicine, 22*, 27-35.

367. Lhermitte, F. (1983). "Utilisation Behavior" and its Relation to Lesions of the Frontal Lobes. *Brain, 106*, 237-255.

368. Lhermitte, F., Pillon, B.& Serdaru, M. (1986). Human Autonomy and the Frontal lobes. Part I: Imitation and Utilization Behavior: A Neuropsychological Study of 75 Patients. *Annals of Neurology, 19 (4)*, 326-334.

369. Lhermitte, M.D. (1986). Human Autonomy and the Frontal Lobes. Part II: Patient Behavior in Complex and Social Situations: The "Environmental Dependency Syndrome". *Annals of Neurology, 19 (4)*, 335-343.

370. Liddle, P.F. (1987). Schizophrenic Syndromes, Cognitive Performance, and Neurological Dysfunction. *Psychological Medicine, 17*, 49-57.

371. Liddle, P.F. (1994). *Volition and Schizophrenia.* In A.S. David& J.C. Cutting (Eds.).*The Neuropsychology of Schizophrenia.* Hillsdale: Lawrence erlbaum Associates, pp.39-48.

372. Liddle, P.F., Friston, K.J., Frith, C.D., Hirsch, S.R., Jones, T.& Frackowiak, R.S.J. (1992). Patterns of Cerebral Blood Flow in Schizophrenia. *British Journal of Psychiatry, 160*, 179-186.

373. Liddle, P.F., Frith, C.D.& Friston, K.J. (1992). Supervisory Mental Processes in Schizophrenia: A Study usin PET. *Schizophrenia Research, 6 (2)*, 148.

374. Liddle, P.F., Friton, K.J., Frith, C.D., Hirsch, S.R., Jones, T.& Frackowiak, R.S.J. (1992). Patterns of Cerebral Blood Flow in Schizophrenia. *British Journal of Psychiatry, 160*, 179-186.

375. Liddle, P.F.& Morris, D.L. (1991). Schizophrenia Syndromes and Frontal Lobe Performance. *British Journal of Psychiatry, 158*, 340-345.

376. Linden, A., Canavan, A.M.G., Lutzenberger, W., et al (1988). Slow Cortical Potentials in Patients with Parkinson's Disease. *Psychophysiology, 25*,

377. Lindvall, O., Björklund, A., Moore, R.Y.& Stenevi, U. (1974). Mesencephalic Dopamine Neurons Projecting to Neocortex. *Brain Research, 81*, 325-331.

378. Lipska, B.K., Jaskiw, G.E.& Weinberger, D.R. (1993). Postpubertal Emergence of Augmented Exploration and Amphetamine Supersensivity After Neonatal Deefferentation of the Rat Ventral Hippocampus. *Neuropsychophrmacology, 9*, 67-75.

379. Low, M.D., Borda, R.P.& Kellaway, P. (1966). "Contingent Negative Variation" in Rhesus Monkeys: An EEG Sign of a Specific Mental Process. *Perception and Motor Skills*, *22*, 443-446.

380. Luciana, M., Depue, R.A., Arbisi, P.& Leon, A. (1992). Facilitation of Working Memory in Humans by a D2 Dopamine Receptor Agonist. *Journal of Cognitive Neuroscience*, *4 (1)*, 58-68.

381. Lukoff, D., Liberman, R.P.& Nuechterlein, K.H. (1986). Symptom Monitoring in the Rehabilitation of Schizophrenic Patients. *Schizophrenia Bulletin*, *12*, 578-602.

382. Lutzenberger, W., Birbaumer, N., Rockstroh, B.& Elbert, T. (1983). Evaluation of Contingencies and Conditional Probabilities. *Archiv für Psychiatrie und Nervenkrankheiten*, *233*, 471-488.

383. Lutzenberger, W., Elbert, T.& Rockstroh, B. (1987). A Brief Tutorial on the Implications of Volume Conduction. *Journal of Neurophysiology*, *1*, 81-90.

384. Lutzenberger, W., Elbert, T., Rockstroh, B.& Birbaumer, N. (1981). Slow Cortical Potentials in Subjects with High or Low Scores on a Questionnaire Measuring Physical Anhedonia and Body Image Distortion. *Psychophysiology*, *18 (4)*, 371-381.

385. Lutzenberger, W., Elbert, T., Rockstroh, B.& Birbaumer, N. (1981). Principal Analysis of Slow Brain Potentials During Six Second Anticipation Intevals. *Biological Psychology*, *13*, 271-279.

386. Lutzenberger, W., Elbert, T., Rockstroh, B.& Birbaumer, N. (1985). *Das EEG*. Heidelberg: Springer.

387. Lütkenhöner, B. (1992). *Möglichkeiten und Grenzen der Neuromagnetischer Quellenanalyse*. Hamburg: Lit Verlag.

388. Macar, F.& Vitton, N. (1982). An Early Resolution of Contingent Negative Variation (CNV) in Time Discrimination. *Electroencephalography and Clinical Neurophysiology*, *54*, 426-435.

389. Malenka, R.C., Angel, R.W., Hampthon, B.& Berger, P.A. (1982). Impaired Central Error-Correcting Behaviour in Schizophrenia. *Archives of General Psychiatry*, *39*, 101-107.

390. Malla, A.K., Norman, R.M.G., Williamson, P., Cortese, L.& Diaz, F. (1993). Three Syndrome Concept of Schizophrenia. A Factor Analytic Study. *Schizophrenia Research, 10*, 143-150.

391. Mantanus, H., Timsith-Berthier, M., Gerono, A.& Von Frenckell, R. (1981). A Correlation Study of Contingent Negative Variation, Reaction Time, and EEG Power Spectrum in Control and Psychopathological Populations. *Biological Psychology, 13*, 227-237.

392. Marceno, S., Coppola, R., Daniel, D.G., Zigun, J.R.& Weinberger, D.R. (1993). Regional Cerebral Blood Flow During the Wisconsin Card Sorting Test in Normal Subjects Studied by Xenon-133 Dynamic SPECT: Comparison of Absolute Values, Percent Distributions Values and Covariance Analysis. *Psychiatry Research, 50*, 177-192.

393. Marczynski, T.J. (1978). *A Parsimonious Model of Mammalian Brain and Event Related Slow Potentials.* In D.A. Otto (Ed.). *Multidisciplinary Perspektives in Event Related Brain Potential Research.* Washington,D.C.: Environmental Protection Agency, pp.626-634.

394. Marczynski, T.J. (1978). *Neurochemical Mechanisms in the Genesis of Slow Potentials: A Review and some Clinical Implications.* In D.A. Otto (Ed.). *Multidisciplinary Perspectives in Event-Related Brain Potential Research.* Washington D.C.: Environmental Protection Agency, pp.25-35.

395. Marczynski, T.J. (1993). *Neurochemical Interpretation of Cortical Slow Potentials as they Relate to Cognitive Processes and a Parsimonious Model of Mammalian Brain.* In W.C. McCallum& S.H. Curry (Eds.).*Slow Potential Changes in the Human Brain.* New York: Plenum Press, pp.253-273.

396. Marneros, A.& Andreasen, N.C. (1992). Positive und Negative Symptomatik der Schizophrenie. *Der Nervenarzt, 63*, 262-270.

397. Marshall, E.J., Syed, G.M.S., Fenwick, P.B.C.& Lishman, W.A. (1993). A Pilot Study of Schizophrenia-Like Psychosis in Epilepsy Using Single-Photon Emission Computerised Tomography. *British Journal of Psychiatry, 163*, 32-36.

398. Mathew, R.J., Duncan, G.C., Weinman, M.L.& Barr, D.L. (1982). Regional Cerbral Blood Flow in Schizophrenia. *Archives of General Psychiatry, 39*, 1121-1124.

399. Mathew, R.J., Wilson, W.H., Tant, S.R., Robinson, L.& Prakash, R. (1988). Abnormal Resting Regional Cerebral Blood Flow Patterns and Their Correlates in Schizophrenia. *Archives of General Psychiatry, 45*, 542-549.

400. Mattes, R., Cohen, R., Berg, P., Canavan, A.G.M.& Hopmann, G. (1991). Slow Cortical Potentials (SCPS) in Schizophrenic Patients during Performance of the Wisconsin Card Sorting Test (WCST). *Neuropsychologia, 29 (2)*, 195-205.

401. Mauguiere, F. (1992). *Evoked Potentials: Diagnostic Uses and Strategies.* In M.. Hoke (Ed.). *Biomagnetism: Clinical Aspects*. Elsevier Science Publishers, pp.121-132.

402. Mazziotta, J.C., Phelps, M.E., Carson, R.E.& Kuhl, D.E. (1982). Tomographic Mapping of Human Cerebral Metabolism: Auditory Stimulation. *Neurology, 32*, 921-937.

403. McCallum, C.W., Papakostopoulos, D., Gombi, R., Winter, A.L., Cooper, R.& Griffith, H.B. (1973). Event-Related Potentials in Human Brain Stem. *Nature, 242*, 465-467.

404. McCallum, W.C. (1986). *How Many Separate Processes Constitute the CNV?* In W.C. McCallum, R. Zappoli& F. Denoth (Eds.).*ERPs Associated with Preparatory and Movement-Related Processes. A Review*. Elsevier, pp.192-196.

405. McCallum, W.C.& Abraham, P. (1973). *The Contingent Negative Variation in Psychosis.* In W.C. McCallum& J.R. Knott (Eds.).*Event-Related Slow Potentials of the Brain.* Electroenc.Clin.Neurphysiol., Suppl. 33, pp.329-335.

406. McCallum, W.C.& Walter, W.G. (1968). The Differential Effects of Distraction on the Contingent Negative Variation in Normal Subjects and Psychiatric Patients. *Electroencephalography and Clinical Neurophysiology, 24*, 593-596.

407. McCarley, R., Faux, S.F., Shenton, M.E., Nestor, P.G.& Adams, J. (1991). Event-Related Potentials in Schizophrenia: Their Biological Clinical Correlates and a New Model of Schizophrenic pathophysiology. *Schizophrenia Research, 4*, 209-231.

408. McCarley, R.W., Faux, S.F., Shenton, M.E., Nestor, P.G.& Holinger, D.P. (1991). Is There a P300 Asymetry in Schizophrenia. *Archives of General Psychiatry, 48*, 380-383.

409. McCarley, R.W., Shenton, M.E., O'Donnell, B.F., et al (1993). Auditory P300 Abnormalities and Left Posterior Superior Temporal Gyrus Volum Reduction in Schizophrenia. *Archives of General Psychiatry, 50*, 190-197.

410. McCarthy, G.& Donchin, E. (1978). Brain Potentials Associated with Structural and Functional Matching. *Neuropsychologia, 16*, 571-585.

411. McCarthy, G.& Wood, C.C. (1985). Scalp Distributions of Event-Related Potentials: An Ambiguity Associated with Analysis of Variance Models. *Electroencephalography and Clinical Neurophysiology, 62*, 203-208.

412. McConaghy, D., Catts, S.V., Michie, P.T., Fox, A., Ward, P.B.& Shelley, A.-M. (1993). P300 Indexes Thought Disorders in Schizophrenics, but Allusive Thinking in Normal Subjects. *The Journal of Nervous and Mental Disease, 181 (3)*, 176-182.

413. McCulloch, J., Savaki, H.E., McCulloch, M.C., Jehle, J.& Sokolof, L. (1982). The Distribution of Alterations in Energy Metabolism in the Rat Brain Produced by Apomorphine. *Brain Research, 243*, 67-80.

414. McGlashan, T.H. (1984). Testing Four Diagnostic Systems for Schizophrenia. *Archives of General Psychiatry, 41*, 141-144.

415. McMillian, T.M., Powell, G.E., Janota, I.& Polkey, C.E. (1987). Relationships between Neuropathology and Cognitive Functioning in Temporal Lobectomy Patients. *Journal of Neurology,Neurosurgery and Psychiatry, 50*, 167-176.

416. Mendez, M.F., Adams, N.L.& Lewandowski, K.S. (1989). Neurobehavioral Changes Associated with Caudate Lesions. *Neurology, 39*, 349-354.

417. Merrin, E.L.& Floyd, T.C. (1994). P300 Responces to Novel Auditory Stimuli in Hospitalized Schizophrenic Patients. *Biological Psychiatry, 36*, 527-542.

418. Michie, P.T., Fox, A.M., Ward, P.B., Catts, S.W.& McConaghy, N. (1990). Event-Related Potential Indices of Selective Attention and Cortical Lateralization in Schizophrenia. *Psychophysiology, 27 (2)*, 209-226.

419. Millner, B.& Petrides, M. (1984). Behavioural Effects of the Frontal Lobe Lesions in Man. *Trends in Neurosciences, 7*, 403-407.

420. Milner, B.& Petrides, M. (1984). Behavioural Effects of Frontal-Lobe Lesions in Man. *Trends in Neurosciences*, 403-407.

421. Mitzdorf, U. (1991). *Physiological Sources of Evoked Potentials*. In C.M.H. Brunia, G. Mulder& M.N. Verbaten (Eds.).*Event-Related Brain Research*. Elsevier, pp.47-57.

422. Mlakar, J., Jensterle, J.& Frith, C.D. (1994). Central Monitoring Defiency and Schizophrenic Symptoms. *Psychological Medicine, 24*, 557-564.

423. Mombour, W., Kockott, G.& Fliege, K. (1975). Die Brief Psychiatric Rating Scale (BPRS) von Overall und Gorham bei Akuten Paranoid-Halluzinatorischen Psychosen. Untersuchungen zu einer Deutschen Übersetzung der BPRS. *Pharmakopsychologie, 8,* 279-288.

424. Morihisha, .M.& McAnulty, G.B. (1985). Structure and Function: Brain Electrical Activity Mapping and Computed Tomography in Schizophrenia. *Biological Psychiatry, 20,* 3-19.

425. Morstyn, R., Duffy, F.H.& McCarley, R.W. (1983). Altered P300 Topography in Schizophrenia. *Archives of General Psychiatry, 40,* 729-734.

426. Moscovitsch, M. (1979). *Information Processing and the Cerebral Hemispheres.* In M.S. Gazzagnia (Ed.). *Handbookof Behavioural Neurology, vol. 2 Neuropsychology.* New York: Plenum,

427. Muller, K.E., Otto, D.A.& Benignus, V.A. (1983). Methodology: Design and Analysis Issues and Strategies in Psychophysical Research. *Psychophysiology, 20 (2),* 212-218.

428. Musalek, M., Podreka, I., Suess, E., et al (1988). Neurophysiological Aspects of Auditory Hallucinations. *Psychopathology, 21,* 275-280.

429. Müller-Preuss, P., Newman, J.D.& Jürgens, U. (1980). Anatomical and Physiological Evidence for a Relationship Between the "Cingular" Vocalization Area and the Auditory Cortex in the Squirrel Monkey. *Brain Research, 202,* 307-315.

430. Nagata, K., Tagawa, K., Shishido, F.& Uemura, K. (1986). *Topographic EEG Correlations of Cerebral Blood Flow and Oxygen Consumption in Patients with Neuropsychological Disorders.* In F.H. Duffy (Ed.). *Topographic Mapping of Brain Electrical activity.* Butterworths, pp.357-370.

431. Nageshi, Y.& Shimokochi, M. (1990). *Distraction Effects on Readiness Potential under a Dual-Task Paradigm.* In C.M.H. Brunia, A.W.K. Gaillard& A. Kok (Eds.).*Psychophsiological Brain Research.* Tilburg University Press, pp.146-149.

432. Nakamura, M., Ozawa, N., Shinba, T.& Yamamoto, K.-. (1993). CNV-like Potentials On the Cortical Surface Associated with Conditioning in Head-Restrained Rats. *Electroencephalography and Clinical Neurophysiology, 88,* 155-162.

433. Neshige, R., Lüders, H.& Shibasaki, H. (1988). Recording of Movement-Related Potentials from Scalp and Cortex in Man. *Brain, 111*, 719-736.

434. Niwa, S.-I., Hiramatsu, K.-I., Fukuda, M., Kameyama, T., Itoh, K.& Hayashida, S. (1992). Information Dysregulation and Event-Related Potentials in Schizophrenia. *Schizophrenia Bulletin, 18 (1)*, 95-105.

435. Nunez, P.L. (1981). *Electric Fields of the Brain: The Neurophysics of the EEG*. New York: Oxford University Press.

436. Nunez, P.L. (1989). Estimation of Large Scale Neocortical Source Activity with EEG Surface Laplacians. *Brain Topography, 2 (1/2)*, 141-154.

437. Nunez, P.L., Silberstein, R.B., Cadusch, P.J., Wijesinghe, R.S., Westdorp, A.F.& Srinivasan, R. (1994). A Theoretical and Experimental Study of High Resolution EEG Based on Surface Laplacians and Cortical Imaging. *Electroencephalography and Clinical Neurophysiology, 90*, 40-57.

438. Nunez, P.L.& Westdorp, A.F. (1994). The Surface Laplacian, High Resolution EEG and Contrversies. *Brain Topography, 6 (3)*, 221-226.

439. O'Connor, K. (1981). Post-Imperative Negativity and Prognosis in Senile Dementia. *Perception and Motor Skills, 52*, 234.

440. O'Toole, D.M.& Iacono, W.G. (1987). An Evaluation of Different Techniques for Removing Eye-Blink Artifact from Visual Evoked Response Recordings. *Psychophysiology, 24 (4)*, 487-497.

441. Ogura, C., Nageishi, Y., Shimokochi, M., Matsubayashi, M.& Kobayashi, T. (1987). *Evaluation of Event-Related Potentials in Schizophrenia Using Principal Components Analysis*. In R. Johnson, J.W. Rohrbaugh& R. Parasuramann (Eds.).*Current Trends in Event-Related Potential Research*. Amsterdam: Elsevier, pp.733-737.

442. Oscar-Berman, M. (1978). The Effects of Dorsolateral-Frontal and Ventrolateral- Orbitofrontal Lesions on Nonspatial Test Performance. *Neuropsychologia, 16*, 259-267.

443. Overall, J.E.& Gorham, D.R. (1962). The Brief Psychiatric Rating Scale. *Psychological Reports, 10*, 799.

444. Overall, J.E.& Hollister, L.E. (1979). Comparative Evaluation of Research Diagnostic Criteria for Schizophrenia. *Archives of General Psychiatry, 36*, 1198-1205.

445. O'Brien, R.G.& Kaiser, M.K. (1985). MANOVA Method for Analyzing Repeated Measures Designs: An Extensive Primer. *Psychological Bulletin, 97 (2)*, 316-333.

446. Pakkenberg, B. (1992). The Volume of the Mediodorsal Thalamic Nucleus in Treated and Untreated Schizophrenics. *Schizophrenia Research, 7*, 95-100.

447. Pallanti, S., Ragazzoni, A., Bartoletti, L., et al (1993). Basic Symptoms and P300 in Young Schizophrenic Patients. *Brain Topography, 5 (3)*, 308.

448. Papakostopoulos, D. (1978). *Electrical Activity of the Brain Associated with Skilled Performance*. In D.A. Otto (Ed.). *Multidisciplinary Perspektives in Event-Related Brain Potential Research*. Washington,D.C.: U.S. Environmental Agency, pp.134-137.

449. Papkostopoulos, D. (1986). *From Action to Interaction*. In W.C. McCallum, R. Zappoli& F. Denoth (Eds.).*ERPs Associated with Preparatory and Movement-Related Processes. A Review*. Elsevier, pp.219-222.

450. Park, S.& Holzman, P.S. (1992). Schizophrenics Show Spatial Working Memory deficits. *Archives of General Psychiatry, 49*, 975-982.

451. Park, S.& Holzmann, P.S. (1993). Assosciation of Working Memory Deficit and Eye Tracking Dysfunction in Schizophrenia. *Schizophrenia Research, 11*, 55-61.

452. Pass, H.L., Klormann, R., Salzmann, L.F., Klein, R.H.& Kaskey, G.B. (1980). The Late Positive Component of the Evoked Response in Acute Schizophrenics During a Test of Sustained Attention. *Biological Psychiatry, 15 (1)*, 9-20.

453. Passingham, R.E. (1985). Prefrontal Cortex and the Sequencing of Movements in Monkeys (Macaca Mulatta). *Neuropsychologia, 23 (4)*, 453-462.

454. Passqual-Marqui, R.D. (1993). The Spherical Spline Laplacian does not Produce Artifactially High Coherences: Comments on Two Articles by Biggins et al. *Electroencephalography and Clinical Neurophysiology, 87*, 62-64.

455. Paty, J., Brenth, P., Tignol, J.& Bourgeois, M. (1978). La Variation Contingente Negative en Psychiatrie. *Ann.Med.-Psychol., 36*, 328-338.

456. Paulman, R.G., Devous, M.D., Gregory, R.R., et al (1990). Hypofrontality and Cognitive Impairment in Schizophrenia: Dynamic Single-Photon Tomography

and Neuropsychological Assessment of Schizophrenic Brain Function. *Biological Psychiatry*, *27*, 377-399.

457. Peralta, V.& Cuesta, M.J. (1991). Schneiderian Versus Bleulerian Schizophrenia and Basic Symptoms. *Psychopathology*, *24*, 151-157.

458. Perrin, F., Pernier, J., Bertrand, O.& Echallier, J.F. (1989). Spherical Spline for Scalp Potential and Current Density Mapping. *Electroencephalography and Clinical Neurophysiology*, *72*, 184-187.

459. Perrin, F., Pernier, J., Bertrand, O., Giard, H.M.& Echallier, J.F. (1987). Mapping of Scalp Potentials by Surface Spline Interpolation. *Electroencephalography and Clinical Neurophysiology*, *66*, 75-81.

460. Pfefferbaum, A.& Ford, J.M. (1988). ERPs to Stimuli Requiring Response Production and Inhibition: Effects of Age, Probability and Visual Noise. *Electroencephalography and Clinical Neurophysiology*, *71*, 55-63.

461. Pfefferbaum, A., Ford, J.M., White, P.M.& Roth, W.T. (1989). P3 in Schizophrenia is Affected by Stimulus Modality, Response Requirements, and Negative Symptoms. *Archives of General Psychiatry*, *46*, 1035-1044.

462. Pfefferbaum, A., Ford, J.M., White, P.M.& Roth, W.T. (1989). P3 Schizophrenia is Affected by Stimulus Modality, Response Requirements, Medication Status, and Negative Symptoms. *Archives of General Psychiatry*, *46*, 1035-1044.

463. Pfefferbaum, A., Ford, J.S., Weller, B.J.& Kopell, B.S. (1985). ERPs to Responce Production and Inhibition. *Electroencephalography and Clinical Neurophysiology*, *60*, 423-434.

464. Pfefferbaum, A., Lim, K.O., Rosenbloom, M.& Zipursky, R.B. (1990). Brain Magnetic Resonance Imaging: Approaches for Investigating Schizophrenia. *Schizophrenia Bulletin*, *16 (3)*, 453-476.

465. Pfefferbaum, A., Wenegrat, B.G., Ford, J.M., Roth, W.T.& Kopell, B.S. (1984). Clinical Applications of the P3 Component of Event-Related Potentials II. Dementia, Depression and Schizophrenia. *Electroencephalography and Clinical Neurophysiology*, *59*, 104-124.

466. Picton, T., Woods, D., Stuss, D.& Campbell, K. (1978). *Methodology and Meaning of human Evoked Potential Scalp Distribution Studies*. In D.A. Otto (Ed.). *Multidisciplinary Perspectives in Event-Related Brain Potential Research*. Washington D.C.: Environmental Protection Agency, pp.515-522.

467. Picton, T.W.& Hillyard, S.A. (1972). Cephalic Skin Potentials in Electroencephalography. *Electroencephalography and Clinical Neurophysiology*, *33*, 419-424.

468. Pilowsky, L.S., Costa, D.C., Ell, P.J., Verhoeff, N.P.L.G, Murray, R.M.& Kerwin, R.W. (1994). D2 Dopamine Receptor Binding in the Basal Ganglia of Antipsychotic-Free Schizophrenic Patients. *British Journal of Psychiatry*, *164*, 16-26.

469. Pirch, J.H. (1978). *Effects of Amphetamine and Pentobarbital on Event Related Slow Potentials in Rats*. In D.A. Otto (Ed.). *Multidisciplinary Perspectives in Event-Related Brain Potential Research*. Washington D.C.: U.S. Environmental Agency, pp.56-60.

470. Pivik, R.T., Bylsma, F.W.& Cooper, P.M. (1988). Dark Condition Normalization of Smooth Pursuit Tracking: Evidence of Cerebellar Dysfunction in Psychosis. *European Archives of Psychiatry and Clinical Neuroscience*, *237*, 334-342.

471. Pope, H.G.& Lipinski, J.F. (1978). Diagnosis in Schizophrenia and Manic-Depressive Illness. *Archives of General Psychiatry*, *35*, 811-828.

472. Posner, M.I., Early, T.S., Reiman, E., Pardo, P.J.& Dhawan, M. (1988). Asymmetries in Hemispheric Control of Attention in Schizophrenia. *Archives of General Psychiatry*, *45*, 814-821.

473. Pritchard, W.S. (1981). Psychophysiology of P300. *Psychological Bulletin*, *89*, 506-540.

474. Pritchard, W.S. (1986). Cognitive Event-Related Potential Correlates of Schizophrenia. *Psychological Bulletin*, *100 (1)*, 43-66.

475. Proulx, G.B.& Picton, T.W. (1980). *The CNV during Cognitive Learning and Extinction*. In H.H. Kornhuber& L. Deecke (Eds.).*Motivation, Motor and Sensory Processes of the Brain. Electrical Potentials, Behavior and Clinical Use. Progress in Brain Research*. Amsterdam: Elsevier, pp.309-313.

476. Pycock, C.J., Kerwin, R.W.& Carter, C.J. (1980). Effect of Lesion of Cortical Dopamine Terminals on Subcortical Receptors in Rats. *Nature*, *286*, 74-77.

477. Radwan, M., Hermesh, H., Mintz, M.& Munitz, H. (1991). Event-Related Potential in Drug-Naive Schizophrenic Patients. *Biological Psychiatry*, *29*, 265-272.

478. Raine, A. (1987). Validation of Schizoid Personality Scales Using Indices of Schizotypal and Borderline Disorder in a Criminal Population. *British Journal of Clinical Psychology, 26*, 305-309.

479. Raine, A. (1988). *Antisocial Behaviour and Social Psychophysiology.* In H.L. Wagner (Ed.). *Social Psychophysiology and Emotion: Theory and Clinical Applications.* Wiley & Sons, pp.231-250.

480. Raine, A. (1991). The SPQ: A Scale for the Assesment of Schizotypal Personality Based on the DSM-III-R Criteria. *Schizophrenia Bulletin, 17(4)*, 555-564.

481. Raine, A. (1992). Sex Differences in Schizotypal Personality in a Nonclinical Population. *Journal of Abnormal Psychology, 101(2)*, 361-364.

482. Raine, A.& Allbutt, J. (1989). Factors of Schizoid Personality. *British Journal of Clinical Psychology, 28*, 31-40.

483. Raine, A., Harrison, G.N., Reynolds, G.P., Sheard, C., Cooper, J.E.& Medley, I. (1990). Structural and Functional Characteristics of the Corpus Callosum in Schizophrenics, Psychiatric Controls, and Normal Controlls. *Archives of General Psychiatry, 47*, 1060-1064.

484. Raine, A., Reynolds, C., Lencz, T., Scerbo, A., Triphon, N.& Kim, D. (1994). Cognitive-Perceptual, Interpersonal, and Disorganized Features of Schizotypal Personality. *Schizophrenia Bulletin, 20(1)*, 191-201.

485. Raine, A., Reynolds, G., Harrison, G., et al (1990). Reduced Frontal Area and Poorer Frontal Functioning in Schizophrenics: An MRI, Neuropsychological and Psychophysiological Examination. *Schizophrenia Research, 3*, 21.

486. Raine, A., Sheard, C., Reynolds, G.P.& Lencz, T. (1992). Pre-Frontal Structural and Functional Deficits Associated With Individual Differences in Schizotypal Personality. *Schizophrenia Research, 7*, 237-247.

487. Raine, A.& Venables, P.H. (1988). Enhanced P3 Evoked Potentials and Longer P3 Recovery Times in Psychopaths. *Psychophysiology, 25(1)*, 30-38.

488. Rawlins, J.N.P., Lyford, G.L., Seferiades, A., Deacon, R.M.J.& Cassaday, H.J. (1993). Critical Determinants of Nonspatial Working Memory Deficits in Rats with Conventional Lesions of the Hippocampus or Fornix. *Behavioral Neuroscience, 107 (3)*, 420-433.

489. Rebert, C., Berry, R.& Merlo, J. (1976). *DC Potential Consequences of Induced Muscle Tension: Effects on Contingent Negative Variation.* In W.C. McCallum& J.R. Knott (Eds.).*The Responsive Brain.* Bristol: Wright, pp.126-131.

490. Rebert, C.S. (1972). Cortical and Subcortical Slow Potentials in the Monkey's Brain During a Preparatory Interval. *Electroencephalography and Clinical Neurophysiology, 33*, 389-402.

491. Rebert, C.S., Diehl, J.J.& Matteucci, M.J. (1993). *The Distribution and Detection of Brain Slow Potentials and their Relationsship to Neuroglial Activities and Higher Mental Functions.* In W.C. McCallum& S.H. Curry (Eds.).*Slow Potential Changes in the Human Brain.* New York: Plenum Press, pp.275-291.

492. Rebert, C.S., Hennessy, M.B.& Donovan, W.J. (1986). *Slow Potentials in Substantia Nigra and other Regions of Monkey Brain during a Cued Reaction Time Task.* In W.C. McCallum, R. Zappoli& F. Denoth (Eds.).*Cerebral Psychophysiology: Studies in Event-Related Potentials.* Amsterdam: Elsevier Science Division, B.V., pp.558-560.

493. Resnick, S., Gur, R.E., Alavi, A., Gur, R.C.& Reivich, M. (1988). Positron Emission Tomography and Subcortical Glucose Metabolism in Schizophrenia. *Psychiatry Research, 24*, 1-11.

494. Reynolds, G.P. (1983). Increased Concentrations and Lateral Asymmetry of Amygdala Dopamine in Schizophrenia. *Nature, 44*, 625-632.

495. Riehle, A.& Requin, J. (1993). The Predictive Value for Performance Speed of Preparatory Changes in Neuronal Activity of the Monkey Motor and Premotor Cortex. *Behavioral Brain Research, 53*, 35-49.

496. Ritter, W.& Ruchkin, D.S. (1994). A Rewiew of Event-Related Potential Components Discovered in the Context of Studying P300. *Annals of the New York Academy of Science, 658*, 1-32.

497. Rizzo, P.A., Albani, G.F., Spadaro, M.& Morocutti, C. (1983). Brain Slow Potentials (CNV), Prolactin, and Schizophrenia. *Biological Psychiatry, 18 (2)*, 175-183.

498. Rizzo, P.A., Amabile, G., Caporali, M., et al (1979). A Longitudinal CNV Study in a Group of Five Bipolar Cyclothymic Patients. *Biological Psychiatry, 14 (4)*, 581-587.

499. Rizzo, P.A., Amabile, G., Caporali, M., Spadaro, M., Zanasi, M.& Morocutti, C. (1978). A CNV Study in a Group of Patients with Traumatic Head Injuries. *Electroencephalography and Clinical Neurophysiology, 45*, 281-285.

500. Rizzo, P.A., Amabile, G., Fiumara, R., et al (1980). Brain Slow Potentials and Hypnosis. *Biological Psychiatry, 15 (3)*, 499-506.

501. Rizzo, P.A., Spadaro, M., Albani, G.F.& Morocutti, C. (1984). Contingent Negative Variation and Schizophrenia: A Long-Term Follow-up Study. *Biological Psychiatry, 19 (12)*, 1719-1725.

502. Roberts, G.W., Colter, N., Lofthouse, R., Bogerts, B., Zech, M.& Crow, T.J. (1986). Gliosis in Schizophrenia: A Survey. *Biological Psychiatry, 21*, 1043-1050.

503. Roberts, L.E., Rau, H., Lutzenberger, W.& Birbaumer, N. (1994). Mapping P300 Waves onto Inhibition: Go/No-Go Discrimination. *Electroencephalography and Clinical Neurophysiology, 92*, 44-55.

504. Robinson, D.L.& Wurz, R.H. (1976). Use of an Extraretinal Signal by Monkey Superior Colliculus Neurons to Distinguish Real From Self-Induced Movement. *Journal of Clinical Neurophysiology, 39*, 852-870.

505. Rochester, S.& Martin, J.R. (1979). *Crazy Talk: A Study of the Discourse of Schizophrenic Speakers*. New York: Plenum Press.

506. Rockstroh, B. (1993). *Regulation of Cortical Excitability in Patients with Epilepsy and its Measurement by Means of Slow Cortical Potentials*. In W.C. McCallum& S.H. Curry (Eds.).*Slow Potential Changes in the Human Brain*. New York: Plenum Press, pp.209-227.

507. Rockstroh, B.& Elbert, T. (1990). *On the Regulation of Excitability in Cerebral Cortex - A Bridge Between EEG and Attention*. In H.-G. Geissler (Ed.). *Psychophysical Explorations of Mental Structures*. Hogrefe & Huber, pp.323-332.

508. Rockstroh, B., Elbert, T., Canavan, A., Lutzenberger, W.& Birbaumer, N. (1989). *Slow Cortical Potentials and Behaviour*. Baltimore: 2nd Ed., Urban & Schwarzenberg.

509. Rockstroh, B., Elbert, T., Lutzenberger, W.& Altenmüller, E. (1991). Effects of Anticonvulsatory Benzodiazepine Clonazepam on Event- Related Brain Potentials in Humans. *Electroencephalography and Clinical Neurophysiology, 78*, 142-149.

510. Rockstroh, B., Elbert, T., Lutzenberger, W.& Birbaumer, N. (1979). Slow Cortical Potentials Under Conditions of Uncontrollability. *Psychophysiology, 16*, 374-380.

511. Rockstroh, B., Elbert, T., Lutzenberger, W.& Birbaumer, N. (1990). Biofeedback Produced Hemispheric Asymmetry of Slow Cortical Potentials and Its Behavioural Effects. *International Journal of Psychophysiology, 9*, 151-168.

512. Rockstroh, B.& McCallum, W.C. (1993). *Theories and Significance of SPs.* In W.C. McCallum& S.H. Curry (Eds.).*Slow Potential Changes in the Human Brain.* New York: Plenum Press, pp.229-234.

513. Rockstroh, B., Müller, M., Wagner, M., Cohen, R.& Elbert, T. (1993). Probing the Nature of the CNV. *Journal of Electroencephalography and Clinical Neurophysiology, 871*, 235-241.

514. Rockstroh, B., Müller, M., Wagner, M., Cohen, R.& Elbert, T. (1994). Event-Related and Motor Responses to Probes in a Forewarned Reaction Time Task in Schizophrenic Patients. *Schizophrenia Research, 13*, 23-34.

515. Rohrbaugh, J.W. (1986). *Introduction.* In W.C. McCallum, R. Zappoli& F. Denoth (Eds.).*Cerebral Psychophsiology: Studies in Event-related Potentials.* Elsevier, pp.189.

516. Rohrbaugh, J.W.& Gaillard, A.W.K. (1983). *Sensory and Motor Aspects of the Contingent Negative Variation.* In A.W.K. Gaillard& W. Ritter (Eds.).*Tutorials in ERP Research: Endogenous Components.* North-Holand Publishing Company, pp.269-310.

517. Roland, P.E. (1987). *Metabolic Mapping of Sensorimotor Integration in the Human Brain.* In CIBA Foundation Symposium 132 (Ed.). *Motor Areas of the Cerebral Cortex.* Wiley, pp.251-268.

518. Romani, A., Merello, S., Gozzoli, L., Zerbi, F., Grazzi, M.& Cosi, V. (1987). P300 and CT Scan in Patients with Chronic Schizophrenia. *British Journal of Psychiatry, 151*, 506-513.

519. Rosch, E. (1977). *Human Categorization.* In N. Warren (Ed.). *Advances in Cross-Cultural Psychology.* London: Academic Press,

520. Rosenkilde, C.E., Bauer, R.H.& Fuster, J.M. (1981). Single Cell Activity in Ventral Prefrontal Cortex of Behaving Monkeys. *Brain Research, 209*, 375-394.

521. Rossi, A., Serio, A., Stratta, P., et al (1994). Planum Temporale Asymmetry and Thought Disorder in Schizophrenia. *Schizophrenia Research, 12*, 1-7.

522. Rossi, A., Stratta, P., Mancini, F., et al (1994). Magnetic Resonance Imaging Findings of Amygdala-Anterior Hippocampus Shrinkage in Male Patients With Schizophrenia. *Psychiatry Research, 52*, 43-53.

523. Roth, W.T. (1977). Late Event-Related Potentials and Psychopathology. *Schizophrenia Bulletin, 3 (1)*, 105-120.

524. Roth, W.T.& Cannon, E.H. (1972). Some Features of the Auditory Evoked Response in Schizophrenics. *Archives of General Psychiatry, 27*, 466-471.

525. Roth, W.T., Duncan, C.C., Pfefferbaum, A.& Timsit-Berthier, M. (1986). *Applications of Cognitive ERPs in Psychiatric Patients*. In W.C. McCallum, R. Zappoli& F. Denoth (Eds.).*Cerebral Psychophysiology: Studies in Event-Related Potentials*. Elsevier Science Publishers, pp.419-438.

526. Roth, W.T., Horvath, T.B., Pfefferbaum, A.& Kopell, B.S. (1980). Event-Related Potentials in Schizophrenics. *Electroencephalography and Clinical Neurophysiology, 48*, 127-139.

527. Roth, W.T., Horvath, T.B., Pfefferbaum, A., Tinklenberg, J.R., Mezzich, J.& Kopell, B.S. (1979). *Late Event-Related Potentials and Schizophrenia*. In H. Begleiter (Ed.). *Evoked Brain Potentials and Behaviour*. New York & London: Plenum Press, pp.499-515.

528. Roth, W.T., Horvath, T.B., Pfefferbaum, A., Tinklenberg, J.R., Mezzich, J.& Kopell, B.S. (1979). *Late Event-Related Potentials in Schizophrenia*. In H. Begleieter (Ed.). *Evoked Brain Potentials and Behavior*. New York: Plenum Press, pp.499-516.

529. Roth, W.T., Kopell, B.S., Tinklenberg, J.R., Darley, C.F., Sikora, R.& Vesecky, T.B. (1975). The Contingent Negative Variation During a Memory Retrieval Task. *Electroencephalography and Clinical Neurophysiology, 38*, 171-174.

530. Roth, W.T., Krainz, J.M., Ford, J.R., Tinklenberg, R.M., Rothbart, R.M.& Kopel, B.S. (1976). Parameters of Temporal Recovery of the Human Auditory Evoked Potential. *Electroencephalography and Clinical Neurophysiology, 40*, 623-632.

531. Roth, W.T., Rothbarth, R.M.& Kopell, B.S. (1978). The Timing of CNV Resolution in a Memory Retrieval Task. *Biological Psychology, 6*, 39-49.

532. Roth, W.T., Tinklenberg, J.R.& Kopell, B.S. (1977). Ethanol and Marihuana Effects on Event-Related Potentials in a Memory Retrieval Paradigm. *Electroencephalography and Clinical Neurophysiology*, *42*, 381-388.

533. Roxborough, H., Muir, W.J., Blackwood, D.H.R., Walker, M.T.& Blackburn, I.M. (1993). Neuropsychological and P300 abnormalities in Schizophrenics and their Relatives. *Psychological Medicine*, *23*, 305-314.

534. Rösler, F. (1991). *Perception or Action: Some Components on Preparatory Negative Potentials.* In C.M.H. Brunia, G. Mulder& M.N. Verbaten (Eds.).*Event-Related Brain Research.* Amsterdam: Elsevier Science Publishers, B.V., pp.116-129.

535. Rösler, F., Heil, M.& Hennighausen, E. (1994).. (UnPub)

536. Rösler, F., Heil, M.& Henninghausen, E. (1994). Exploring Memory Functions By Means of Brain Electrical Topography. *In Preparation*,

537. Rösler, F., Sutton, S., Johnson, R., et al (1986). *Endogenous ERP Components and Cognitive Constructs. A Review.* In W.C. McCallum, R. Zappoli& F. Denoth (Eds.).*Cerebral Psychophysiology: Studies in Event-Related Potentials*. Elsevier Science Publishers, pp.51-92.

538. Rubin, P., Holm, S., Friberg, L., et al (1991). Altered Modulation of Prefrontal and Subcortical Brain Activity in Newly Diagnosed Schizophrenia and Schizophreniform Disorder. *Archives of General Psychiatry*, *48*, 987-995.

539. Rubin, P., Karle, A., Moller-Madsen, S., et al (1993). Computerised Tomography in Newly Diagnosed Schizophrenia and Schizophreniform Disorder. *British Journal of Psychiatry*, *163*, 606-612.

540. Rutter, D.R. (1985). Language in Schizophrenia: The Structure of Monologues and Conversations. *British Journal of Psychiatry*, *146*, 399-404.

541. Sagawa, K., Kawakatsu, S., Shibuya, I., et al (1990). Correlation of Regional Cerebral Blood Flow With Performance on Neuropsychological Tests in Schizophrenic Patients. *Schizophrenia Research*, *3*, 241-246.

542. Saitoh, O., Niwa, S.-I., Hiramatsu, K.-I., Kameyama, T., Rymar, K.& Itoh, K. (1984). P300 in Siblings of Schizophrenic Probands. *Advances in Biological Psychiatry*, *15*, 46-59.

543. Salisbury, D.F., O'Donnell, B.F., McCarley, R.W., Nestor, P.G., Faux, S.F.& Smith, R.S. (1994). Parametric Manipulations of Auditory Stimuli

Differentially Affect P3 Amplitude in Schizophrenics and Controls. *Psychophysiology, 31*, 29-36.

544. Sasaki, H.& Gemba, H. (1982). Development and Changes of Cortical Field Potentials During Learning Processes of Visually Initiated Hand Movements in the Monkey. *Experimental Brain Research, 48*, 429-437.

545. Sasaki, K. (1979). *Cerebro-Cerebellar Interconnection in Cats and Monkeys*. In J. Massion& K. Sasaki (Eds.).*Cerebr-Cerebellar Interaction*. Amsterdam: Elsevier, pp.105-124.

546. Sasaki, K., Gemab, H., Hashimoto, S.& Mizuno, N. (1979). Influences of Cerebellar Hemispherectomy on Slow Potentials in the Motor Cortex Preceding Self-Paced Hand Movement in the Monkey. *Neuroscience Letters, 15*, 23-28.

547. Sasaki, K.& Gemba, H. (1986). Electrical Activity in the Prefrontal Cortex to No-Go Reaction of Conditioned Hand Movement with Colour Discrimination in the Monkey. *Experimental Brain Research, 64*, 603-606.

548. Sasaki, K.& Gemba, H. (1991). *Cortical Potentials Associated with Voluntary Movements in Monkeys*. In C.H.M. Brnia, G. Mulder& M.N. Verbaten (Eds.).*Event-Related Brain Research*. Elsevier Science Research, pp.80-96.

549. Sawaguchi, T.& Goldman-Rakic, P.S. (1991). D1 Dopamine Reports in the Prefrontal Cortex: Involvement in Working Memory. *Science, 251*, 947-951.

550. Sawaguchi, T., Matsumara, M.& Kubota, K. (1988). Dopamine Enhances the Neuronal Activity of Spatial Short Term Memory Task in the Primate Prefrontal Cortex. *Neuroscience Research, 5*, 465-473.

551. Schandry, R. (1991). *Origins of the Main Human Biosignals*. In R. Weitkunat (Ed.). *Digital Bioscanning Processing*. Elsevier Science Publishers B.V., pp.1-12.

552. Scherg, M. (1990). *Fundamentals of Dipole Source Potential Analysis*. In F. Grandori, M. Hoke& G.L. Romani (Eds.).*Auditory Evoked Magnetic Fields and Electric Potential*. Basel: S.Karger, pp.40-69.

553. Scherg, M. (1992). Functional Imaging and Localisation of Electromagnetic Brain Activity. *Brain Topography, 5(2)*, 103-111.

554. Scherg, M.& Berg, P. (1991). Use of Prior Knowledge in Brain Electromagnetic Source Analysis. *Brain Topography, 4(2)*, 143-150.

555. Scherg, M.& Ebersole, J.S. (1993). Models of Brain Sources. *Brain Topography, 5(4)*, 419-423.

556. Scherg, M.& Picton, T. (1991). *Separation and Identification of Event-Related Potential Components by Brain Electric Source Analysis*. In C.H.M. Brunia, G. Mulder& M.N. Verbaten (Eds.).*Event-Related Brain Research*. Elsevier, pp.24-37.

557. Schreiber, H., Lang, H., Lang, M., et al (1983). Frontal Hemispheric Differences in the Bereitschaftspotential Associated with Writing and Drawing. *Human Neurobiology, 2*, 197-202.

558. Schweitzer, L. (1982). Evidence of Right Cerebral Hemisphere Dysfunction in Schizophrenic Patients with Left Hemisphere Overactivation. *Biological Psychology, 17 (6)*, 655-673.

559. Schweitzer, L., Becker, E.& Welsch, H. (1978). Abnormalities of Cerebral Lateralization in Schizophrenia Patients. *Archives of General Psychiatry, 35*, 982.

560. Seidman, L.J., Yurgelen-Todd, D., Kremen, W.S., et al (1994). Relationship of Prefrontal and Temporal Lobe MRI Measures to Neuropsychological Performance in Chronic Schizophrenia. *Biological Psychiatry, 35*, 235-246.

561. Shagass, C., Ornitz, E.M., Sutton, S.& Tueting, P. (1978). *Event-Related Potentials and Psychopathology*. In E. Callaway, P. Tueting& S. Koslow (Eds.).*Event-Related Potentials in Man*. New York: Academic Press, pp.443-496.

562. Shallice, T. (1990). *From Neuropsychology to Mental Structure*. Cambridge/New York: 2nd Ed., Cambridge University Press.

563. Shenton, M.E., Kirkinis, R.& Jolez, F.A. (1992). Abnormalities of the Left Temporal Lobe and Thought Disorder in Schizophrenia. A Quantitative Magnetic Resonance Imaging Study. *New England Journal of Medicine, 327*, 604-612.

564. Sheppard, G., Gruzelier, J., Manchanda, R.& Hirsch, S.R. (1983). 15-0 Positron Emission Tomographic Scanning in Predominantly Never Treated Acute Schizoprenic Patients. *Lancet, II*, 1448-1452.

565. Shibasaki, H., Barrett, G., Neshige, R.& Tomoda, H. (1986). Volitional Movement Is Not Preceded by Cortical Slow Negativity in Cerebellar Dentate Lesions in Man. *Brain Research, 368*, 361-365.

566. Siegel, B.V., Buchsbaum, M.S., Bunney, W.E., et al (1993). Cortical-Striatal-Thalamic Circuits and Brain Glucose Metabolic Activity in 70 Unmedicated Male Schizophrenic Patients. *American Journal of Psychiatry, 150*, 1325-1336.

567. Simons, R.F. (1982). Physical Anhedonia and Future Psychopathology: An Electrocortical Continuity. *Psychophysiology, 19 (4)*, 433-441.

568. Simons, R.F. (1986). *Can the Autonomic Nervous System Contribute to Our Understanding of Event-Related Slow Brain Potentials?* In W.C. McCallum, P. Zappoli& R. Denoth (Eds.).*ERPs Associated with Preparatory and Movement Related Processes. A Review*. Elsevier, pp.206-212.

569. Simons, R.F., Balaban, M.T., Macy, M.H.& Graham, F.K. (1986). Heart Rate, Blink, and ERPs to Modality Defined Targets and Novels. *Psychophysiology, 23*,

570. Simons, R.F.& Russo, K.R. (1987). Event-Related Potentials and Continouos Performance in Subjects with Physical Anhedonia. *Journal of Psychophysiology, 1*, 401-410.

571. Simson, R., Vaughan, H.G.& Ritter, W. (1977). The Scalp Topography of Potentials in Auditory and Visual Go- Nogo Tasks. *Electroencephalography and Clinical Neurophysiology, 43*, 864-875.

572. Singh, J.& Knight, R.T. (1990). Frontal Lobe Contribution to Voluntary Movements in Humans. *Brain Research, 531*, 45-54.

573. Singh, J., Knight, R.T., Rosenlicht, N., Kotun, J.M., Beckley, D.J.& Woods, D.L. (1992). Abnormal Premovement Brain Potentials in Schizophrenia. *Schizophrenia Research, 8*, 31-41.

574. Skinner, J.E. (1978). *A Neurophysiological Model for Regulation of Sensory Input to Cerebral Cortex*. In D.A. Otto (Ed.). *Multidisciplinary Perspectives in Event-Related Brain Potential Research*. Washington,D.C.: Environmental Protection Agency, pp.615-625.

575. Skinner, J.E.& Yingling, C.D. (1976). Regulation of Slow Potential Shifts in Nucleus Reticularis Thalami by teh Mesencephalic Reticular Formation and the Frontal Granular Cortex. *Electroencephalography and Clinical Neurophysiology, 40*, 288-296.

576. Skinner, J.E.& Yingling, C.D. (1977). *Central Gating Mechanisms that Regulate Event-Related Potentials and Behaviour*. In J. Desmedt (Ed.).

Attention,Voluntary Contraction and Event-Related Cerebral Potentials. Basel: Karger, pp.30-69.

577. Small, J.G.& Small, I.F. (1971). Contingent Negative Variation. *Archives of General Psychiatry*, *25*, 550-554.

578. Snyder, A.Z. (1991). Dipole Source Localisation in the Study of EP Generators: A Critique. *Electroencephalography and Clinical Neurophysiology*, 321-325.

579. Sperry, R.W. (1950). Neural Basis of the Spontaneous Optokinetic Response Produced by Visual Inversion. *Journal of Comparative and Physiological Psychology*, *43*, 482-489.

580. Spitzer, M. (1993). The Psychopathology, Neuropsychology, and Neurobiology of Associative and Working Memory in Schizophrenia. *European Archives of Psychiatry and Clinical Neuroscience*, *243*, 57-70.

581. Squire, L.R.& Zola-Morgan, S. (1991). The Medial Temporal Lobe Memory System. *Science*, *253*, 1380-1385.

582. Srebo, R.& Purdy, P.D. (1990). Localization of Visually Evoked Cortical Activity Using Magnetic Resonance Imaging and Computerized Tomography. *Vision Research*, *30(3)*, 351-358.

583. Stamm, J., Stapelton, J.M.& Halgren, E. (1984). Human Medial Lobe Potentials Evoked in Memory and Language Tasks. *Electroencephalography and Clinical Neurophysiology*, *63*, 145-163.

584. Stamm, J.S. (1984). *Performance Enhancement with Cortical Negative Slow Shifts in Monkey and Man*. In B. Rockstroh, T. Elbert, W. Lutzenberger& N. Birbaumer (Eds.).*Self Regulation of the Brain and Behavior*. Heidelberg/Berlin: Springer, pp.199-215.

585. Stephens, J.H., Ota, K.Y., Carpenter, W.T.& Shaffer, J.W. (1980). Diagnostic Criteria for Schizophrenia: Prognostic Implications and Diagnostic Overlap. *Psychiatry Research*, *2*, 1-12.

586. Stevens, J. (1986). *Applied Multivariate Statistics for the Social Sciences*. Hillsdale,N.J.: Lawrence Erlbaum Publishers.

587. Stevens, J.R. (1982). Neuropathology of Schizophrenia. *Archives of General Psychiatry*, *39*, 1131-1139.

588. Strandburg, R.J., Marsh, J.T., Brown, W.S., Asarnow, R.F.& Guthrie, D. (1984). Event-Related Potential Concomitants of Information Processing Dysfunction in Schizophrenic Children. *Electroencephalography and Clinical Neurophysiology, 57*, 236-253.

589. Strauss, J.S.& Gift, T.E. (1977). Choosing an Approach for Diagnosing Schizophrenia. *Archives of General Psychiatry, 34*, 1248-1253.

590. Stuss, D.T.& Benson, D.F. (1994). *The Frontal Lobes and Control of Cognition and Memory*. In Perecman (Ed.). *The Frontal Lobes Revisited*. IRBN Press, pp.141-158.

591. Suddath, R.L., Christison, G.W., Torrey, E.F.& Weinberger, D.R. (1990). Cerebral Anatomical Abnormalities in Monozygotic Twins Discordant to Schizophrenia. *New England Journal of Medicine, 322*, 789-794.

592. Summerfelt, A.T., Alphs, L.D., Wagman, A.M.I., Funderburk, F.R., Hierholzer, R.M.& Strauss, M.E. (1991). Reduction of Perseverative Errors in Patients With Schizophrenia Using Monetary Feedback. *Journal of Abnormal Psychology, 100 (4)*, 613-616.

593. Sutton, S.& Ruchkin, D.S. (1984). *The Late Positive Complex: Advances and New Problems*. In R. Karrer, J. Cohen& P. Tueting (Eds.).*Brain and Information: Event-Related Potentials*. Anals of the New York Academy of Science (Vol.425),

594. Suzuki, M., Yuasa, S., Minabe, Y., Murata, M.& Kurachi, M. (1993). Left Superior Temporal Blood Flow Increases in Schizophrenic and Schizophreniform Patients with Auditory Hallucination: A Longitudinal Case Study Using 123-I-IMP SPECT. *European Archives of Psychiatry and Clinical Neuroscience, 242*, 257-261.

595. Szechtman, H., Nahmias, C., Garnett, E.S., et al (1988). Effect of Neuroleptics on Altered Cerebral Glucose Metabolism in Schizophrenia. *Archives of General Psychiatry, 45*, 523-532.

596. Tam, S.-Y.& Roth, R.H. (1985). Selective Increase in Dopamine Metabolism in the Prefrontal Cortex by the Anxiogenic Beta-Carboline FG 7142. *Biochemical Pharamcology, 34 (9)*, 1595-1598.

597. Tamminga, C.A., Thaker, G.K., Buchanan, R.W., et al (1992). Limbic System Abnormalities Identified in Schizophrenia Using Positron Emission Tomography With Fluorodeoxyglucose and Neocortical Alterations With Deficit Syndrome. *Archives of General Psychiatry, 49*, 522-530.

598. Taylor, P.J. (1987). *Hemispheric Lateralization and Schizophrenia*. In H. Helmchen& F.A. Henn (Eds.).*Biological Perspectives of Schizophrenia*. John Wiley and Sons Ltd., pp.213-235.

599. Tecce, J.J. (1978). *A CNV Rebound Function: Preliminary Report*. In D.A. Otto (Ed.). *Multidisciplinary Perspectives in Event-Related Brain Potentials*. Washington D.C.: U.S. Environmental Protection Agency, pp.222-225.

600. Tecce, J.J.& Cole, J.O. (1976). *The Distraction-Arousal Hypothesis, CNV, and Schizophrenia*. In D.J. Mostofsky (Ed.). *Behavioral Control and Modification of Physiological Activity*. London: Englewood, Cliffs, N.J. Prentice Hall, pp.162-220.

601. Tecce, J.J., Gardos, G., Cole, J.O.& Bowers, P.A. (1979). A CNV Typology of Chronic Schizophrenics and Response to Drug Treatment. *Psychopharamacology Bulletin, 15 (1)*, 39-41.

602. Tecce, J.J., Orzak, M.H.& Mirsky, A.F. (1978). *Abesence of CNV Reboumd in Psychosurgery Patients*. In D.A. Otto (Ed.). *Multidisciplinary Perspectives in Event-related Brain Potential Research*. Washington D.C.: Environmental Protection Agency, pp.383-386.

603. Tecce, J.J., Savignano-Bowman, J.& Cole, J.O. (1978). *Drug Effects on Contingent Negative Variation and Eyeblinks: The Distraction-Arousal Hypothesis*. In M.A. Lipton, A. DiMascio& K.F. Killam (Eds.).*Psychopharmacology: A Generation of Progress*. New York: Raven Press, pp.745-758.

604. Thaker, G., Kirkpatrick, B., Buchanan, R., Ellsberry, R., Lahti, A.& Tamminga, C. (1989). Oculomotor Abnormalities and Their Clinical Correlates in Schizophrenia. *Psychopharamacology Bulletin, 25(3)*, 491-497.

605. Thompson, J.W., Newton, P., Pocock, P.V., Crow, H., McCallum, W.C.& Papakostopoulos, D. (1978). *Preliminary Study of Pharmacology of Contingent Negative Variation in Man*. In D.A. Otto (Ed.). *Multidisciplinary Perspektives in Event-Related Brain Potential Research*. Washington D.C.: Environmental Protection Agency, pp.51-55.

606. Thomson, J.W., Newton, P., Pococ, P.V., et al (1978). *Preliminary Study of Pharmacology of Contingent Negative Variation in Man*. In D.A. Otto (Ed.). *Multidisciplinary Perspectives in Event-Related Potential Research*. Wahington D.C.: U.S. Environmental Protection Agency, pp.51-55.

607. Timsit, M., Koninckx, N., Dargent, J., Fontaine, O.& Dongier, M. (1970). Variations Contingentes Négatives en Psychiatrie. *Electroencephalography and Clinical Neurophysiology, 28*, 41-47.

608. Timsit, M., Schoenen, J.& Timsit-Bertier, M. (1986). *A Psychophysiological Approch to Psychosomatic Headache.* In W.C. McCallum, R. Zappoli& F. Denoth (Eds.).*Cerebral Psychophysiology: Studies in Event-Related Potentials.* Elsevier Science Publishers B.V., pp.455-457.

609. Timsit-Berthier, M. (1983). *Implications and Applications of SPs.* In W.C. McCallum& S.H. Curry (Eds.).*Slow Potential Change in the Human Brain.* New York: Plenum Press, pp.163-164.

610. Timsit-Berthier, M., Ansseau, M., Mantanus, H.& Legros, J.-J. (1986). *Concurrent Use of Contingent Negative Variation and Dexamethasone Supression Test in Major Depression.* In C. Shagass, R.C. Josiassen& R.A. Roemer (Eds.).*Brain Electrical Potentials and Psychopathology.* Elsevier Science Publishers B.V., pp.245-258.

611. Timsit-Berthier, M., de Thier, D.& Timsit, M. (1987). Electrophysiological and Psychological Aspects of the Derealization State Induced by Nitrous Oxide in Nine Control Subjects. *Advances in Biological Psychiatry, 16*, 90-101.

612. Timsit-Berthier, M., Delanuoy, J.& Gerono, A. (1978). *Morphological Analyses of the CNV in Psyhiatry: Comparison of Resolution Mode and Cumulative Curve Methods.* In D.A. Otto (Ed.). *Multidisciplinary Perspectives in Event-Related Brain Potential Research.* Washington D.C.: Environmental Protection Agency, pp.389-391.

613. Timsit-Berthier, M., Delaunoy, J., Koninckx, N.& Rousseau, J.C. (1973). Slow Potential Changes in Psychiatry. I. Contingent Negative Variation. *Electroencephalography and Clinical Neurophysiology, 35*, 355-361.

614. Timsit-Berthier, M., Delaunoy, J.& Rousseau, J. (1976). *Some Problems and Tentative Solutions to Questions Raised by Slow Potential Changes in Psychiatry.* In W.C. McCallum& J Knott (Eds.).*The Responsive Brain.* Bristol: Wright, pp.138-143.

615. Timsit-Berthier, M., Delauony, J.& Rousseau, J.C. (1973). Slow Potential Changes in Psychiatry. II. Motor Potential. *Electroencephalography and Clinical Neurophysiology, 35*, 363-367.

616. Timsit-Berthier, M., Gerono, M., Rousseau, J., et al (1984). *An International Pilot Study of CNV in Mental Illness - Second Report.* In R. Karrer,

J. Cohen& P. Tueting (Eds.).*Brain and Information*. New York: The New York Academy of Science, pp.629-637.

617. Timsit-Berthier, M., Mantanus, H., Ansseau, M., Doumont, A.& Legros, J.J. (1983). Methodological Problems Raised by Contingent Negative Variations Interpretation in Psychopathological Conditions. *Advances in Biological Psychiatry, 13*, 80-92.

618. Timsit-Berthier, M., Mantaus, H., Ansseau, M., Dumont, A.& Legros, J.J. (1983). *Methodological Problems Raised by Contingent Negative Variation Interpretation in Psychopathological Conditions*. In C. Perris, D. Kemali& M. Koukkou-Lehmann (Eds.).*Neurophysiological Correlates of Normal Cognition and Psychopathology*. Basel: Karger, pp.80-92.

619. Timsit-Berthier, M., Rousseau, J.C.& Delaunoy, J. (1971). Réactivité de l'onde d'attente et des ondes négatives post- imperatives. *Rev.EEG.Neurophys., 1(2)*, 245-248.

620. Timsit-Berthier, M.& Timsit, M. (1981). Toward a Neurochemical Interpretation of CNV in Psychiatry. *Advances in Biological Psychiatry, 6*, 165-172.

621. Timsith-Berthier, M., Delaunoy, J., Xhensel, B.& Timsit, M. (1978). *Reliability of Contingent Negative Variation in Psychopathology*. In D.A. Otto (Ed.). *Multidisciplinary Perspectives in Event-Related Brain Potential Research*. Washington D.C.: Environmental Protection Agency, pp.373-375.

622. Timsith-Berthier, M., Mantanus, H., Ansseau, M., Devoitille, M., Mas, A.D.& Legros, J.-J. (1987). *Contingent Negative Variation in Major Depressive Patients*. In R. Johnson, J.W. Rohrbaught& R. Parasuraman (Eds.).*Current trends in Event-Related Potential Research*. Elsevier Science Publishers B.V., pp.762-771.

623. Timsith-Berthier, M., Mantanus, H., Marassiaux, P., et al (1986). *CNV and Dopamine Receptor Reactivity: Correlations with the Apomorphine Test*. In W.C. McCallum, R. Zappoli& F. Denoth (Eds.).*Cerebral Psychophysiology: Studies in Event-Related Potentials*. Elsevier Science Publishers, pp.403-405.

624. Tomer, R., Mintz, M., Levy, A.& Myslobodsky, M. (1981). Smooth Pursuit Pattern in Schizophrenic Patients During Cognitive Tasks. *Biological Psychology, 16(2)*, 131-143.

625. Tomer, R., Mintz, M., Levy, A.& Myslobodsky, M. (1981). Smooth Pursuit Pattern in Schizophrenic Patients During Cognitive Task. *Biological Psychiatry, 16(2)*, 131-143.

626. Tsuda, T. (1982). Contingent Negative Variation in Parkinsonism. *Tokushima Journal of Experimental Medicine, 29*, 87-94.

627. van Boxtel, G.J.M.& Brunia, C.H.M. (1990). *The Effects of Precuing Isometric Contraction Speed On the Contingent Negative Variation and Reaction Time.* In C.H.M. Brunia, A.K.W. Gaillatd& A. Kok (Eds.).*Psychophysiological Brain Research.* Tilburg: Tilburg University Press,

628. van Boxtel, G.J.M.& Brunia, C.H.M. (1994). Motor and Non-Motor Components of the Contingent Negative Variation. *International Journal of Psychophysiology, 17*, 269-279.

629. van Boxtel, G.J.M., van den Boogart, B.& Brunia, C.H.M. (1993). The Contingent Negative Variation in a Choice Reaction Time Task. *Journal of Psychophysiology, 7*, 11-23.

630. van den Bosch, R.J. (1983). Contingent Negative Variation and Psychopathology: Frontal- Central Distribution, and Association with Performance Measures. *Biological Psychiatry, 18 (6)*, 615-634.

631. van den Bosch, R.J. (1984). Contingent Negative Variation: Components and Scalp Distribution in Psychiatric Patients. *Biological Psychiatry, 19 (7)*, 963-972.

632. van den Bosch, R.J., Rozendaal, N.& Mol, JM.F.A. (1988). Slow Potential Correlates of Frontal Functions, Psychosis, and Negative Symptoms. *Psychiatry Research, 2*, 201-208.

633. Vasey, M.W.& Thayer, J.F. (1987). The Continuing Problem of False Positives in Repeated Measures ANOVA in Psychophysiology: A Multivariate Solution. *Psychophysiology, 24 (4)*, 479-486.

634. Verhey, F., Lamers, T.& Emonds, P. (1984). *A Second Baseline in Relating ERP and Measured Psychopathology.* In R. Karrer& J. Cohen (Eds.).*Brain and Information.* New York: The New York Academy of Science, pp.638-644.

635. Verhey, F., Lamers, T., Timsit-Berthier, M., et al (1986). *An International Pilot study of CNV in Mental Illness. III.* In W.C. McCallum, R. Zappoli& F. Denoth (Eds.).*Cerebral Psychophysiology: Studies in Event-Related Potentials.* Elsevier Science Publishers B.V., pp.458-460.

636. Verhey, F.M.H., Lamers, T.B.& Emonds, P.M.J. (1980). *Intercorrelations of CNV Variables*. In H.H Kornhuber& L. Deecke (Eds.).*Motivation, Motor and Sensory Processes of the Brain. Electrical Potentials, Behavior and Clinical Use. Progress in Brain Research*. Amsterdam: Elsevier, pp.531-534.

637. Verleger, R. (1988). Event-Related Potentials and Cognition: A Critique of the Context Updating Hypothesis and an Alternative Interpretation of P3. *Behavioral Brain Research, 11 (3)*, 343-327.

638. Verleger, R.& Cohen, R. (1978). Effects of Certanity, Modality Shift and Guess Outcome on Evoked Potentials and Reaction Times in Chronic Schizophrenics. *Psychological Medicine, 8*, 81-93.

639. Volkow, N.D., Wolf, A.P., Van Gelder, P., et al (1987). Phenomenological Correlates of Metabolic Activity in 18 Patients With Chronic Schizophrenia. *American Journal of Psychiatry, 144 (2)*, 151-158.

640. von Holst, E.& Mittelstaedt, H. (1950). Das Reafferenzprinzip (Wechselwirkungen zwischen Zentralnervensystem und Periferie). *Naturwissenschaften, 37*, 464-476.

641. Wagner, M., Kurz, G.& Engel, R.R. (1989). Normal P300 in Acute Schizophrenics during a Continuous Performance Test. *Biological Psychiatry, 25*, 1-3.

642. Wagner, M., Rendtorff, N., Kathamnn, N.& Engel, R.R. (1994). CNV, PINV, And Probe-Evoked Potentials in Schizophrenics (In Press)

643. Wagner, M., Rendtorff, N., Kathmann, N.& Engel, R.R. (1993). CNV, PINV, and Probe-Evoked Potentials in Schizophrenics, Submitted. *Journal of Psychophysiology, 7*, 172-173.

644. Wagner, M., Rendtorff, N., Kathmann, N.& Engell, R.R. (1994). Sensitivity of the CNV to Irrelevant Novels in Schizophrenics, Alcoholics and Normal Controls. *Journal of Psychophysiology, 7*, 172-173.

645. Walker, E.& McGuire, M. (1982). Intra- and Interhemispheric Information Processing in Schizophrenia. *Psychological Bulletin, 92 (3)*, 701-725.

646. Walter, W.G. (1964). The Contingent Negative Variation. An Electrical Sign of Significance of Association in the Human Brain. *Science, 146*, 434.

647. Walter, W.G. (1968). *Electric Signs of Expectancy and Decision in the Human Brain*. In H.L. Oestreiceher& D.R. Moore (Eds.).*Cybernetic Problems in Bionics*. New York: Gordon & Breach, pp.361-396.

648. Warkentin, S., Nilson, A., Risberg, J.& Carlson, S. (1989). Abscence of Frontal Lobe Activation in Schizophrenia. *Journal of Cerebral Blood Flow and Metabolism*, *9*, 354.

649. Warkentin, S., Nilsson, A., Risberg, J., et al (1990). Regional Cerebral Blood Flow in Schizophrenia: Repeated Studies During a Psychotic Episode. *Psychiatry Research*, *35*, 27-38.

650. Watanabe, M. (1986). Prefrontal Unit Activity During Delayed Conditional Go/No-Go Discrimination in the Monkey. I. Relation to the Stimulus. *Brain Research*, *382*, 1-14.

651. Watanabe, M. (1986). Prefrontal Unit Activity During Delayed Conditional Go/No-Go Discrimination in the Monkey. II. Relation to Go/No-Go Responses. *Brain Research*, *382*, 15-27.

652. Weinberg, H. (1973). The Contingent Negative Variation: Its Relation to Feedback and Expectant Attention. Related Slow Potentials of the Brain: Their Relations to Behavior. *Electroencephalography and Clinical Neurophysiology*, *33*, 219-228.

653. Weinberger, D.R. (1987). Implications of Normal Brain Development for the Pathogenesis of Schizophrenia. *Archives of General Psychiatry*, *44*, 660-669.

654. Weinberger, D.R. (1991). *Anteromedial Temporal-Prefrontal Connectivity: A Functional Neuroanatomical System Implcated in Schizophrenia*. In B.J. Anteromedial Tempral-PrefronCarroll& J.E. Barrett (Eds.).*Psychopathology and the Brain*. New York: Raven Press, pp.25-42.

655. Weinberger, D.R. (1994). *Schizophrenia as a Neurodevelopmental Disorder: A Review of the Concept*. In S.R. Hirsch& D.R. Weinberger (Eds.).*Schizophrenia*. London: Blackwood Press, pp.2-76.

656. Weinberger, D.R. (1994). *Schizophrenia as a Neurodevelopmental Disorder: A Review of the Concept*. In S.R. Hirsch& D.R. Weinberger (Eds.).*Schizophrenia*. London (in press): Blackwood Press, pp.1-73.

657. Weinberger, D.R.& Berman, K.F. (1988). Speculation on the Meaning of Cerebral Metabolic Hypofrontality in Schizophrenia. *Schizophrenia Bulletin*, *14 (2)*, 157-168.

658. Weinberger, D.R., Berman, K.F.& Daniel, D.G. (1991). *Prefrontal Cortex Dysfunction in Schizophrenia.* In H.S. Levin, H.M. Eisenberg& A.L. Benton (Eds.).*Frontal Lobe Function and Dysfunction.* New York: Oxford University Press, pp.275-287.

659. Weinberger, D.R., Berman, K.F.& Illowsky, B.P. (1988). Physiological Dysfunction of Dorsolateral Prefrontal Cortex in Schizophrenia / III. A New Cohort and Evidence for a Monoaminergic Mechanism. *Archives of General Psychiatry, 45,* 609-615.

660. Weinberger, D.R., Berman, K.F., Suddath, R.& Torrey, E.F. (1992). Evidence of Dysfunction of a Prefrontal-Limbic Network in Schizophrenia: A Magnetic Resonance Imaging and Regional Cerebral Blood Flow Study of Discordant Monozygotic Twins. *American Journal of Psychiatry, 149 (7),* 890-897.

661. Weinberger, D.R., Berman, K.F.& Zec, R.F. (1986). Physiologic Dysfunction of Dorsolateral Prefrontal Cortex in Schizophrenia / I. Regional Cerebral Blood Flow Evidence. *Archives of General Psychiatry, 43,* 114-124.

662. Weinrich, M., Wise, S.P.& Mauritz, K.H. (1984). A Neurophysiological Study of the Premotor Cortex in the Rhesus Monkey. *Brain, 107,* 385-414.

663. Westphal, K.P., Grözinger, B., Diekmann, V., et al (1990). Slower Theta Activity Over the Midfrontal Cortex in Schizophrenic Patients. *Acta Psychiatrica Scandinavica, 81,* 132-138.

664. Westphal, K.P., Neher, K.D., Grözinger, B., Diekmann, V.& Kornhuber, H.H. (1980). *Differences Between Schizophrenic Patients and Normal Controls in Bereitschaftspotential, Alpha Activity and Other EEG Signs.* In W.C. McCallum, R. Zappoli& F. Denoth (Eds.).*Cerebral Psychophysiology: Studies in Event-Related Potentials.* Amsterdam: Elsevier, pp.464-465.

665. Westphal, K.P., Neher, K.D., Grözinger, B., Diekmann, V.& Kornhuber, H.H. (1986). *Differences Between Schizophrenic Patients and Normal Controlls in Bereitschaftspotential, Alpha Activity and Other EEG Signs.* In W.C. McCallum, R. Zappoli& F. Denoth (Eds.).*Studies in Event-Related Potentials.* Amsterdam: Elsevier, pp.464-465.

666. Wilkinson, R.T.& Spence, M.T. (1973). Determinants of the Post-Stimulus Resolution of Contingent Negative Variation (CNV). *Electroencephalography and Clinical Neurophysiology, 35,* 503-509.

667. Winn, P. (1994). Schizophrenia Research Moves to the Prefrontal Cortex. *Trends in Neurosciences, 17 (7)*, 265-272.

668. Winn, P. (1994). Schizophrenia Works Moves to the Prefrontal Cortex. *Trends in Neurosciences, 17 (7)*, 7-94.

669. Winocur, G. (1992). A Comparison of Normal Old Rats and Young Adult Rats with Lesions to the Hippocampus or Prefrontal Cortex on a Test of Matching-To-Sample. *Neuropsychologia, 30 (9)*, 769-781.

670. Wise, P.S. (1991). *A Brief History of Basal Ganglia.* In B.J. Carroll& J.E. Barrett (Eds.).*Psychopathology and the Brain.* New York: Raven Press Ltd., pp.45-63.

671. Wolkin, A., Angrist, B., Wolf, A., et al (1988). Low Frontal Glucose Utilization in Chronic Schizophrenia: A Replication Study. *American Journal of Psychiatry, 145 (2)*, 251-253.

672. Wolkin, A., Jaeger, J., Brodie, J.D., et al (1985). Persistence of Cerebral Metabolic Abnormalities in Chronic Schizophrenia as Determined by Positron Emission Tomography. *American Journal of Psychiatry, 142 (5)*, 564-571.

673. Wong, P.K.H. (1991). *Introduction to Brain Topography.* New York: Plenum Press.

674. Woods, B.T.& Yurgelen-Todd, D. (1991). Brain Volume Loss in Schizophrenia: When Does it Occur and is it Progressive ? *Schizophrenia Research, 5*, 202-204.

675. Wurtz, R.H. (1966). Steady Potential Correlates of Intracranial Reinforcement. *Electroencephalography and Clinical Neurophysiology, 20*, 59-67.

676. Yingling, C.D.& Skinner, J.E. (1976). Selective Regulation of Thalamic Sensory Relay Nuclei by Nucleus Reticularis Thalami. *Electroencephalography and Clinical Neurophysiology, 41*, 476-482.

677. Young, A.H., Blackwood, D.H.R., Roxborough, H., McQueen, J.K., Martin, M.J.& Kean, D. (1991). A Magnetic Resonance Imaging Study of Schizophrenia: Brain Structure and Clinical Symptoms. *British Journal of Psychiatry, 158*, 158-164.

678. Zappoli, R., Papini, M., Pasquinelli, A., Martinelli, M.G.& Guerri, S. (1978). *CNV in Patients After Psychosurgery Affecting Thalamocortical*

Pathways to Prefrontal Cortex. In D.A. Otto (Ed.). *Multidisciplinary Perspectives in Event-Related Potential Research.* Washington D.C.: Environmental Protection Agency, pp.376-382.

679. Zappoli, R., Versari, A., Arnetoli, G., et al (1993). The Effects of Prefrontal Association Cortical Areas Excision or Degeneration on Some Bit-Mapped ERPs (CNV Complex). *Brain Topography, 5 (3),* 314.

680. Zappoli, R., Versari, A., Arnetoli, G., et al (1992). The Effects of Deafferenting Prefrontal Association Cortical Areas On Neurocognitive Event-Related Potentials (CNV Complex). *Journal of Psychophysiology, 6(3),* 267.

681. Zappoli, R., Versari, A., Bartelli, M., Gengomi, P.F., Mascalchi, M.& Poggiolini, D. (1993). Neurocognitive Potentials, Cortico-Cortical Homohemispheric Two- Way Connections and Prefrontal Cortex Ablation or Degeneration. *International Journal of Psychophysiology, 14 (2),* 156.

682. Zola-Morgan, S.& Squire, L.R. (1985). Medial Temporal Lessions in Monkeys Impair Memory On a Variety of Tasks Sensitive to Human Amnesia. *Behavioral Neuroscience, 99,* 22-34.

683. Zubin, J. (1950). Symposium on Statistics for the Clinician. *Journal of Clinical Psychology, VI (1),* 1-7.

684. Zubin, J.& Spring, B. (1977). Vulnerability - A New View of Schizophrenia. *Journal of Abnormal Psychology, 86 (2),* 103-126.

Appendix A Additional Statistical Results not Listed in the Text

Appendix A.1 Multivariate Analysis of Variance (MANOVA) Results of the Slow Negative Potential Shifts of Experiment A

Table A.1.1 Experimental Effects on iCNV and tCNV

		iCNV		tCNV	
Effect	df1, df2	F	p	F	p
M	1,36	4.17	**.05**	4.96	**.03**
*M * G*	*1,36*	*1.34*	*.25*	*0.47*	*.50*
A	1,36	1.22	.30	0.06	.81
*A * G*	*1,36*	*1.41*	*.24*	*4.91*	*.06*
H	1,36	7.88	**.008**	8.88	**.005**
*H * G*	*1,36*	*0.87*	*.36*	*.01*	*.91*
E	4,33	15.44	**.0001**	21.0	**.0001**
*E * G*	*4,33*	*1.25*	*.31*	*2.51*	*.06*
M * A	1,36	1.89	.18	0.01	.92
*M * A * G*	*1,36*	*0.07*	*.80*	*0.83*	*.37*
M * H	1,36	0.10	.75	0.10	.75
*M * H * G*	*1,36*	*0.86*	*.35*	*1.63*	*.21*
A * H	1,36	1.56	.22	1.89	.18
*A * H * G*	*1,36*	*0.01*	*.93*	*0.08*	*.93*
M * E	4,33	5.00	**.003**	3.53	**.02**
*M * E * G*	*4,33*	*0.76*	*.56*	*0.44*	*.78*
A * E	4,33	0.27	.97	0.11	.98
*A * E * G*	*4,33*	*0.45*	*.77*	*1.09*	*.38*
H * E	4,33	0.46	.76	1.42	.25
*H * E * G*	*4,33*	*0.70*	*.60*	*0.63*	*.65*
M * A * H	1,36	0.07	.80	0.12	.73
*M * A * H * G*	*1,36*	*0.99*	*.33*	*1.38*	*.25*
M * A * E	4,33	1.43	.24	2.57	.06
*M * A * E * G*	*4,33*	*0.38*	*.82*	*1.05*	*.39*
M * H * E	4,33	3.13	**.03**	1.06	.39
*M * H * E * G*	*4,33*	*1.25*	*.31*	*1.37*	*.27*
A * H * E	4,33	1.85	.14	1.42	.25
*A * H * E * G*	*4,33*	*0.20*	*.94*	*0.51*	*.73*
M * A * H * E	4,33	1.42	.25	3.15	**.03**
*M * A* H * E * G*	*4,33*	*0.90*	*.48*	*0.53*	*.71*
Group	1,36	0.00	1.0	0.0	1.0

M*=Provisional Memory factor,* ***A****=Ambiguity factor,* ***H****=Hemisphere factor,* ***E****=Electrode factor,* ***G****=Group factor; Results of multivariate ANOVA*

Table A.1.2 Experimental Effects on PINV-1 and PINV-2

		PINV-1		PINV-2	
Effect	df1, df2	F	p	F	p
M	1,36	11.51	**.002**	7.43	**.01**
*M * G*	*1,36*	*5.19*	***.03***	*4.36*	***.04***
A	1,36	21.16	**.0001**	36.96	**.0001**
*A * G*	*1,36*	*2.88*	*.10*	*2.55*	*.12*
H	1,36	0.40	.53	0.22	.64
*H * G*	*1,36*	*18.23*	***.0001***	*19.59*	***.0001***
E	4,33	13.09	**.0001**	15.87	**.0001**
*E * G*	*4,33*	*4.44*	***.006***	*4.53*	***.005***
M * A	1,36	1.79	.19	2.51	.12
*M * A * G*	*1,36*	*1.99*	*.17*	*2.41*	*.13*
M * H	1,36	0.92	.34	0.60	.44
*M * H * G*	*1,36*	*1.14*	*.29*	*3.11*	*.09*
A * H	1,36	1.32	.26	2.38	.13
*A * H * G*	*1,36*	*8.87*	***.005***	*10.25*	***.003***
M * E	4,33	1.99	.12	1.50	.22
*M * E * G*	*4,33*	*1.04*	*.40*	*0.73*	*.58*
A * E	4,33	19.14	**.0001**	18.91	**.0001**
*A * E * G*	*4,33*	*8.17*	***.0001***	*5.28*	***.002***
H * E	4,33	1.02	.41	1.66	.18
*H * E*	*4,33*	*2.27*	*.08*	*2.44*	*.07*
M * A * H	1,36	2.84	.10	3.39	.07
*M * A * H * G*	*1,36*	*0.28*	*.60*	*0.31*	*.58*
M * A * E	4,33	0.52	.72	0.28	.89
*M * A * E*	*4,33*	*0.98*	*.43*	*1.21*	*.32*
M * H * E	4,33	0.67	.62	0.47	.76
*M * H * E * G*	*4,33*	*2.17*	*.09*	*2.64*	***.05***
A * H * E	4,33	7.60	**.0002**	8.23	**.0001**
*A * H * E*	*4,33*	*1.18*	*.34*	*1.03*	*.41*
M * A * H * E	4,33	1.63	.19	2.02	.11
*M * A* H * E * G*	*4,33*	*0.96*	*.44*	*0.92*	*.46*
Group	1,36	4.40	**.04**	4.54	**.04**

M*=Provisional Memory factor,* ***A****=Ambiguity factor,* ***H****=Hemisphere factor*
E*=Electrode factor,* ***G****=Group factor; Results of multivariate ANOVA*

Appendix A.2 ANOVA Results for the Data of Experiment A Using Different Baselines

Table A.2.1 Experimental Effects on PINV-1 and PINV-2 /1 s Pretrial Baseline

		PINV-1		PINV-2	
Effect	df1, df2	F	p	F	p
M	1,36	12.00	.001	8.93	.005
*M * G*	*1,36*	*0.52*	*.48*	*0.32*	*.58*
A	1,36	9.47	.004	15.30	.0004
*A * G*	*1,36*	*0.28*	*.60*	*0.38*	*.54*
H	1,36	5.12	.03	2.28	.0004
*H * G*	*1,36*	*14.60*	***.0005***	*15.52*	***.0004***
E	4,144	21.61	.0001	25.15	.0001
*E * G*	*4,144*	*1.48*	*.23*	*1.80*	*.17*
M * A	1,36	0.92	.34	1.35	.25
*M * A * G*	*1,36*	*0.00*	*.99*	*0.03*	*.86*
M * H	1,36	0.84	.36	0.69	.41
*M * H * G*	*1,36*	*0.21*	*.65*	*0.00*	*.99*
A * H	1,36	3.66	.06	4.82	.03
*A * H * G*	*1,36*	*5.96*	***.02***	*6.16*	***.02***
M * E	4,144	6.99	.0002	7.94	.0002
*M * E * G*	*4,144*	*1.23*	*.30*	*1.27*	*.29*
A * E	4,144	7.14	.0001	9.69	.0001
*A * E * G*	*4,144*	*2.91*	***.04***	*3.34*	***.02***
H * E	4,144	0.46	.73	0.76	.52
*H * E * G*	*4,144*	*6.24*	***.0004***	*6.80*	***.0002***
M * A * H	1,36	2.17	.15	1.84	.18
*M * A * H * G*	*1,36*	*2.57*	*.12*	*2.31*	*.14*
M * A * E	4,144	1.11	.35	0.99	.39
*M * A * E * G*	*4,144*	*0.54*	*.63*	*0.69*	*.54*
M * H * E	4,144	2.25	.07	2.23	.07
*M * H * E * G*	*4,144*	*0.67*	*.61*	*0.58*	*.67*
A * H * E	4,144	2.50	.05	2.86	.03
*A * H * E * G*	*4,144*	*0.89*	*.46*	*0.92*	*.44*
M * A * H * E	4,33	1.38	.25	1.18	.32
*M * A* H * E * G*	*4,33*	*0.14*	*.96*	*0.08*	*.98*
Group	1,36	1.05	.31	1.24	.27

__M__=Provisional Memory factor, __A__=Ambiguity factor, __H__=Hemisphere factor, __E__=Electrode factor, __G__=Group factor; univariate ANOVA; Huynh-Feldt corrected p-values

Table A.2.1(suppl.) Results of Within-Group ANOVA for PINV-1 and PINV-2 / 1 s Pretrial-Baseline

	PINV-1					PINV-2			
	dfs	Patients		Controls		Patients		Controls	
M	1,19	7.95,	.01	4.19	.06	5.14	.03	3.80	.07
H	1,19	29.58	.0001	0.88	.36	22.44	.0002	2.20	.15
A * H	1,19	0.16	.70	8.57	.009	0.05	.83	6.98	.003
A * E	4.72	2.32	.10	7.45	.0002	3.61	.03	8.83	.0001
H * E	4,72	4.33	.01	2.80	.04	3.70	.03	3.83	.009

for abbreviations see **table A.2.1**

Table A.2.2 Experimental Effects on PINV-1 and PINV-2/IS-LPC-Baseline

		PINV-1		PINV-2	
Effect	df1, df2	F	p	F	p
M	1,36	9.36	.004	7.30	.01
*M * G*	*1,36*	*6.41*	***.02***	*5.65*	***.02***
A	1,36	21.16	.0001	13.13	.0009
*A * G*	*1,36*	*3.75*	***.06***	*4.78*	***.03***
H	1,36	10.67	.002	18.42	.0001
*H * G*	*1,36*	*4.20*	***.05***	*3.01*	*.09*
E	4,144	29.75	.0001	26.96	.0001
*E * G*	*4,144*	*6.71*	***.0009***	*6.70*	***.0009***
M * A	1,36	0.30	.59	0.49	.49
*M * A * G*	*1,36*	*0.00*	*.97*	*0.01*	*.94*
M * H	1,36	0.03	.87	0.01	.94
*M * H * G*	*1,36*	*1.35*	*.25*	*2.22*	*.14*
A * H	1,36	2.04	.16	2.81	.10
*A * H * G*	*1,36*	*6.77*	***.01***	*7.87*	***.008***
M * E	4,144	2.56	.07	2.39	.08
*M * E * G*	*4,144*	*1.87*	*.15*	*1.63*	*.19*
A * E	4,144	14.90	.0001	17.61	.0001
*A * E * G*	*4,144*	*2.96*	*.06*	*3.25*	***.05***
H * E	4,144	4.12	.008	4.47	.005
*H * E * G*	*4,144*	*8.09*	***.0001***	***7.57***	***.0001***
M * A * H	1,36	4.52	.04	4.00	.05
*M * A * H * G*	*1,36*	*4.79*	***.04***	***4.36***	***.04***
M * A * E	4,144	1.01	.38	0.93	.41
*M * A * E * G*	*4,144*	*1.07*	*.36*	*0.86*	*.45*
M * H * E	4,144	3.49	.02	3.36	.02
*M * H * E * G*	*4,144*	*0.85*	*.48*	*1.03*	*.39*
A * H * E	4,144	5.31	.003	4.78	.004
*A * H * E * G*	*4,144*	*0.73*	*.52*	*0.71*	*.54*
M * A * H * E	4,33	1.92	.12	1.69	.17
*M * A* H * E * G*	*4,33*	*0.68*	*.58*	*0.56*	*.66*
Group	1,36	2.47	.12	2.46	.13

for abbreviations see **table A.2.1**

Table A.2.2 (suppl.) Results of Within-Group ANOVA for PINV-1 and PINV-2/IS-LPC-Baseline

		PINV-1				PINV-2			
	dfs	Patients		Controls		Patients		Controls	
M	1,19	9.25	.007	0.45	.51	7.52	.01	0.19	.67
A	1,19	3.52	.08	21.53	.0002	1.09	.31	15.99	.0008
H	1,19	0.73	.40	14.32	.001	----	----	----	----
E	4,72	36.66	.0001	6.01	.003	33.17	.0001	5.80	.003
A * H	1,19	0.75	.40	7.55	.01	0.71	.41	9.01	.008
H * E	4,72	3.79	.02	7.84	.0001	3.09	.04	8.04	.0001
M * A * H	1,19	4.93	.04	0.02	.90	4.35	.05	0.05	.83
A * E	4,72	----	----	----	----	3.27	.06	18.76	.0001

for abbreviations see table A.2.1

Appendix A.3 ANOVA Results for the Data of Experiment B Using the Pretrial Baseline

Table A.3.1 Experimental Effects on PINV-1, PINV-2, and PINV-3 /Pretrial-Baseline

		PINV-1		PINV-2		PINV-3	
Effect	df1, df2	F	p	F	p	F	p
C	2,44	5.31	.01	12.66	.0001	11.68	.0001
*C * G*	*2,44*	*3.04*	*.07*	*1.68*	*.20*	*1.90*	*.16*
L	2,44	0.83	.44	0.72	.49	1.74	.19
*L * G*	*2,44*	*3.77*	***.03***	*3.69*	***.03***	*5.88*	***.005***
A	2,44	10.78	.0004	5.15	.02	6.88	.006
*A * G*	*2,44*	*1.19*	*.31*	*1.23*	*.30*	*0.92*	*.39*
C * L	4,88	4.53	.006	2.50	.07	3.22	.03
*C * L * G*	*4,88*	*2.04*	*.12*	*0.41*	*.75*	*0.51*	*.67*
C * A	4,88	1.80	.15	1.16	.33	1.06	.38
*C * A * G*	*4,88*	*0.74*	*.55*	*0.76*	*.53*	*0.46*	*.74*
L * A	4,88	0.22	.91	1.15	.33	1.72	.18
*L * A * G*	*4,88*	*1.80*	*.14*	*1.45*	*.24*	*0.80*	*.49*
C * L * A	8,172	0.72	.65	0.80	.58	0.78	.59
*C*L*A*G*	*8,172*	*1.63*	*.13*	*2.53*	***.02***	*1.75*	*.11*
G	1,22	5.90	.02	6.07	.02	7.89	.01

__C__=Condition factor; __L__=Laterality factor; __A__=Anteriority factor; __G__=Group factor; PINV-1=1.0-1.5 s, PINV-2=2.0-2.5 s, PINV-3=2.5-3.0 s (each after IS onset; F=univariate F-values, p= Huynh-Feldt corrected p values; average mastoids referene, baseline is tCNV;

Table A.3.1 (suppl.) Results of Within-Group ANOVA for PINV-1, PINV-2, and PINV-3/Pretrial-Baseline

	PINV-1				PINV-2				PINV-3			
	Patients		Controls		Patients		Controls		Patients		Controls	
L	3.30	.06	0.68	.52	1.62	.22	3.71	.04	1.38	.27	9.40	.001

Patients: df1, df2=2,20; Controls: df1,df2=2,24; for abbreviations see **table A.3.1**

Appendix A.4 ANOVA Results for the Data of Experiment D Excluding Two Left-Handed Subjects in the High-Scoring Group

Table A.4.1 Experimental Effects on PINV-1 to PINV-4

		PINV-1		PINV-2		PINV-3		PINV-4	
Effect	df	F	p	F	p	F	p	F	p
C	1,33	3.36	**.07**	9.93	**.004**	12.20	**.001**	8.13	**.008**
C * G	1,33	0.05	.83	0.01	.92	0.37	.55	0.50	.48
Cr	1,33	0.33	.57	2.49	.12	11.70	**.002**	19.92	**.0001**
Cr * G	1,33	0.00	.95	0.43	.52	1.16	.29	1.47	.23
H	1,33	0.64	.43	4.16	**.05**	8.36	**.007**	12.86	**.001**
H * G	1,33	8.54	**.006**	8.84	**.006**	6.41	**.02**	3.54	**.06**
Sr	2,66	39.68	**.0001**	29.77	**.0001**	14.99	**.0001**	8.84	**.0004**
Sr* G	2,66	1.42	.25	1.74	.19	1.60	.21	1.47	.24
C * Cr	1,33	7.52	**.01**	6.58	**.02**	13.51	**.0008**	14.15	**.0007**
C * Cr * G	1,33	1.36	.25	1.10	.30	2.29	.14	0.29	.59
C * H	1,33	1.14	.29	2.76	.11	3.46	.07	5.36	**.03**
C * H * G	1,33	4,74	**.04**	5.04	**.03**	4.44	**.04**	3.13	.09
Cr * H	1,33	0.40	.53	1.83	.19	3.37	.08	4.41	.04
Cr * H * G	1,33	0.30	.59	0.61	.44	0.86	.36	0.71	.40
C * Sr	2,66	0.72	.46	1.71	.20	0.14	.86	0.82	.44
C * Sr * G	2,66	0.69	.47	1.11	.33	0.67	.52	1.17	.32
Cr * Sr	2,66	5.42	**.01**	9.36	**.001**	13.60	**.0001**	13.31	**.0001**
Cr * Sr * G	2,66	0.27	.74	0.50	.56	1.05	.35	1.71	.19
H * Sr	2,66	0.31	.73	0.88	.42	1.54	.22	1.85	.17
H * Sr * G	2,66	0.32	.72	0.75	.47	0.67	.51	0.41	.65
C * Cr * H	1,33	0.87	.36	1.39	.25	4.30	.05	6.66	**.01**
C * Cr * H * G	1,33	0.00	.96	1.11	.30	1.46	.24	1.58	.22
C * Cr * Sr	2,66	5.68	**.005**	4.17	**.02**	5.82	**.005**	6.00	**.005**
C * Cr * Sr * G	2,66	2.54	.09	3.08	**.05**	2.90	.06	3.95	**.03**
C * H * Sr	2,66	5.01	**.01**	4.26	**.02**	3.01	.06	0.63	.53
C * H * Sr * G	2,66	1.04	.36	0.74	.48	0.59	.56	0.27	.76
Cr * H * Sr	2,66	0.68	.68	0.58	.56	1.15	.32	1.65	.20
Cr * H * Sr * G	2,66	0.84	.44	1.68	.19	1.70	.19	1.49	.23
C * Cr * H * Sr	2,66	0.28	.69	0.34	.67	1.05	.34	1.25	.29
C * Cr * H * Sr * G	2,66	0.36	.63	1.17	.31	1.03	.35	0.96	.38
G	1,33	0.09	.77	0.04	.85	0.16	.69	0.41	.52

__C__=Condition factor, __CR__=Coronal Row factor, __H__=Hemisphere factor, __SR__=Sagittal Row factor, __G__=Group factor, univariate ANOVA, Huynh-Feldt corrected p-values

Appendix B Scalp Amplitude and Scalp Laplacian Maps of Experiments C and D not Presented in Chapter 8.

Appendix B.1 Maps of Experiment C

Figure B.1.1 Scalp Amplitude Maps of iCNV, tCNV, and PINV in Healthy Controls

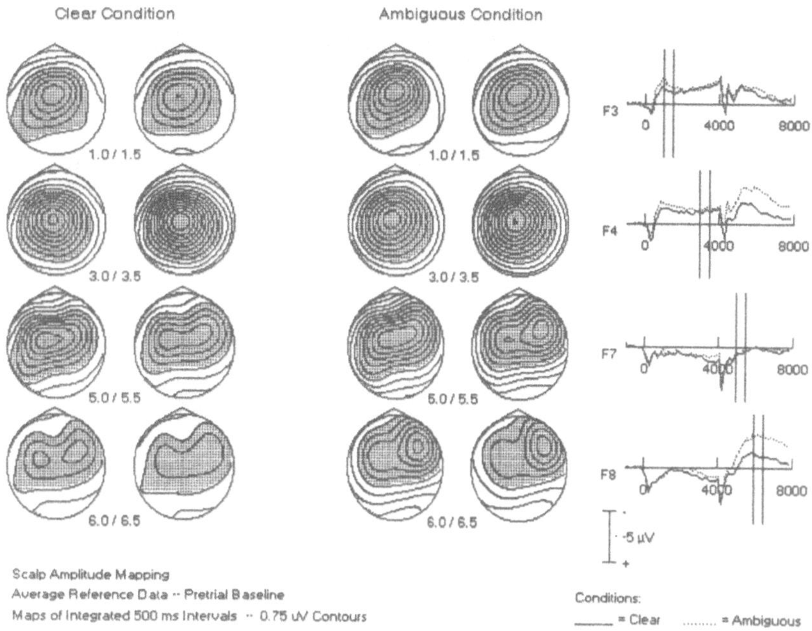

Figure B.1.2 Scalp Amplitude Maps of iCNV, tCNV, and PINV in Schizophrenic Patients

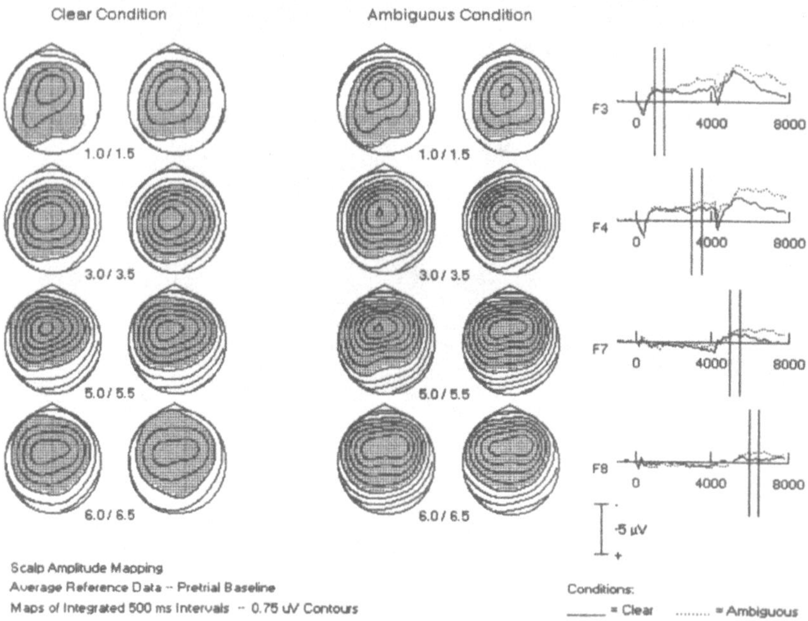

Figure B.1.3 Scalp Laplacian Maps of iCNV, tCNV, and PINV in Healthy Controls

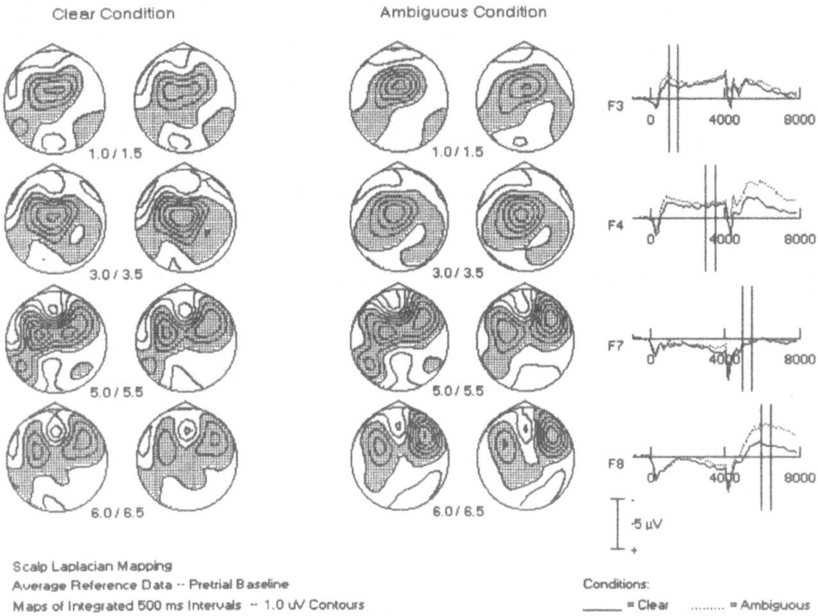

Figure B.1.4 Scalp Laplacian Maps of iCNV, tCNV, and PINV in Schizophrenic Patients

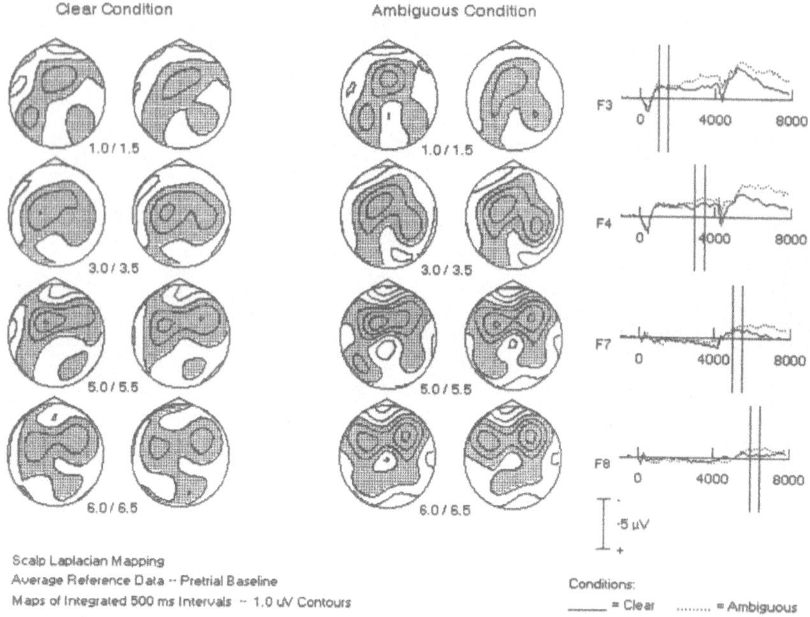

Figure B.1.5 Scalp Amplitude Maps of the tCNV-Referenced PINV in Healthy Controls

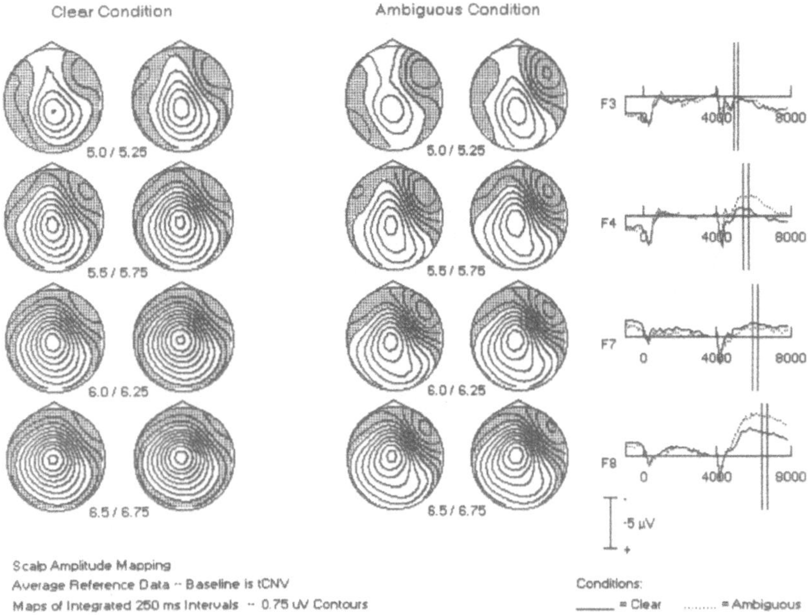

Figure B.1.6 Scalp Amplitude Maps of the tCNV-Referenced PINV in Schizophrenic Patients

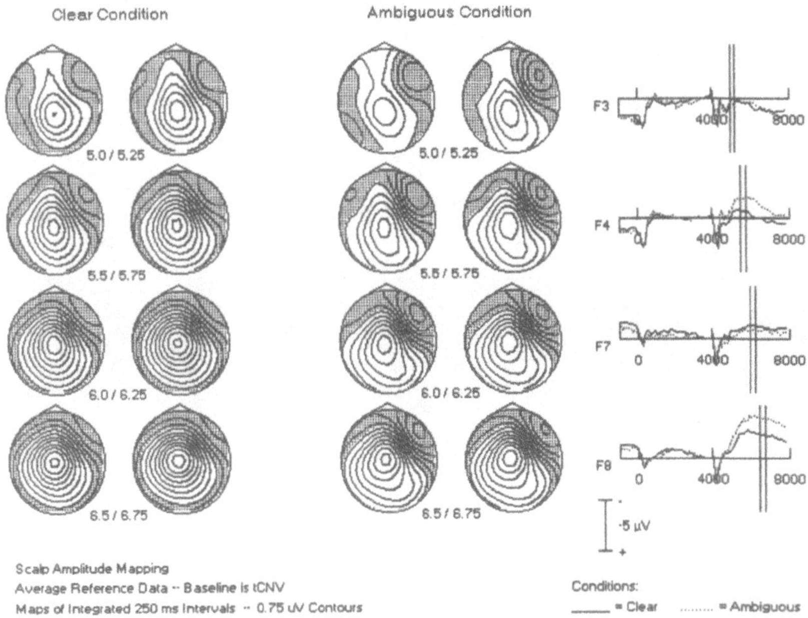

Figure B.1.7 Cortical Amplitude Maps of the tCNV-Referenced PINV in Healthy Controls

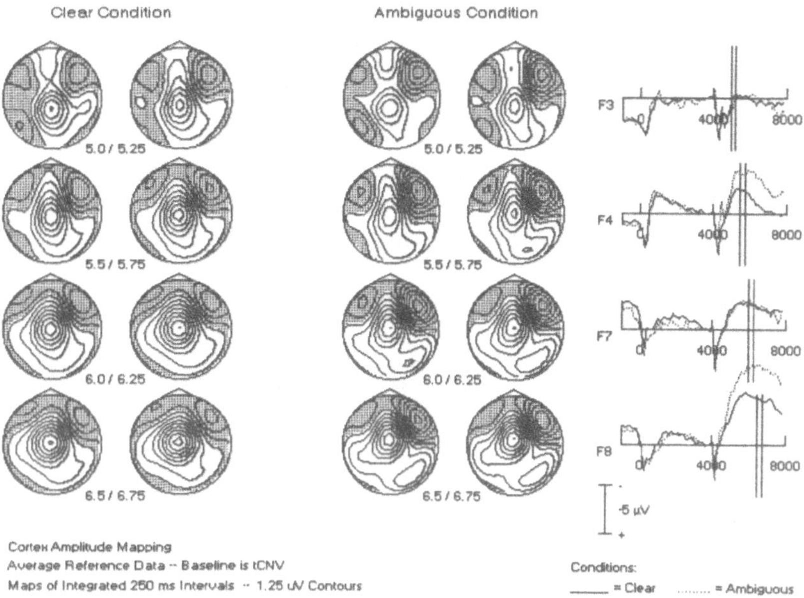

Figure B.1.8 Cortical Amplitude Maps of the tCNV-Referenced PINV in Schizophrenic Patients

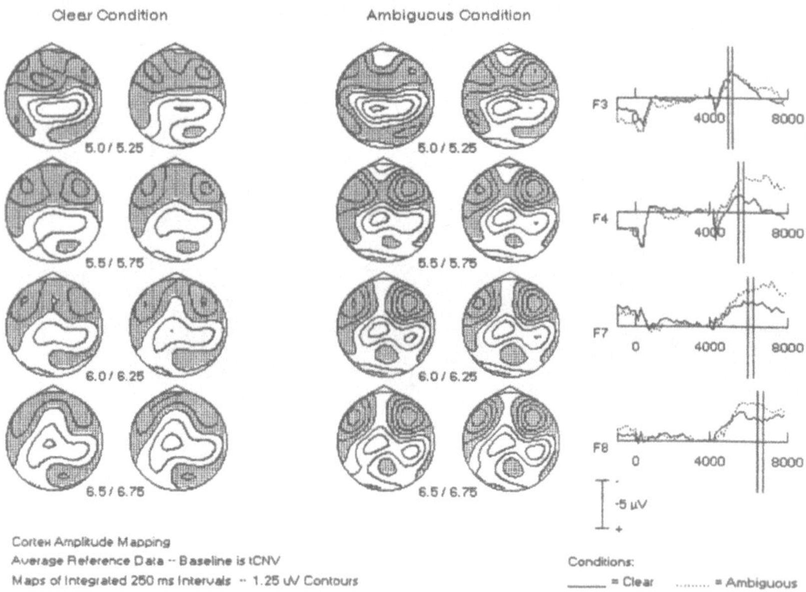

Appendix B.2 Maps of Experiment D

Figure B.2.1 Scalp Amplitude Maps of iCNV, tCNV, and PINV in the Low-Scoring Group

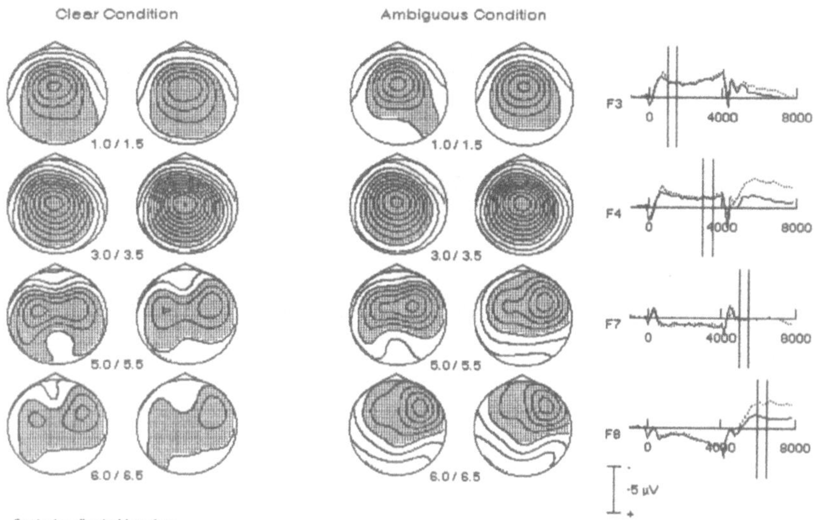

Scalp Amplitude Mapping
Average Reference Data -- Pretrial Baseline
Maps of Integrated 500 ms Intervals -- 0.75 uV Contours

Conditions:
_____ = Clear .... = Ambiguous

Figure B.2.2 Scalp Amplitude Maps of iCNV, tCNV, and PINV in the High-Scoring Group

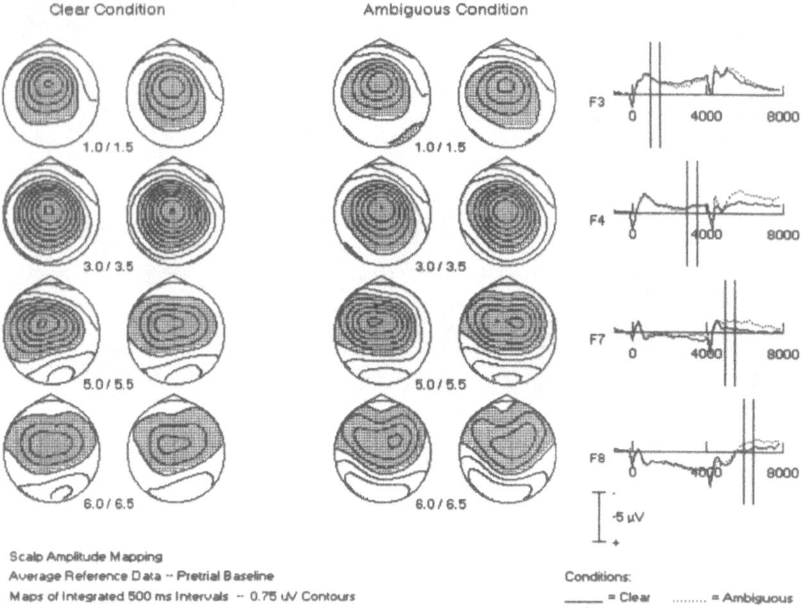

Figure B.2.3 Scalp Laplacian Maps of iCNV, tCNV, and PINV in the Low-Scoring Group

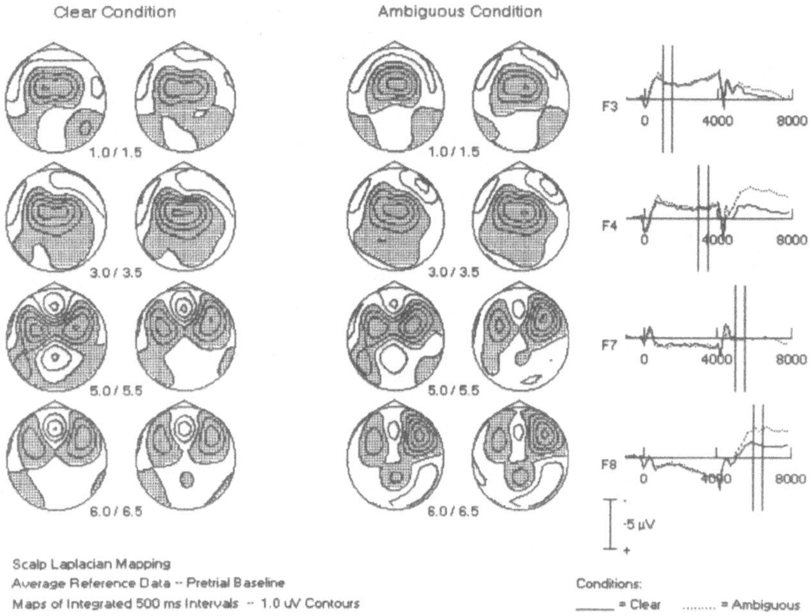

Figure B.2.4 Scalp Laplacian Maps of iCNV, tCNV, and PINV in the High-Scoring Group

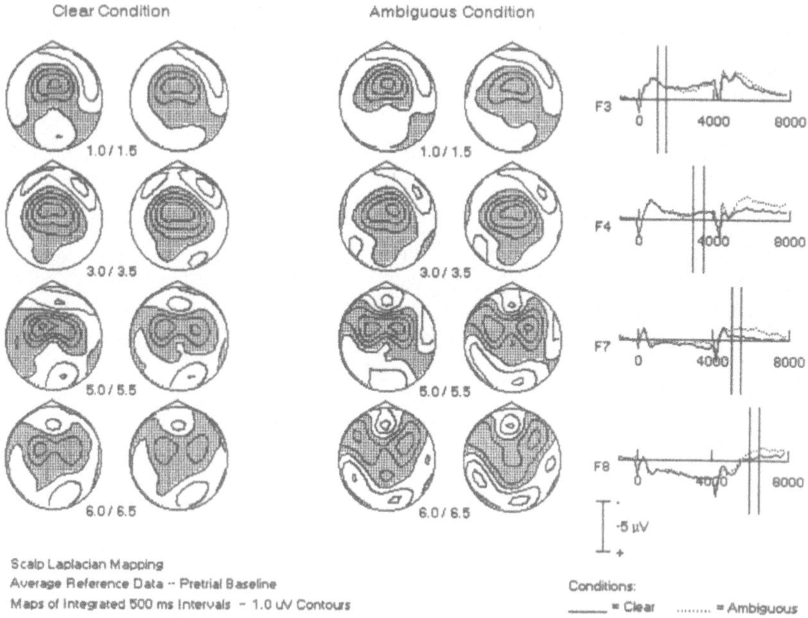

Figure B.2.5 Scalp Amplitude Maps of the tCNV-Referenced PINV in the Low-Scoring Group

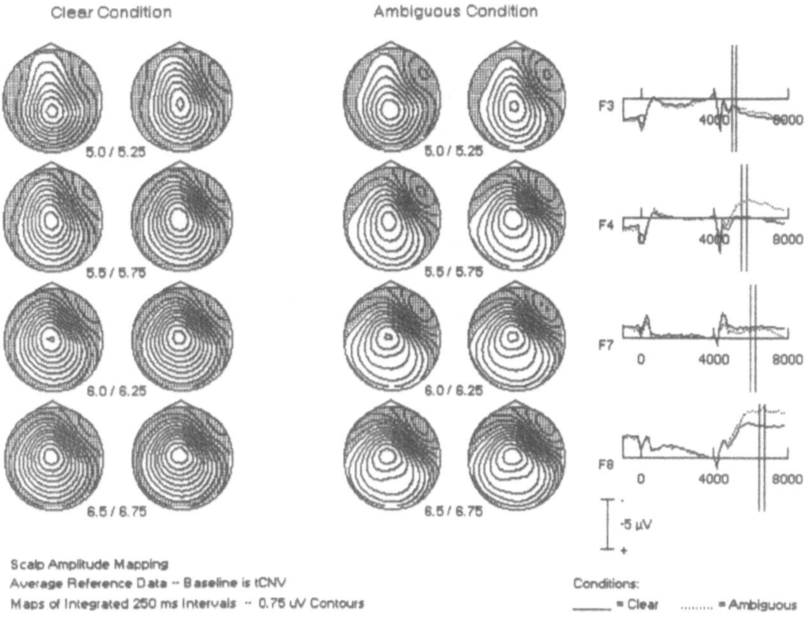

Figure B.2.6 Scalp Amplitude Maps of the tCNV-Referenced PINV in the High-Scoring Group

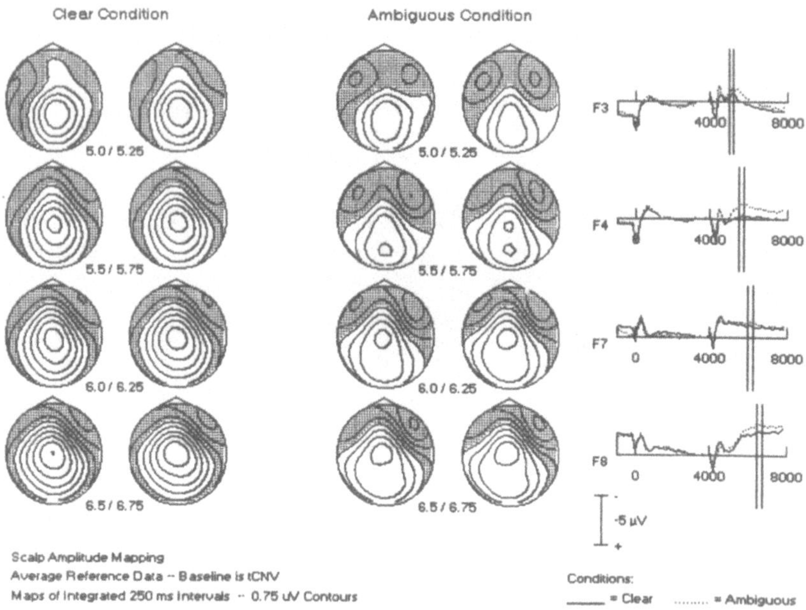

Figure B.2.7 Scalp Laplacian Maps of the tCNV-Referenced PINV in the Low-Scoring Group

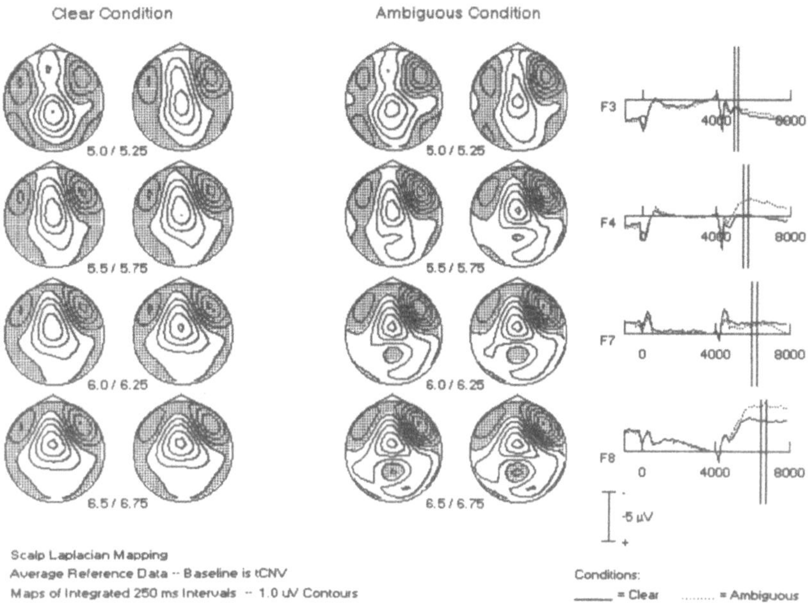

Figure B.2.8 Scalp Laplacian Maps of the tCNV-Referenced PINV in the High-Scoring Group

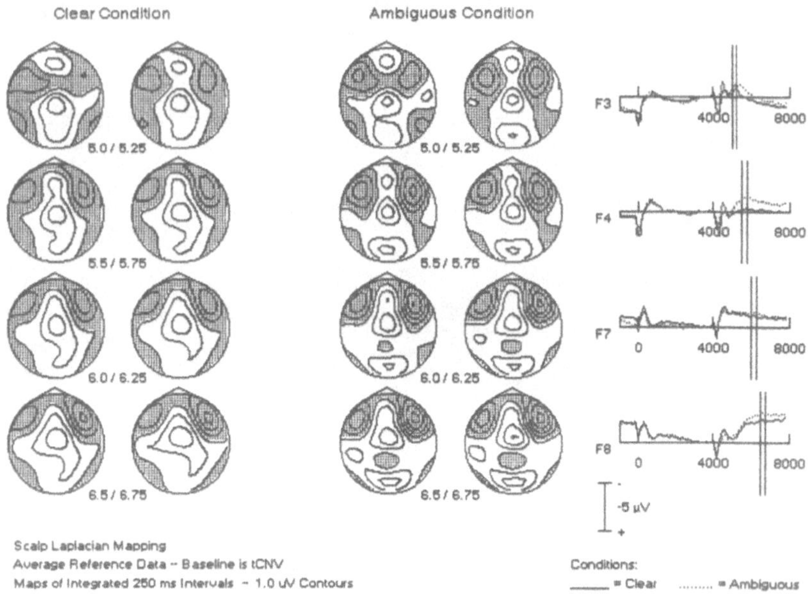

Appendix C The German Translation of the Schizotypal Personality Questionnaire (SPQ; Raine, 1991)

Methods

The SPQ by Raine (1991) was translated into German by the author. Verbal translations were prefered, and American/English idioms were replaced by comparable German idioms. The German translation of the SPQ, the SPQ-G, was filled in by 683 students from all faculties (Biology, Physics, Chemistry, Psychology, Philosophy, etc.) of the University of Konstanz. Together with the SPQ-G, a 50 item-version of the EPQ, the EPQ-KR (Ruch, 1989, 1992) as well as further 36 items, which are not relevant for the present purpose, were administered. The questionnaire was filled in completely by N=665 students (mean age 22.4±3.7 years; 424 male (63.8%), 241 female (36.2%)).

The following analyses were accomplished: (1) mean, standard deviation, curtosis, skewness and difficulty for the SPQ-G total score and the SPQ-G subscales, (2) product moment correlations for the SPQ-G subscales and the EPQ-KR scales; (3) explorative principle component analyses (PCAs) with varimax rotation for (a) all SPQ-G subscales and (b) the SPQ-G subscales and the EPQ-KR scales; (4) reliabilities (Cronbach´s α) and stability coefficients[226] for the SPQ-G total score, SPQ-G subscales, and the EPQ-KR scales; (5) mean differences between responding during the first and the second assessment (repeated measures ANOVA); (6) sex differences in SPQ-G subscales and EPQ-KR scales (MANOVA with subsequent univariate F-Tests).

Results & Discussion

Descriptive statistics Means and standard deviations of the SPQ-G total score and its subscales are documented in **table C.1**. The mean SPQ-G total score is 21.6±10.1 with a tendency for positive skewness in the sample. The scale difficulty of .28 appears adequate for the purpose of this questionnaire.

Subscale difficulties range between .19 and .42, with the more difficult subscales having greater deviations in curtosis and skewness. It should be noted that the easiest subscale, US, contains most frequently expressions like “sometimes“ or “ occasionally“.

Furthermore, the German sample comprised more male than female subjects, while the reverse held for the American and British samples. Hence, the results reported here may also be explained by the sample composition and sex differences in subscale means (see below).

For the samples of Raine (1991) SPQ total scores of 26.9 and 26.3 were reported. Gruzelier et al. (1995) obtained a SPQ total score of 19.3 using a 66

---

[226] N=40, with intervals ranging between about 2 (75%) to 6 months (25%).

item-version of the SPQ[227]. The lower SPQ-G total score in the German, compared to the American, sample may be explained by the high proportion of students of "hard" sciences (Physic, Chemistry, Biology) in our sample, while Raine exclusively tested students enrolled in introductory Psychology courses. Independent of the "true" mental states, the readiness to affirm items with "esoteric-spiritual" or "magic-irrational" contents may be reduced in students of subjects other than Psychology.

Table C.1 Descriptive Statistics for the SPQ-G (N = 665)

SPQ	M	D	SD	C	S
IR	3.2	.36	2.2	-.54	.48
SA	2.2	.27	2.0	.31	.92
MT	1.4	.20	1.8	.67	1.25
UPE	2.4	.27	2.0	-.10	.70
EB	2.2	.31	2.1	-.65	.73
NCF	1.7	.19	1.7	1.79	1.34
US	3.8	.42	2.7	-.94	.37
CA	2.2	.27	1.7	.27	.79
S	2.4	.30	1.7	.18	.74
Total	21.6	.28	10.1	-.18	.49

M: Mean; D: subscale difficulty (mean subscale score/maximum subscale score); SD: Standard Deviation; C: Curtosis, S: Skewness; IR: Ideas of Reference, SA: Social Anxiety; MT: Magical Thinking; UPE: Unusual Perceptual Experiences; EB: Exccentic Behavior; NCF: No Close Friends; US: Unuasual Speech; CA: Constricted Affect; S: Suspiciousness; Total: Total Score from 74 items; N=665

The internal consistencies (Cronbach´s α) of the SPQ-G total score and the subscales are similar to those reported by Raine (1991), except for S (see **table C.2**). The SPQ-G total score ($r=.88$) and the SPQ-G subscales ($.74 \leq r \leq .93$), except for S ($r=.69$), have good 2-6 months retest reliabilities, which are comparable to those of the original version and the stability coefficients of the EPQ-KR scales (L: $r=.84$; N: $r=.94$; E: $r=.94$; P: $r=.68$). The repeated measures ANOVA for the comparison of the first and the second assessment reveals a significant main effect "Retest" ($F(1,39)=7.56$, $p<.01$) and a significant interaction "Retest x Subscale" ($F(8,32)=2.71$, $p<.01$), indicating a lower level of responding during the second as compared to the first assessment, except for the US and EB subscales (see **table C.2**). It should be noted that the first assessment was an anonymous screening of large subject groups, while the second assessment was related to an individual laboratory testing situation for about 75% of the subjects.

227 This yields a SPQ total score of 21.6 if this score is " extrapolated" to a 74 item version.

Table C.2 Means for the First and Second Assessment and Retest-Realibilities (N=40)

	Tot.	S	US	EB	IR	NCF	CA	SA	UPE	MT
M 1st Assessment	29.6	3.4	4.7	3.2	3.9	3.0	3.1	3.5	3.1	1.7
M 2nd Assessment	26.4	2.7	5.0	3.3	3.2	2.7	2.8	2.9	2.6	1.2
r1.2	.88	.69	.80	.89	.74	.93	.87	.76	.78	.88
Cronbach´s α	.88	.62	.79	.81	.67	.64	.62	.73	.65	.77

Subscale abbreviations are explained in table C.1

Table C.3 Correlations between the SPQ-G subscales and the EPQ-KR scales

SPQ-G	IR	UPE	MT	S	NCF	EB	SA	CA	US	Total
IR	1.00	**0.43**	0.31	**0.50**	-0.07	0.34	0.10	-0.01	0.32	**0.60**
UPE		1.00	**0.50**	0.31	-0.01	0.28	0.19	0.01	0.35	**0.62**
MT			1.00	0.18	-0.11	0.14	0.02	-0.07	0.12	**0.41**
S				1.00	0.16	0.36	0.30	0.23	0.33	**0.66**
NCF					1.00	0.18	**0.46**	**0.56**	0.11	**0.41**
EB						1.00	0.13	0.18	**0.42**	**0.63**
SA							1.00	0.41	0.21	**0.54**
CA								1.00	0.22	**0.46**
US									1.00	**0.67**
EPQ-KR	IR	UPE	MT	S	NCF	EB	SA	CA	US	Total
L	0.14	0.07	0.00	0.05	-0.15	0.06	-0.09	-0.10	0.11	0.08
E	0.20	0.16	0.18	0.04	**-0.44**	0.06	-0.32	**-0.43**	0.07	-0.14
N	0.28	0.32	0.17	**0.45**	0.22	0.30	**0.47**	0.25	0.35	**0.56**
P	0.01	-0.09	-0.03	0.13	0.10	-0.04	0.13	0.13	-0.06	0.00

Subscale abbreviations are explained in table C.1; correlations ≥ .40 are printed in bold letters

In **table C.3** the correlations between the SPQ-G subscales and the EPQ-KR scales are documented. Correlations greater than .40 are found for the subscale pairs IR & UPE, IR & S, and UPE & MT, the subscale triple NCF & SA & CA, and the subscale pair US & EB. All SPQ-G subscales are substantially correlated with the SPQ-G total score. However MT, NCF, CA, and SA are weakly correlated with the SPQ-G total score ($.41 \leq r \leq .54$) when compared to the correlations between the SPQ-G total score and the remaining subscales ($.60 \leq r \leq .67$). The SPQ-G total score correlates positively with N ($r=.56$), but is unrelated to E ($r=-.14$) and P($r=.00$).

Since N is considered to indicate a non-specific vulnerability for the experience of negative affect (Montag &Levin, 1992; Clarke et al., 1994; Andresen, 1995) its substantial correlation with the SPQ-G total scores is not surprising. Positive correlations between N and, for instance, the

"Schizophrenism" scale were reported by Neilsen & Petersen (1976) and Claridge & Hewitt (1987).

By contrast, P was not related to the SPQ-G total score, confirming that "Psychoticism" and Schizotypal Personaliy are different constructs (Raine & Manders, 1988; Bentall et al., 1989; Andresen, 1995).

The first Principle Component Analysis (PCA-1) reveals 2 factors of the Schizotypal Personality with Eigenvalues greater than 1.0 (see **table C.4**). Factor 1 comprises the subscales S, IR, MT, UPE, EB, and US and explains 32.2% of the total variance. Concerning its subscale composition this first factor is identical with the first factor reported by Raine et al. (1992), and shall, therefore, be called the "cognitive-perceptual factor" of schizotypy. Factor 2 comprises the NCF, CA, and SA subscales, explains 21.5% of the total variance, and is, of course, identical with the "interpersonal factor" of schizotypy reported by Raine et al. (1992).

For the second PCA, PCA-2, three factors were deliberately extracted in order to provide a forced presentation of the 3-factor LISREL solution and the 3-factor PCA solution reported by Raine et al. (1994) and Gruzelier et al. (1995), respectively. PCA-2 resulted in a splitting of the "cognitive-perceptual factor" from PCA-1 and is similar to the results of Raine and Gruzelier in this respect. However, the subscale composition of remaining first and the new third factor, which explains 10.7% of the total variance, is dissimilar to that reported by the two authors: Instead of US and EB, the subscales UPE and MT pertain to the new third factor, which describes a "magical-illusionary" component of the schizotypal personality. The first factor, now comprising the S, IR, EB, and US subscales, shall be called the "paranoid-disorganized factor" of schizotypy.

Table C.4 Loadings and Communalities (h^2) of the SPQ-G Subscales on the (1) Cognitive-Perceptual and (2) Interpersonal Varimax Factors of PCA-1 and on the (1) Disorganized-Paranoid, (2) Interpersonal, and (3) Magical-Illusionary Varimax Factors of PCA-2

PCA-1	US	EB	S	IR	UPE	MT	NCF	CA	SA
factor 1	**.59**	**.58**	**.63**	**.77**	**.76**	**.61**	-.06	.02	.19
factor 2	.28	.28	.34	-.06	-.04	-.25	**.83**	**.81**	**.70**
h^2	.43	.41	.51	.60	.58	.44	.69	.66	.53
PCA-2	US	EV	AW	RI	UW	MD	KEF	EA	SA
factor 1	**.72**	**.78**	**.65**	**.63**	.33	.04	.06	.17	.10
factor 2	.13	.10	.24	-.12	.06	-.06	**.84**	**.79**	**.77**
factor 3	.08	.01	.23	**.44**	**.77**	**.87**	-.12	-.12	.21
h^2	.54	.61	.53	.61	.71	.76	.72	.67	.64

Subscale abbreviations are explained in table C.1; loadings ≥ .40 are printed in bold letters

Including the EPQ-KR scales into a third PCA, PCA-3, yields 4 factors with Eigenvalues larger than 1.0, explaining 25.6%, 19.0%, 8.9%, and 7.7% of the total variance (see **table C.4**). While N has high loadings on the "paranoid-disorganized" factor and E negative loadings on the "interpersonal factor", the "magical-illusionary" factor remains unaltered in PCA-3 without factorial relations with the EPQ-KR scales. P and L compose a fourth "antisocial-nonconformable" factor.

Altogether, the PCA results confirm the distinction between "positive" and "negative" schizotypal traits that has been reported by others (e.g. Siever & Gunderson, 1983; Kendler er al., 1983, 1991; Raine et al., 1992). The forced three-factor solution obtained with PCA-2 yielded a splitting of the "positiv" schizotypy factor that differed from that reported by Raine et al. (1994). It should, however, be noted that the corresponding factors reported by Raine et al. (1994) were highly correlated (r=.71, r=.75).

Table C.5 Sex Differences in SPQ-G Subscales and EPQ-KR Scales

	RI	UW	MD	US	EV	AW	KEF	SA	EA	Tot.	L	E	N	P
M(m)	2.9	2.2	1.1	3.9	**2.4**	2.3	**1.9**	2.1	**2.4**	21.2	8.8	6.1	4.9	**7.6**
M(f)	**3.7**	**2.9**	**2.1**	3.8	1.7	2.5	1.5	**2.4**	1.9	22.4	8.7	**6.7**	**5.4**	7.3
F	19.3	19.6	44.1	0.2	17.7	0.9	6.8	4.2	12.8	2.1	0.8	9.8	4.6	4.1
p	***	***	***	n.s.	***	n.s.	**	*	***	n.s.	n.s.	**	*	***

Subscale abbreviations are reported in table C.1; M(m)=mean of the male subjects; M(f)=mean of the female subjects; if sex differences are significant, the mean of the group with the larger subscale mean is printed in bold letters; F=F-term (ANOVA), p=significance of F; *=p<.05;**=p<.01;***p=p<.001; n.s.=non-significant

While male and female subjects exhibited no differences in the SPQ-G total score (t-test), sex differences were found for the SPQ-G subscales (Wilks $F(9,655)=13.74$, $p<.0001$) and the EPQ-KR scales (Wilks $F(4,661)=7.95$, $p>.0001$; see **table C.5**). Female subjects reported significantly more ideas of reference (IR), unusual perceptual experiences (UPE), magical thinking (MT), and social anxiety (SA); male subjects reported more constricted affect (CA) and had fewer close friends or confidents (NCF). Suspiciousness (S) and eccentric behavior (EB) did not distinguish the sexes. Finally, male subjects reported to be less extraverted, less neurotic, and more "psychoticistic" than female subjects.

These results largely confirm those reported by Raine et al.(1992) and emphasize the importance of sex differences in the fine-grained patterns of schizotypal personality. The SPQ-G total score, however, does not reveal sex differences.

Conclusions

Altogether, these analyses indicate that the SPQ-G is psychometrically similar to the SPQ. Hence, results of research using the SPQ-G can be integrated into the Anglo-American literature on Schizotypal Personality.

PSYCHOPHYSIOLOGIE IN LABOR UND FELD

Herausgeber: Michael Myrtek und Jochen Fahrenberg

Band 1 Michael Marwitz: Psychophysiologische Aspekte der normotonen und der labil-hypertonen Blutdruckregulation. 1997.

Band 2 Beatrice Cadalbert: Die Psychophysiologie des niedrigen Blutdrucks: Kreislaufregulation, Lebensgewohnheiten und Beschwerden. 1997.

Band 3 Christoph Klein: The Post-Imperative Negative Variation in Schizophrenic Patients and Healthy Subjects. 1997.

Tilo Held

Schizophreniebehandlung in der Familie
Eine kontrollierte Studie zur Wirksamkeit familiärer Verhaltenstherapie bei der Rückfallprophylaxe schizophrener Erkrankungen

Frankfurt/M., Berlin, Bern, New York, Paris, Wien, 1995. 157 S., zahlr. Tab. u. Graf.
Europäische Hochschulschriften: Reihe 6, Psychologie. Bd. 505
ISBN 3-631-48398-8 br. DM 54.--*

Die "psychoedukativen Therapieansätze" der Schizophrenie stellen sich immer deutlicher als ein wesentlicher Durchbruch bei der Behandlung dieser Erkrankung heraus. Die Einbeziehung der Familien in ein strukturiertes Behandlungskonzept erweist sich dabei als dem traditionellen individual-therapeutischen Ansatz deutlich überlegen. Dieses Buch zeigt, wie im Rahmen einer 2-Jahresstudie die Vorteile dieser Behandlungsform bezüglich der Behandlungseffizienz, der sozialen Integration und der Kostenwirksamkeit zutagetreten.
Aus dem Inhalt: Ein Überblick über die neuen, empirisch überprüften Therapieverfahren der Schizophrenie. Eine Studie, die belegt, wie sich auch im europäischen Kontext die therapeutischen, sozialen und wirtschaftlichen Vorteile der "psychoedukativen Familienbetreuung" manifestieren. Hinweise auf die praktische Anwendung dieser Therapieform

Peter Lang Europäischer Verlag der Wissenschaften
Frankfurt a.M. • Berlin • Bern • New York • Paris • Wien
Auslieferung: Verlag Peter Lang AG, Jupiterstr. 15, CH-3000 Bern 15
Telefon (004131) 9402121, Telefax (004131) 9402131
- Preisänderungen vorbehalten - *inklusive Mehrwertsteuer

FSC
www.fsc.org
MIX
Papier | Fördert
gute Waldnutzung
FSC® C083411

Kolibri 360 GmbH
Ferdinand-Jühlke-Straße 7
99095 Erfurt, Deutschland
produktsicherheit@kolibri360.de

Druck:
CPI Druckdienstleistungen GmbH
im Auftrag der
Kolibri 360 GmbH
Ferdinand-Jühlke-Str. 7
99095 Erfurt